2012 International Symposium on System on Chip

(SoC 2012)

Tampere, Finland
10 – 12 October 2012

IEEE Catalog Number: CFP12554-PRT
ISBN: 978-1-4673-2895-1

**Copyright © 2012 by the Institute of Electrical and Electronic Engineers, Inc
All Rights Reserved**

Copyright and Reprint Permissions: Abstracting is permitted with credit to the source. Libraries are permitted to photocopy beyond the limit of U.S. copyright law for private use of patrons those articles in this volume that carry a code at the bottom of the first page, provided the per-copy fee indicated in the code is paid through Copyright Clearance Center, 222 Rosewood Drive, Danvers, MA 01923.

For other copying, reprint or republication permission, write to IEEE Copyrights Manager, IEEE Service Center, 445 Hoes Lane, Piscataway, NJ 08854. All rights reserved.

***This publication is a representation of what appears in the IEEE Digital Libraries. Some format issues inherent in the e-media version may also appear in this print version.**

IEEE Catalog Number: CFP12554-PRT
ISBN 13:   978-1-4673-2895-1

Additional Copies of This Publication Are Available From:

Curran Associates, Inc
57 Morehouse Lane
Red Hook, NY 12571 USA
Phone: (845) 758-0400
Fax: (845) 758-2633
E-mail: curran@proceedings.com
Web: www.proceedings.com

2012 International Symposium on System on Chip (SoC 2012)

Tampere, Finland
10-12 October 2012

IEEE Catalog Number: CFP12554-POD
ISBN: 978-1-46732-895-1

Table of Contents

Application-Aware Spinlock Control using a Hardware Scheduler in MPSoC Platforms   1

Diandian Zhang[1], Li Lu[1], Jeronimo Castrillon[1], Torsten Kempf[1], Gerd Ascheid[1], Rainer Leupers[1], Bart Vanthournout[2]

[1]Institute for Communication Technologies and Embedded Systems (ICE), RWTH Aachen University, Germany, [2]Synopsys Inc., Leuven, Belgium

A Multi-banked Shared-L1 Cache Architecture for Tightly Coupled Processor Clusters   7

Mohammad Reza Kakoee[1], Vladimir Petrovic[2], Luca Benini[1]

[1]DEIS, University of Bologna, [2]Elsys Eastern Europe

An Automated Framework for the Simulation of Mapping Solutions on Heterogeneous MPSoCs   12

Antonio Miele[1], Christian Pilato[1], Donatella Sciuto[1]

[1]Politecnico di Milano

Instrumentation-Driven Model Detection for Datafow Graphs   18

Ilya Chukhman[1], William Plishker[1], Shuvra S. Bhattacharyya[1]

[1]University of Maryland, College Park

Thermal/Performance Trade-off in Network-on-Chip Architectures   26

Davide Zoni[1], Simone Corbetta[1], William Fornaciari[1]

[1]Politecnico di Milano

A Double Data Rate 8T-Cell SRAM Architecture for Systems-on-Chip   34

Saleh M. Abdel-Hafeez[1], Mohammad Shatnawi[1], Ann Gordon-Ross[2]

[1]Jordan University of Science and Technology, [2]University of Florida

Scalability Analysis of Release and Sequential Consistency Models in NoC based Multicore Systems   38

Abdul Naeem[1], Axel Jantsch[1], Zhonghai Lu[1]

[1]Royal Institute of Technology (KTH), Sweden

Resource-shared Custom Instruction Generation under Performance/Area Constraints   45

Di Wu[1], Junwhan Ahn[2], Imyong Lee[2], Kiyoung Choi[2]

[1]SAP Labs Korea, Seoul, Korea, [2]Seoul National University, Seoul, Korea

Comparative Analysis of Dynamic Task Mapping Heuristics in Heterogeneous NoC-based MPSoCs   51

Leandro Möller[1], Leandro Soares Indrusiak[2], Luciano Ost[3], Fernando Moraes[4], Manfred Glesner[1]

[1]TU Darmstadt, [2]University of York, [3]LIRMM, [4]PUCRS

A Hybrid Chip Interconnection Architecture with a Global Wireless Network Overlaid on Top of a Wired Network-on-Chip   55

Ling Wang[1], Zhen Wang[1], Yingtao Jiang[2]

[1]Harbin Institute of Technology, [2]University of Nevada, Las Vegas

Statistical Timing Characterization   59

Nadine Azemard[1], Zeqin Wu[1], Philippe Maurine[1], Gilles Ducharme[2]

[1]LIRMM, University of Montpellier II, Montpellier, France, [2]Dept. Math, University of Montpellier II, Montpellier, France

Hierarchical Control Flow Matching for Source-level Simulation of Embedded Software   63

Kun Lu[1], Daniel Müller-Gritschneder[1], Ulf Schlichtmann[1]

[1]Technical University of Munich

PowerMemo: A Power Profiling Tool for Mobile Devices in an Emulated Wireless Environment   68

Shiao-Li Tsao[1], Chih-Chen Kao[1], Ilter Suat[1], Yuchen Kuo[1], Yi-Hsin Chang[1], Cheng-Kun Yu[1]

[1]National Chiao Tung University

Architecture Efficiency of Application-Specific Processor: a 170Mbit/s 0.644mm2 Multi-standard Turbo Decoder   73

Rachid Al-Khayat[1], Amer Baghdadi[1], Michel Jézéquel[1]

[1]Institut Mines-Telecom, Telecom Bretagne, CNRS Lab-STICC

Improving Logic-to-Memory Ratio in an Embedded Multi-Processor System via Code Compression 80

Roberto Airoldi[1], Piia Saastamoinen[1], Jari Nurmi[1]

[1]Tampere University of Technology

Effects of Scaling a Coarse-Grain Reconfigurable Array on Power and Energy Consumption 84

Waqar Hussain[1], Tapani Ahonen[1], Jari Nurmi[1]

[1]Department of Computer Systems, Tampere University of Technology

CRAVE: An Advanced Constrained RAndom Verification Environment for SystemC 89

Finn Haedicke[1], Hoang M. Le[1], Daniel Große[1], Rolf Drechsler[12]

[1]University of Bremen, [2]DFKI

Asynchronous Parallel MPSoC Simulation on the Single-chip Cloud Computer 96

Christoph Roth[1], Simon Reder[1], Gökhan Erdogan[1], Oliver Sander[1], Gabriel M. Almeida[1], Harald Bucher[1], Jürgen Becker[1]

[1]Karlsruhe Institute of Technology

Ultra-Low Latency NoC testing via Pseudo-Random Test Pattern Compaction 104

Hervé Tatenguem[1], Alessandro Strano[1], Vineeth Govind[2], Jaan Raik[2], Davide Bertozzi[1]

[1]University of Ferrara, [2]Tallinn Institute of Technology

A flexible platform architecture for Gbps Wireless Communication 110

Jeroen Declerck[1], Prabhat Avasare[1], Miguel Glassee[1], Amir Amin[1], Erik Umans[1], Andy Dewilde[1], Praveen Raghavan[1], Martin Palkovic[1]

[1]IMEC, Belgium

Efficient VLSI Architectures of QPP Interleavers for LTE Turbo Decoders 116

Martin Broich[1], Tobias G. Noll[1]

[1]EECS - RWTH Aachen University

Tiny and Application-Specific Programmable Processor for BCH Decoding 122

Anthony Van Herrewege[1], Ingrid Verbauwhede[1]

[1]KU Leuven

Dataflow-Based Reconfigurable Architecture for Streaming Applications 126

Anja Niedermeier[1], Jan Kuper[1], Gerard Smit[1]

[1]University of Twente

Enhancing Cache Coherent Architecture with access patterns for embedded manycore systems 130

Jussara Marandola[1], Stephane Louise[2], Loïc Cudennec[2], Jean-Thomas Acquaviva[2], David A. Bader[1]

[1]Georgia Tech computing center, [2]CEA, LIST

System-level Software Performance Simulation Considering Out-of-order Processor Execution 137

Roman Plyaskin[1], Thomas Wild[1], Andreas Herkersdorf[1]

[1]TU Munich, Germany

Coarse and Fine-Grained Monitoring and Reconfiguration for Energy-Efficient NoCs 144

Liang Guang[1], Ethiopia Nigussie[1], Juha Plosila[1], Jouni Isoaho[1], Hannu Tenhunen[2]

[1]University of Turku, Finland, [2]Royal Institute of Technology, Sweden

2012 International Symposium on System on Chip (SoC)

Tampere, Finland, October 11-12, 2012

Welcome

Welcome to the System-on-Chip event. SoC is an annual symposium held at Tampere, Finland - The SoC City. It builds on the tradition of a series of SoC events organized annually since 1999. The mission of SoC is to provide a forum that is fully and comprehensively dedicated to SoC issues.

The event is based on a balanced mixture of

- world-class invited presentations
- exhibition of state-of-the-art commercial technology
- contributed paper track (since 2003)
- industrial paper track
- panel discussion
- tutorial course (since 2001)

The main organizer for the event is Tampere University of Technology Department of Computer Systems in cooperation with timely research activities of the field. Since 2003 the event has technical co-sponsorship by the Circuits and Systems Society of IEEE.

On these pages you will find information on SoC 2012.
For more information, please see SoC homepage or email Prof. Jari Nurmi (Jari dot Nurmi at TUT dot FI).

Foreword

International Symposium on System-on-Chip 2012 is the 14th annual SoC event in Tampere, it builds on a tradition started back in 1999. In addition to the invited lectures, commercial exhibit and vendor programme, the conference is open for contributions from researchers on this broad but focused field. The symposium also features a panel discussion on topics of high interest to the SoC community. This will reflect the theme of the year, Reconfigurable Circuits and Systems. The mission of SoC 2012 is to provide a forum that is fully and comprehensively dedicated to SoC. We enjoy the privilege to have IEEE Circuits and Systems Society as our technical co-sponsor.

The event was the first to use solely "SoC" as its name and focus. Later on, many counterparts have emerged worldwide, adopting this magnificent abbreviation in their names. Still, it is the major international SoC event in the Northern Europe, equally appreciated by the companies and academics in Europe but also increasingly in Americas and Far East. This is also reflected in the spectrum of countries where the papers presented in SoC 2012 originate from, they come from 13 countries all over the world (first author). We think that a very interesting thing is that even when the four top countries are Germany, Italy, France and Finland, also countries a bit further away like Taiwan, Korea, China and USA are represented. Thanks to all contributors for their submissions, whether or not exceeding the publication threshold this time.

We would like to acknowledge the sponsorship received from Nokia Corporation and IEEE Finland Section. Even more than that, we are especially pleased about the presence of numerous Nokia representatives in the event, which has also become a tradition. We also thank the exhibitors for their support.

We are grateful to the technical program committee members and other reviewers of the submitted papers, with their help we could provide valuable feedback to the authors to improve the quality of the Proceedings. Last but definitely not least, we extend our thanks to the invited speakers of this year. Traditionally, the backbone of the event has been formed by the invited talks. We believe that with the selected four distinguished people from the academy and industry, the event will be in a good shape. They all approach the theme of the year from different viewpoints.

Thanks also to our steering committee comprising Prof. Jan Rabaey, Prof. Heinrich Meyr, Prof. Hannu Tenhunen, and Dr. Fabio Campi, chaired by the permanent general chair Prof. Jari Nurmi.

This year we are co-located with another major event, Embedded Systems Week, consisting of three conferences (CASES, CODES+ISSS, EMSOFT), half a dozen smaller workshops, and several tutorials. We wish that the SoC participants will capitalize on it and enjoy the increased networking opportunities.

Welcome to Tampere, the SoC City!

Jari Nurmi
General Chair

Jarmo Takala
Program Chair

Olli Vainio, Jussi Raasakka
Proceedings Chairs

Steering Committee

- Jari Nurmi, TUT, Finland (chairman)
- Jan Rabaey, UC Berkeley, USA
- Heinrich Meyr, RWTH Aachen / CoWare, Germany
- Hannu Tenhunen, KTH, Sweden / UTU, Finland / INPG, France / Fudan University, China
- Fabio Campi, ST Microelectronics, Italy

Retired Steering Committee Members

- Mika Kuulusa, Nokia, Finland

General Chair

- Prof. Jari Nurmi, Tampere University of Technology, Finland

Proceedings Co-chairs

- Prof. Olli Vainio, Tampere University of Technology, Finland
- Jussi Raasakka, Tampere University of Technology, Finland

Exhibit and Sponsor Chair

- Tapani Ahonen, Tampere University of Technology, Finland

Technical Program Committee Chair

- Prof. Jarmo Takala, Tampere University of Technology, Finland

Technical Program Committee

- Andrea Acquaviva, University of Verona, Italy
- Brian Bailey, independent consultant, USA
- Heikki Berg, Nokia, Finland
- Koen Bertels, TU Delft, The Netherlands
- Davide Bertozzi, University of Ferrara, Italy
- Shuvra S. Bhattacharyya, University of Maryland, USA
- Abdelhafid Bouhraoua, KFUPM, Saudi-Arabia
- Claudio Brunelli, Nokia, Finland
- Peeter Ellervee, TU Tallinn, Estonia
- M. A. Al Faruque, Siemens, USA
- William Fornaciari, Politecnico di Milano, Italy
- Kees Goossens, TU Eindhoven, The Netherlands
- Lasse Harju, ST-Ericsson, Finland
- Hannu Heusala, University of Oulu, Finland
- Heikki Hurskainen, Microteam, Finland
- Jouni Isoaho, University of Turku, Finland
- Tariq Jamil, Sultan Qaboos University, Oman
- Murali Jayapala, IMEC, Belgium
- Tuomas Järvinen, ST-Ericsson, Finland
- Kimmo Kuusilinna, Nokia, Finland
- Vesa Lahtinen, Renesas Mobile, Finland
- Rainer Leupers, RWTH Aachen, Germany
- Oz Levia, independent consultant, USA
- Samy Meftali, LIFL/INRIA, France

- Dragomir Milojevic, ULB/IMEC, Belgium
- Fernando Moraes, PUCRS, Brazil
- Tobias Noll, RWTH Aachen, Germany
- Gianluca Palermo, Politecnico di Milano, Italy
- Juha Plosila, University of Turku, Finland
- Yang Qu, Renesas Mobile, Finland
- Tero Rissa, Nokia, Finland
- Stefan Rusu, Intel, USA
- Marco Santambrogio, Politecnico di Milano, Italy
- Olli Silven, University of Oulu, Finland
- Gerard J. M. Smit, University of Twente, The Netherlands
- Wonyong Sung, Seoul National University, Korea
- Lionel Torres, LIRMM, France
- Seppo Virtanen, University of Turku, Finland
- Steve Wilton, UBC, Canada

External Reviewers

(in addition to Technical Program Committee members and chairmen)

- Roberto Airoldi
- Francescantonio Della Rosa
- Vladimir Guzma
- Ismo Hänninen
- Pekka Jääskeläinen
- Erno Salminen
- Jarno Vanne

SOC 2012: Program

Thursday 11.10.2012

Registration

8:30 - 10:00

Coffee

9:00 - 10:00

- Rondo

Opening

10:00 - 10:15

- Studio

Invited1

10:15 - 11:00

- Studio

ROMA: Reconfigurable Operator Based Architecture for Multimedia Applications
Emmanuel Casseau
IRISA

Advanced Platform Architectures

11:00 - 11:40

- Studio

Application-Aware Spinlock Control using a Hardware Scheduler in MPSoC Platforms
Diandian Zhang[1], Li Lu[1], Jeronimo Castrillon[1], Torsten Kempf[1], Gerd Ascheid[1], Rainer Leupers[1], Bart Vanthournout[2]
[1]Institute for Communication Technologies and Embedded Systems (ICE), RWTH Aachen University, Germany, [2]Synopsys Inc., Leuven, Belgium

A Multi-banked Shared-L1 Cache Architecture for Tightly Coupled Processor Clusters
Mohammad Reza Kakoee, Vladimir Petrovic, Luca Benini
DEIS, University of Bologna

Lunch

11:40 - 13:00

Invited2

13:00 - 13:45

- Studio

Rethinking FPGAs: And-Inverter Cones Challenge LUT's Supremacy
David Novo
EPFL

System-level Design Flow and Methodology

13:50 - 14:50

- Studio

An Automated Framework for the Simulation of Mapping Solutions on Heterogeneous MPSoCs
Antonio Miele, Christian Pilato, Donatella Sciuto
Politecnico di Milano

Instrumentation-Driven Model Detection for Datafow Graphs
Ilya Chukhman, William Plishker, Shuvra Bhattacharyya
University of Maryland College Park

Thermal/Performance Trade-off in Network-on-Chip Architectures
Davide Zoni, Simone Corbetta, William Fornaciari
Politecnico di Milano

Posters and Coffee

14:50 - 15:30

- Rondo

A Double Data Rate 8T-Cell SRAM Architecture for Systems-on-Chip
Saleh Abdel-Hafeez[1], Mohammad Shatnawi[1], Ann Gordon-Ross[2]
[1]Jordan University of Science and Technology, [2]University of Florida

Scalability Analysis of Release and Sequential Consistency Models in NoC based Multicore Systems
Abdul Naeem, Axel Jantsch, Zhonghai Lu
Royal Institute of Technology (KTH), Sweden

Resource-shared Custom Instruction Generation under Performance/Area Constraints
Di Wu[1], Junwhan Ahn[2], Imyong Lee[2], Kiyoung Choi[2]
[1]SAP Labs Korea, [2]Seoul National University

Comparative Analysis of Dynamic Task Mapping Heuristics in Heterogeneous NoC-based MPSoCs
Leandro Moller[1], Leandro Soares Indrusiak[2], Luciano Ost[3], Fernando Moraes[4], Manfred Glesner[1]
[1]TU Darmstadt, [2]University of York, [3]LIRMM, [4]PUCRS

A Hybrid Chip Interconnection Architecture with a Global Wireless Network Overlaid on Top of a Wired Network-on-Chip
Ling Wang[1], Zhen Wang[1], Yingtao Jiang[2]
[1]Harbin Institute of Technology, [2]University of Nevada, Las Vegas

Statistical Timing Characterization
Zeqin Wu[1], Philippe Maurine[2], Nadine Azemard[1], Gilles Ducharme[3]
[1]LIRMM, [2]UM2 _ LIRMM, [3]UM2 - I3M

Hierarchical Control Flow Matching for Source-level Simulation of Embedded Software
Kun Lu, Daniel Müller-Gritschneder, Ulf Schlichtmann
Technical University of Munich

PowerMemo: A Power Profiling Tool for Mobile Devices in an Emulated Wireless Environment
Shiao-Li Tsao, Chih-Chen Kao, Ilter Suat, Yuchen Kuo, Yi-Hsin Chang, Cheng-Kun Yu
National Chiao Tung University

Architecture Efficiency of Application-Specific Processor: a 170Mbit/s 0.644mm2 Multi-standard Turbo Decoder
Rachid Al-Khayat, Amer Baghdadi, Michel Jezequel
Institut Mines-Telecom; Telecom Bretagne; CNRS Lab-STICC

Improving Logic-to-Memory Ratio in an Embedded Multi-Processor System via Code Compression
Roberto Airoldi, Piia Saastamoinen, Jari Nurmi
Tampere University of Technology

Effects of Scaling a Coarse-Grain Reconfigurable Array on Power and Energy Consumption
Waqar Hussain, Tapani Ahonen, Jari Nurmi
Department of Computer Systems, Tampere University of Technology

Verification and Testing

15:30 - 16:30

- Studio

CRAVE: An Advanced Constrained RAndom Verification Environment for SystemC
Finn Haedicke[1], Hoang Le[1], Daniel Grosse[1], Rolf Drechsler[2]
[1]University of Bremen, [2]University of Bremen and DFKI

Asynchronous Parallel MPSoC Simulation on the Single-chip Cloud Computer
Christoph Roth, Simon Reder, Gökhan Erdogan, Oliver Sander, Gabriel M. Almeida, Harald Bucher, Jürgen Becker
Karlsruhe Institute of Technology

Ultra-Low Latency NoC testing via Pseudo-Random Test Pattern Compaction.
Hervé Tatenguem[1], Alessandro Strano[1], Vineeth Govind[2], Jaan Raik[2], Davide Bertozzi[1]
[1]University of Ferrara, [2]Tallinn Institute of Technology

Panel Discussion

16:30 - 17:45

- Studio

Banquet

19:00 - 22:00

Friday 12.10.2012

Invited3

9:00 - 9:45

- Studio

Xilinx 3D Architecture
Jari Keskinen
Xilinx

Posters and Coffee

9:45 - 10:40

- Rondo

Application-Specific Architectures

10:40 - 12:00

- Studio

A flexible platform architecture for Gbps Wireless Communication
Jeroen Declerck, Prabhat Avasare, Miguel Glassee, Amir Amin, Erik Umans, Andy Dewilde, Praveen Raghavan, Martin Palkovic
IMEC, Belgium

Efficient VLSI Architectures of QPP Interleavers for LTE Turbo Decoders
Martin Broich and Tobias G. Noll
EECS - RWTH Aachen University

Tiny Application-Specific Programmable Processor for BCH Decoding
Anthony Van Herrewege and Ingrid Verbauwhede
KU Leuven

Dataflow-Based Reconfigurable Architecture for Streaming Applications
Anja Niedermeier, Jan Kuper, Gerard J.M. Smit
University of Twente

Lunch

12:00 - 13:20

Techniques for Manycore System on Chips

13:20 - 14:20

- Studio

Enhancing Cache Coherent Architecture with access patterns for embedded manycore systems
Jussara Marandola[1], Stephane Louise[2], Loic Cudennec[2], Jean-Thomas Acquaviva[2], David Bader[1]
[1]Georgia Tech computing center, [2]CEA, LIST

System-level Software Performance Simulation Considering Out-of-order Processor Execution
Roman Plyaskin, Thomas Wild, Andreas Herkersdorf
TU Munich, Germany

Coarse and Fine-Grained Monitoring and Reconfiguration for Energy-Efficient NoCs
Liang Guang[1], Ethiopia Nigussie[1], Juha Plosila[1], Jouni Isoaho[1], Hannu Tenhunen[2]
[1]University of Turku, Finland, [2]Royal Institute of Technology, Sweden

Posters and Coffee

14:20 - 15:00

- Rondo

Invited4

15:00 - 15:45

- Studio

Partially Reconfigurable ASIPs
Gerd Ascheid
RWTH Aachen University

Closing

15:45 - 16:00

- Studio

Social Event

17:30 - 22:00

Invited talk abstracts and biographies

ROMA: Reconfigurable Operator Based Architecture for Multimedia Applications
Prof. Emmanuel Casseau, IRISA, France

In multimedia applications, video and image processing is one of the challenges embedded systems have to face. Such applications are typically computationally intensive with control statements and designers have to cope with power and performance stringent requirements. The ROMA project proposes to develop both a design methodology and a reconfigurable processor able to adapt its computing structure to video and image processing applications. The reconfigurable processor is in charge of implementing parts of the code corresponding to loops and frequently executed computation code fragments that can be accelerated and/or which are good candidates to save power. It is built around a pipeline of coarse grain reconfigurable operators exhibiting interesting performance/power trade-off. The operators are designed such that their granularity matches the domain-specific computation patterns. Flexibility is obtained through these operators which can be configured for the function they implement and the width of the data. We have also developed a design flow to configure the processor. From the application source code, the software framework identifies the different computation patterns as well as their successive arrangements and completes transformations for the processor mapping.

Bio:

Emmanuel Casseau received the M.S. degree in Electrical Engineering in 1990 and the Ph.D degree in Electrical and Computer Engineering from the University of West Brittany, France, in 1994. From 1994 to 1996 he was a research engineer at ENST Bretagne, a graduate engineering school in France, where he developed high-speed Viterbi decoder architectures for turbo-code VLSI implementations. From 1996 to 2006 he was an Associate Professor in the Electronic Department at the University of South Brittany, France, where he led the IP project of the LESTER Laboratory. He his currently a Professor in IRISA/INRIA (French National Institute for Research in Computer Science and Control), University of Rennes1, France. His research interests include system design, high-level synthesis, SoCs design methodologies and reconfigurable architectures for multimedia applications.

Rethinking FPGAs: And-Inverter Cones Challenge LUT's Supremacy
Dr. David Novo, EPFL, Switzerland

Look-Up Tables (LUTs) are universally used in FPGAs as the fundamental unchallenged unit of functional reconfiguration. A k-input LUT can implement any k-input logic function, and thus, mapping a whole circuit onto an FPGA is a relatively straightforward covering problem. Complex circuits, however, require many LUTs connected by a flexible interconnect network, which ends up dominating circuit delay. Increasing the number of LUT inputs to cover larger parts of a circuit could reduce the communication overhead. However, it also entails an exponential increase in LUT area and power. For this reason, LUTs with more than 4-6 inputs have rarely been used in practical FPGAs. Still, the flexibility of current FPGAs comes at a huge price: the ratio of reconfigurable implementations and their ASIC counterparts is on the order of 3-5x in critical path delay, 14-7x in dynamic power consumption and 35-18x in silicon area.

In the quest for new elements to bridge the efficiency gap, we argue in this talk that other elementary logic blocks can provide a better compromise between hardware complexity and flexibility. Inspired by recent trends in synthesis and verification, we explore blocks based on And-Inverter Graphs (AIGs): their hardware complexity grows only linearly in the number of inputs, they sport the potential for multiple independent outputs, and the delay grows only logarithmically with the number of inputs. Of course, these new blocks are extremely less flexible than LUTs; yet, we show (i) that effective mapping algorithms are possible, (ii) that, due to their simplicity, poor utilization is less of an issue than with LUTs, and (iii) that a combination with a few LUTs can still be beneficial in extreme unfortunate cases. Our first results indicate that this new logic block alone, or combined with some LUTs in hybrid FPGAs, can reduce delay by 27 and 32% on average, respectively. At the same time, the area is reduced by some 16% on average. Yet, in this initial attempt we have explored only a few design points, and we think that these results could still be improved by a more systematic exploration.

Bio:

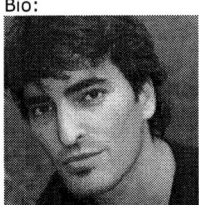

David Novo is Post-doctoral researcher at the EPFL School of Computer and Communication Sciences, where he joined the Processor Architecture Laboratory (LAP) in November 2010. Previously, he conducted his doctoral research at the Wireless Group in the Interuniversity Microelectronics Centre (IMEC), receiving the Ph.D in Engineering from the Katholieke Universiteit Leuven (KUL) in 2010.
David was recipient of the Best Paper Award at the 20th ACM/SIGDA International Symposium on Field-Programmable Gate Arrays (FPGA) in 2012 and nominated at the IEEE Workshop on Signal Processing Systems (SiPS) in 2005. In 2012, he has been awarded with the EU Marie Curie Intra-European Fellowship for Career Development. In 2011, he has been Guest Editor of a Special Issue on Quantization of VLSI Digital Signal Processing Systems, which appeared in February 2012 on the EURASIP Journal on Advances in Signal Processing. He is also author of more than 30 international papers distributed in the areas of signal processing, computer-aided design and computer architectures. David is currently working on hardware and software techniques for increasing computation efficiency in next-generation computers.

Xilinx 3D Architecture
Jari Keskinen, Xilinx, Finland

As system architects seek to achieve ever higher levels of integration and performance whilst minimising costs and remaining within strict power budgets, they find limitations with traditional solutions such as PCBs or ASICs. Whilst the 3D IC is not yet a fully developed technology, there are products in production today which can help to solve these issues in some applications. This session will review a technology called Stacked Silicon Interconnect (SSIT), an enabler for 3D ICs, in which a number of manufacturing techniques have been brought together to create both homogeneous and heterogeneous All Programmable 3D ICs. The session will also explore how these 3D ICs address the challenges faced by system architects today.

Bio:

Jari Keskinen received his M.Sc. degree in Computer Engineering and Applied Electronics in 1993 from Tampere University of Technology. Today he is a Staff Field Application Engineer at Xilinx and works closely with systems architects and hardware designers in telecommunications and industrial markets to solve performance, power, cost and integration challenges with state-of-the-art 28nm programmable FPGAs, SoCs and 3D ICs. Prior to joining Xilinx (1994-2000) Jari worked as an Application Engineer at Mentor Graphics (FINLAND) Oy. After graduation he worked as Research Scientist in Computer Systems laboratory in Tampere University of Technology.

Partially Reconfigurable ASIPs
Prof. Gerd Ascheid, RWTH Aachen University, Germany

Highest throughput and energy efficiency is achieved with dedicated hardware, highest flexibility is provided by processors. Processors with application specific instruction sets (ASIP) trade some flexibility for more efficiency and throughput. On the other hand, coarse grained reconfigurable arrays (CGRA) introduce some hardware flexibility at the cost of efficiency and throughput. Combining both approaches allows building systems with the flexibility of programmable architecture yet achieving throughput and energy efficiency close to dedicated hardware (ASIC) solutions. The talk will discuss architectural options for the CGRA and its integration into the processor architecture, implementation issues and a design flow both for the pre- and post-silicon design phases. Following a general review of these points, specific examples for reconfigurable ASIPs will be discussed in some depth, in particular, an ASIP for MIMO processing in mobile communications.

Bio:

Gerd Ascheid received his Diploma and PhD degrees in Electrical Engineering from RWTH Aachen University. In 1988 he started as a co-founder CADIS GmbH which successfully brought the system simulation tool COSSAP to the market. From 1994-2003 Gerd Ascheid was Director / Senior Director with Synopsys, Inc. In 2002 he was a co-founder of LisaTek whose processor design tools are now part of the Synopsys product portfolio. Since April 2003 Gerd Ascheid heads the Institute for Integrated Signal Processing Systems at RWTH Aachen University. He is also coordinator of the UMIC (Ultra-high speed Mobile Information and Communication) Research Centre at RWTH Aachen University. His research interest is in wireless communication algorithms and application specific integrated platforms, in particular, for mobile terminals.

Technical Co-Sponsor

Financial Sponsors

IEEE Finland Section

Accepted Papers by Country

(Includes contributed scientific papers and invited papers)

Belgium	2
China	1
Finland	4
France	4
Germany	7
Italy	4
Netherlands	1
Republic of Korea	1
Sweden	1
Switzerland	1
Taiwan	1
United Kingdom	1
United States	2
Total	30

Accepted Papers by Type

Scientific - oral	15
Scientific - poster	11
Invited	4
Total	30

Application-Aware Spinlock Control using a Hardware Scheduler in MPSoC Platforms

Diandian Zhang*, Li Lu*, Jeronimo Castrillon*, Torsten Kempf*,
Gerd Ascheid*, Rainer Leupers* and Bart Vanthournout†
*Institute for Communication Technologies and Embedded Systems (ICE), RWTH Aachen University, Germany
†Synopsys Inc., Leuven, Belgium

Abstract—Spinlocks are a common technique in Multi-Processor Systems-on-Chip (MPSoCs) to protect shared resources and prevent data corruption. Without a priori application knowledge, the control of spinlocks has high randomness which can degrade the system performance significantly. This paper presents a centralized control mechanism of spinlocks by using a hardware scheduler called OSIP, that increases system performance by utilizing application-specific information. A complete spinlock control flow, starting from integrating high-level user-defined information down to a low-level realization of the control, is introduced. Two case studies demonstrate the high efficiency of this mechanism.

I. Motivation

Due to the advantages of high performance and power efficiency, Multi-Processor Systems-on-Chip (MPSoCs) are nowadays widely used, especially in the embedded domain. Representative examples are TI OMAP processors [1] and the ARM MPCore technology [2]. In MPSoCs, applications are typically partitioned into tasks and mapped onto different processing elements (PEs). In order to protect shared resources like memories or I/O peripherals and to prevent data corruption, mutual exclusion needs to be guaranteed. Spinlocks are a widely used technique which ensures that only one task is able to access a shared resource at one time.

In literature, many studies have been carried out on the implementation of spinlocks. Most of them focus on the fairness of spinlock acquisition or the reduction of transaction costs due to repeatedly requesting the locks. However, application knowledge is typically not considered. This leads to a random distribution of acquiring spinlocks. In many cases, this could result in a lower parallelism of task execution and consequently worsen the system performance. A simple example is depicted in Figure 1, in which two PEs compete for getting the spinlock. In the example, two different execution sequences of tasks are possible due to acquiring the lock in a different order. By comparing the execution time, assigning the lock to PE_2 first (Figure 1(b)) appears to be a better choice, which achieves a higher performance.

Despite the simplicity of the example, it shows that a smart control of spinlocks based on application knowledge can increase the system performance. This particular problem is addressed by this work and follows the three main questions:

- What application knowledge is needed?
- How to convey the knowledge to the spinlock control?
- How to utilize the knowledge for the spinlock control?

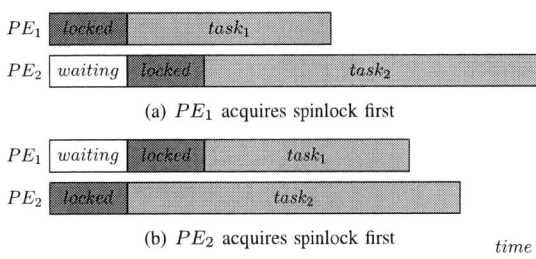

Fig. 1. Impact of spinlock acquisition order on the performance

The remainder of this paper is organized as follows. Section II introduces related work. In section III, a brief overview of OSIP-based MPSoCs, on which this work is based, is given. The main contribution of the work is described in section IV, which answers the three questions above. The OSIP, a hardware scheduler originally designed for task scheduling, is re-used for controlling spinlocks, provided with the user-defined application information. Two case studies – a synthetic application and a multimedia application (H.264 video decoding) – are discussed in section V. Section VI summarizes the work and gives an outlook.

II. Related Work

A commonly used technique to address mutual exclusion is to use spinlocks. Many implementations have been proposed for spinlocks in literature. The simplest approach is the *test-and-set* lock, which repeatedly tries to replace the lock flag with *true* in order to acquire the lock. Although the implementation is rather simple, it introduces heavy traffic load due to continuous updates of the flag, specially in cache coherent systems. One improvement is suggested by using the *test and test-and-set* lock [3], in which updates are only made when the lock is assumed to be available. Another important improvement is introduced in [4] by adding a certain delay (backoff) between two unsuccessful trials.

Having the *test-and-set* lock, the grants to the requests can be very unevenly distributed. In contrast, a queuing lock provides better fairness, in which the PEs acquire the lock in turns (FIFO fairness) [4]–[7].

However, in all above-memtioned approaches, the acquisition of the locks features high randomness, since the grant decision solely depends on the time stamp when the locks are required, regardless of the fairness. As shown in section I, a proper control of spinlocks would most likely have a positive impact on system performance. Priority-based spinlocks [8] [9] are a way of implementing this, in which the lock is granted by comparing the priorities of the requests in the queue. In [8], which process gets the grant as the next, is determined by looping over the queue at the time the lock is released. In contrast, in [9] the queue is maintained in the priority order during the time that the lock is acquired. In both cases, the grant decisions are distributively made by PEs, which potentially increases their workload.

In this work, the spinlocks in the system are controlled in a centralized way, which has similarity with the approach presented in [10]. However, the main distinctions of our approach are made by:

- (re-)using a programmable hardware scheduler called OSIP to flexibly control the spinlocks for different applications,
- enabling the programmer to specify application information, which can be, but is not limited to priorities, to make the most suitable decisions for spinlocks.

III. OVERVIEW OF OSIP-BASED MPSoCS

The efficiency of task scheduling and mapping in MPSoCs has a huge impact on the system performance. For accelerating these operating system (OS) operations, research attempts to shift, at least partially, the traditional software implementation of the OS to hardware. The OSIP-based MPSoCs introduced in [11] employ an Application-Specific Instruction-set Processor (ASIP) [12] called OSIP as the system scheduler to improve the scheduling efficiency.

The advantages of OSIP-based MPSoCs can be highlighted from two perspectives. First, OSIP provides an efficient and yet flexible solution to task scheduling based on the ASIP concept. It combines the advantages of pure hardware schedulers such as [13] [14] and software approaches like [1]. Second, a complete programming model is defined for OSIP-based MPSoCs which greatly eases system programming.

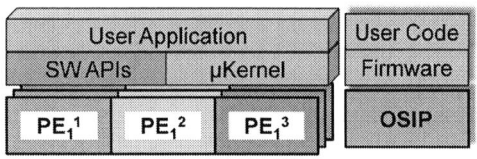

Fig. 2.   OSIP software layers

The programming model consists of a set of software APIs, shown in Figure 2. On the one hand, the APIs provide an interface to the user application to support high-level services such as *system configuration*, *task creation*, *task suspension* etc. On the other hand, the API layer also abstracts the hardware details of the communication between PEs and OSIP.

OSIP is a core component of the system. Provided with the task information by the PEs, it takes runtime decisions of task scheduling and mapping. To fully exploit the special hardware features of the OSIP processor, a firmware containing low-level functions and basic scheduling algorithms such as *FIFO*,

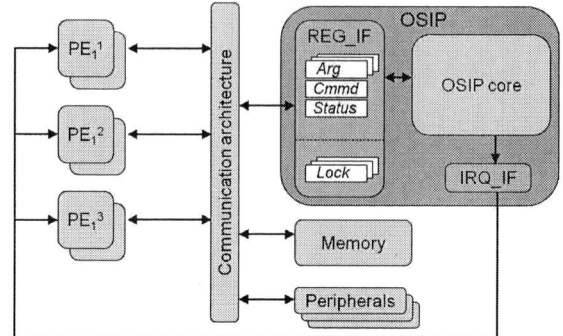

Fig. 3.   OSIP-based system

round-robin, *priority-based* is provided. Upon the firmware, the software programmer has the flexibility to construct own scheduling algorithms optimized for the intended application.

As shown in Figure 3, OSIP comprises three main components: the OSIP core, being an ASIP, a register interface (REG_IF) and an interrupt interface (IRQ_IF). Through the interrupt interface the OSIP core generates interrupts to the PEs for triggering task execution. More complex is the behavior residing in the register interface, in which the information between the PEs and OSIP is exchanged and maintained.

The most important registers in the register interface are listed in the figure. They are grouped into OSIP-core-related and -unrelated registers. In the first group, information and requests for task scheduling are transferred between the PEs and the OSIP core through the *command (Cmmd)* and *argument (Arg)* registers. The *status* register indicates whether the OSIP core is currently busy at handling a command or idle.

The second group of registers contains a set of spinlocks to protect shared resources like shared memories or peripherals. No interaction exists between these locks and the OSIP core. A simple hardware control logic based on *compare-and-swap* is used to grant the spinlock requests of the PEs. The control flow is illustrated in Figure 4(a). If a task requires a lock, it spins the corresponding lock register from the interface by calling the API function *SpinLockAcquire()* (step 1a). Upon receiving the request, the control logic checks the lock status (step 1b). Whenever the lock is available, it grants the lock to the requester and marks the lock as unavailable in the meantime. Otherwise, it returns a *failed* signal. After the task gets the lock and finishes the critical section, it sends a release signal to the interface (step 2a) by calling the API function *SpinlockRelease()*. The control logic sets the lock to available again (step 2b). However, in spinlock-intensive applications, this control mechanism shows high randomness, which could lead to suboptimal performance.

IV. APPLICATION-AWARE SPINLOCK CONTROL

This section introduces an extended control mechanism in OSIP-based systems to optimize the management of spinlocks.

A. Key concept

The key idea is to allow the programmer to control the spinlocks based on application knowledge, instead of letting the system do the job in an arbitrary way. In this context,

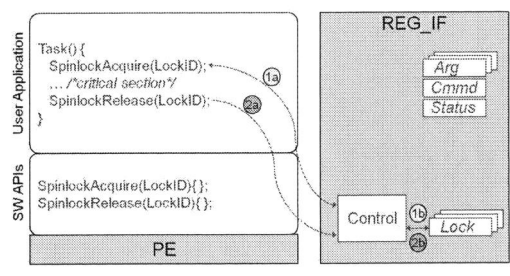
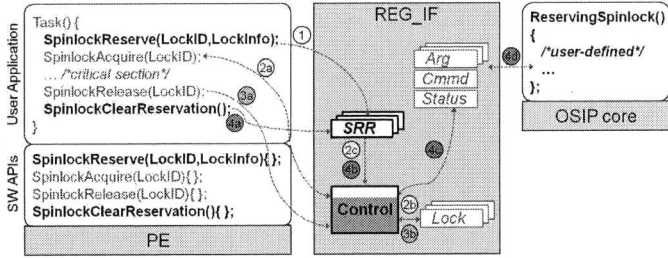

(a) Original spinlock flow   (b) Improved spinlock flow with reservation

Fig. 4. Spinlock flow in OSIP-based systems

application knowledge specifically refers to spinlock-related information, typically the task execution time and the blocking time by spinlocks. In addition, blocking the tasks in a different order could influence resolving task dependencies, and consequently the task parallelism. Therefore, this should be considered as well. The information will then be provided to OSIP, e.g. abstracted as a priority, demonstrating how urgently the lock is required by a task for the sake of maximizing task parallelism. Having the control information, OSIP is able to take proper runtime decision for granting the spinlock requests.

Due to the programmability of OSIP, the programmer is given the flexibility to define the semantic of the spinlock control information, e.g. priority-based, weighted fair queuing, etc. It can even be specified in a way, that some PEs always have a higher priority than the others when requesting a certain lock. The software running on the OSIP core can be adapted to understand the semantic.

To support this application-aware spinlock control, the system needs to be extended both from the application side and the OSIP side. On the application side, the programmer should be given the opportunity to provide control information, when necessary. On the OSIP side, the spinlock control mechanism should be enhanced, not only at the register interface, but also in the user-defined algorithms running on the OSIP core.

Implementing spinlock control algorithm in software on the OSIP core is meant to support different semantics of the control information, as implied above. Naturally, this increases the workload of OSIP, in addition to normal task scheduling and mapping. However, the efficiency of OSIP enables it to afford this extra workload. Related reports on the OSIP efficiency can be found in [11] and [15].

In order to make a good decision on spinlocks, the request information from different PEs should be considered together by OSIP. For this purpose, the PEs send a reservation request to OSIP before trying to acquiring the lock. In the reservation request, the spinlock control information is contained. During the time the spinlock is unavailable, i.e. being held by a PE, several reservation requests can be collected. Then the OSIP core will decide which PE/PEs should get the spinlock as the next and make reservations for it/them. In this way, the randomness of granting spinlock requests is greatly reduced.

The details of the extended spinlock control mechanism are explained below, illustrated by Figure 4(b). The extensions are highlighted in bold in the figure.

B. APIs for spinlock reservation

The spinlock reservation does not try to acquire the lock immediately, but is meant to inform OSIP how its request to the lock should be handled by transferring the lock control information. Therefore, the reservation request is sent to OSIP before the acquisition request starts.

In order to simplify adding the control information into the application, an API called *SpinlockReserve()* is introduced. It contains two parameters: *LockID* and *LockInfo*, specifying the required spinlock ID and user-defined control information. In fact, it is not always necessary to do the reservation for all spinlocks. Instead, only applying reservations for the most critical ones would be sufficient. For minimizing the communication overhead, both parameters are combined into a 32-bit word before they are transferred to the OSIP register interface. Complementarily, another API *SpinlockClearReservation()* is defined to cancel the reservation.

These two APIs should be always used in pair to enclose spinlock acquisition and release. A simple example is given in function *Task()* in Figure 4(b). In practice, this API pair can even be used to enclose multiple spinlock acquisition/release pairs, if desired, as long as they require the same lock.

C. Extended spinlock control in OSIP

The APIs for reserving spinlocks require an extension in the OSIP register interface to keep the reservation information and interact with the OSIP core for enhanced control of locks.

Four stages are defined for the extended spinlock control in OSIP: spinlock information collection, spinlock decision, release and reservation. The first three stages are purely controlled by the hardware in the interface, while the last stage needs a collaboration between the hardware interface and the software control of the OSIP core.

1) Information collection stage: Upon receiving a spinlock reservation request from a PE, the reservation information will be written into an internal register at the interface, which is named Spinlock Reservation Register (SRR) (Figure 4(b), step 1). Each PE has a dedicated SRR, where its request is stored.

As implied by the API arguments, the SRR contains the spinlock ID and control information. In addition, two flags *ReservationFlag* and *ReservedFlag* are used in SRR to specify whether to make a reservation for the lock and whether the lock has been reserved for the PE, respectively. In total, 14 bits are needed in the current OSIP system: 8 bits for the lock ID (the current system supports 256 spinlock registers), 2 bits for both flags and the remaining 4 bits for the control information.

2) Decision stage: In this stage, the interface decides whether a requesting task gets the spinlock. The general control flow is depicted in Figure 5. The difference between this and the original approach lies in the decision whether the lock will be acquired immediately by a request when it is free.

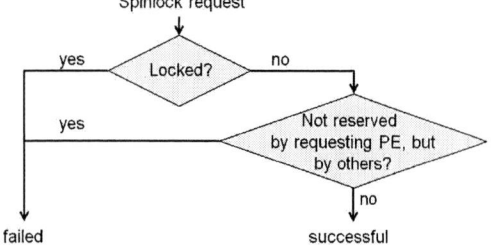

Fig. 5.   Flow of granting a spinlock request

When receiving a spinlock request (step 2a), the interface first checks the status of the lock (step 2b). If it is locked, the request will fail and the PE has to make another try, which is the same in the original approach. The difference occurs when the lock is available. In this case, a further check needs to be done, which examines whether the required lock has been reserved for the PE (step 2c). If no reservation exists for the requesting PE but another PE has the reservation, the request will still fail. In other words, the request will be successful, only when the requesting PE has got the reservation (no matter whether the other PEs also have got it), or the lock currently has not been reserved for any PEs. This implies that the tasks which have got the lock reserved will have a better opportunity to get it. With this, the programmer is able to control the spinlock to a certain extent.

3) Release stage: Spinlocks are released in this stage (step 3a and 3b), which is done the same as in the original approach.

4) Reservation stage: Obviously the spinlock reservation plays an essential role for making a proper decision on granting the request. Certainly, choosing a reservation criterion is one important aspect that the programmer has to consider and implement on the OSIP core. Another critical aspect is when the OSIP core should be triggered to make reservations for the PEs. If triggered too frequently (e.g. it is always triggered whenever there is a reservation request), the core will often unnecessarily stay at a busy state. This would impair the primary purpose of OSIP, namely the normal task scheduling and mapping. On the other hand, if the reservation is only rarely done, its effectiveness will be reduced.

With these considerations, in the current implementation the OSIP core is triggered only when the interface receives a reservation cancellation signal as a compromise. At this point, the flags of the corresponding SRR are cleared (step 4a). Furthermore, the interface will check if the lock is still reserved by other PEs (step 4b). If so, the interface will ignore the signal for reducing the load to OSIP. Only when no further reservations exist for the lock and OSIP is in the idle state, a

successful triggering happens (step 4c and 4d).

When triggering the OSIP core, only the SRRs that contain the same released ID will be considered by the core when making the reservation decisions. The spinlock control information of these SRRs is transferred to the OSIP core by re-using the *command* and *arguments* registers introduced in section III (step 4c). A new OSIP command is defined for the reservation, which is generated at the interface directly. Since the number of SRRs is often larger than that of the argument registers, the control information of SRRs needs to be re-arranged in the registers. To be exact, the bits of the first argument register are used to mask the SRR IDs which will be considered. The lock control information of the SRRs is stored into the rest of the argument registers in a halfbyte-aligned way, ordered by the SRR ID. This is the reason why the control information field of each SRR is currently limited to four bits.

After the OSIP core is triggered, the spinlock reservation is executed as a normal command (step 4d). The OSIP status will be set to busy to prevent further commands. The reservation algorithm is user-defined and decides which PE(s) get(s) the reservation, based on the specified spinlock control information. The interface will update the *reserved* flag of the corresponding SRR(s). The reservation for a lock is not necessarily limited to one PE. In fact, making a reservation for multiple PEs has the advantages of reducing the workload of OSIP and accelerating the response to the lock request.

D. Hardware overhead

For supporting spinlock reservation, the OSIP register interface has to be enhanced with additional hardware. As shown above, a set of 14-bit SRRs are introduced, as many as PEs. Also the spinlock control logic needs to be extended. However, since the most computationally intensive part for making reservation decisions is performed in software, the additional control logic (mainly for storing/extracting information into/from SRRs and equivalence comparison of lock IDs) is almost negligible, both from the area and timing perspective. For a 12-PE-system, the largest system considered in this work, the additional hardware has an area of 2.4 Kgate, achieving a maximum clock frequency of 1.6 GHz (synthesized with Synopsys Design Compiler for a 90 nm standard cell library, supply voltage 1.0 V, temperature 25 °C). As a comparison, the OSIP core has an area of 41 Kgate and achieves 613 MHz under the same synthesis condition [11].

The hardware logic for spinlocks (stage 1–3) at the interface can theoretically be completely implemented by software running on the OSIP core. However, this would on the one hand introduce too much workload to the core. On the other hand, the behavior in these three stage requires very low flexibility, which naturally calls for a pure hardware implementation.

V. RESULTS

In this section two applications are studied, on which the spinlock reservation approach is applied. The first application is a synthetic application, in which the task size is parameterizable. It is chosen to see how the proposed approach performs

978-1-4673-2895-1/12 $31.00 © 2012 IEEE

for different task sizes. The second application is a real-life application - H.264 video decoding.

The system simulation platform is created using Synopsys Platform Architect, which contains several instruction-accurate ARM926EJS processor models, the OSIP, a shared memory and several peripherals. All components are connected by a multi-layer AHB bus, enhanced by a cache system with coherence control [15]. The clock frequency is 200 MHz.

A. Synthetic application

The task graph of the synthetic application is illustrated in Figure 6. Three tasks exist in the application: *data-producing* ($T_{produce}$), *task-generation* (T_{gen}) and *data-consuming* ($T_{consume}$). The first two tasks are executed on one processor. They produce data into the shared memory and repeatedly generate $T_{consume}$. The generated $T_{consume}$ tasks are executed on the remaining processors, which consume the produced data and output the result. In the case study, the size of $T_{consume}$ is parameterized to 2.5, 15 and 27.5 Kcycle, representing small, moderate and large tasks, respectively. The number of the consumer processors is also configurable.

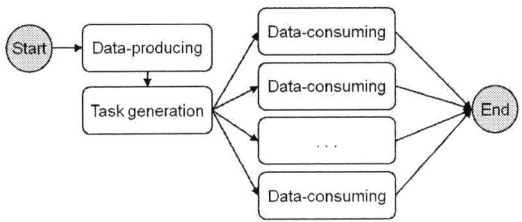

Fig. 6.   Task graph of synthetic application

In the system, $T_{consume}$ and T_{gen} often compete for the same lock for protecting the used standard C library functions. T_{gen} needs a *malloc()* for creating tasks while $T_{consume}$ needs a lock for outputting results to the I/O. Since the GNU C compiler for the ARM processor used in the system internally issues *malloc()* when calling *printf()*, the same lock has to be used for protecting them. Note that it is not always the case when using other compilers. However, this does not influence the generality of this approach.

In general, T_{gen} is much faster than $T_{consume}$. However, if T_{gen} is frequently blocked by $T_{consume}$ due to the competition for the lock (which is unfortunately often the case, since T_{gen} has to compete with multiple $T_{consume}$), the task generation speed will be greatly affected. This has the consequence that the system performance will be reduced because some consumer processors will mostly or even always stay at an idle state for lack of $T_{consume}$ and waste cycles.

In this case, simply introducing a higher task priority to T_{gen} does not help much. A task with a higher priority does not guarantee an earlier lock acquisition, if the grant to the lock request is performed in a random way. Instead, controlling the spinlock with additional application knowledge would lead to more improvement. Figure 7 shows the performance improvement by making a spinlock reservation for T_{gen}. With

this, once the spinlock gets reserved for T_{gen}, T_{gen} will always be the next one that will get the lock after it is released.

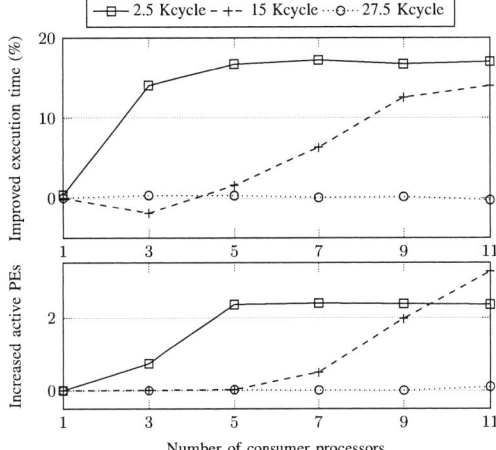

Fig. 7.   Synthetic application: Performance improvement

Two trends can be observed in the upper part of the figure, in which the reduced execution time due to spinlock reservation is presented for different system configurations in terms of different $T_{consume}$ sizes and number of consumer processors. First, the applications with small tasks benefit more from spinlock reservation than those with large tasks. Second, large systems benefit more than small systems. In both cases, the competition for the lock occurs more frequently. This makes the optimization opportunities with spinlock reservation much larger. In contrast, for systems with a small number of processors or large tasks, the performance improvement is only limited. Sometimes it even becomes slightly worse due to additional API overhead and OSIP workload.

The improved spinlock control leads directly to more active consumers in the system, shown in the lower part of the figure. Without unnecessarily waiting for the lock, tasks are now generated more smoothly, which involves more consumers into task execution. This explains the reduced execution time of the application. For small tasks, in most configurations two more consumer processors become active. For moderate tasks in the 11-consumer-system, an even higher average number of increased active PEs is achieved, which lies at 3.3.

B. H.264

A real and more complex application from the multimedia domain – H.264 is chosen as a second case study. It is implemented following the 2D-wave concept [16].

In the application three kernel tasks can be highly parallelized for the macroblocks (MBs) of the video frames: *inverse quantization (IQT)*, *discrete cosine transform (DCT)* and *intra-frame prediction*. The IQT/DCT can be processed for each MB independently, while the prediction for an MB has a dependence on the IQT/DCT of the same MB and on the predictions for the previous neighbouring MBs. If the dependency is not efficiently resolved, the prediction for the MBs will be done at a low parallelism degree, which results in a low frame rate, given in frames per second (fps).

978-1-4673-2895-1/12 $31.00 © 2012 IEEE

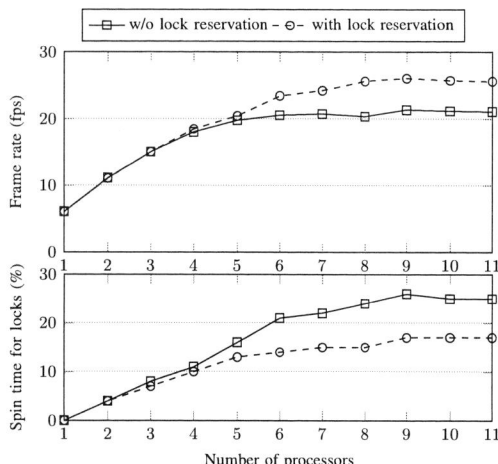

Fig. 8. H.264: Performance improvement

Similarly to the synthetic application, the memory allocation for creating new tasks (both for IQT/DCT and prediction) are often blocked by the I/O accesses for outputting MBs. Therefore, spinlock reservations are made for creating these tasks. Furthermore, the reservation for prediction is assigned with a higher priority, since the parallelism of prediction tasks plays a more important role in improving the performance in our current implementation.

The performance improvement with spinlock reservation is highlighted in the upper part of Figure 8 by comparing the frame rate. Starting from a 6-processor-system, the benefit of reserving the spinlock can be easily observed, reflected by a strongly increased frame rate. Up to a speedup of 1.26 can be achieved. The low improvement for small systems with less processors results from the lower competition for the lock, in which case the optimization opportunities are limited.

An inside look at the reason for the speedup of frame rate is made in the lower part of the figure. Instead of coarsely comparing the number of active PEs, as done for the synthetic application, the polling time for the lock is analyzed. The average spin time of trying to acquire the lock is significantly reduced by reserving the lock, which effectively improves the utilization of the PEs. As a side effect, the communication traffic is also reduced.

VI. SUMMARY AND OUTLOOK

This work introduces a centralized spinlock control mechanism, in which the spinlock is reserved, prior to the attempt of acquisition. The reservation information is user-defined, based on application knowledge. A simple spinlock reservation API pair supports an easy integration of the information into the spinlock control flow. The OSIP, originally designed for task scheduling and mapping, is extended and re-utilized to support this mechanism for making application-aware spinlock decisions. The efficiency and programmability of the OSIP core enable the employment of this mechanism in practice. While the former advantage leaves enough spare time for OSIP to control spinlocks in addition to task scheduling, the latter one provides the programmer a high flexibility to make application-specific decisions. The performance improvement in the presented case studies shows the efficiency of the proposed mechanism in spinlock-intensive applications. Although only experimented on the OSIP-based platforms, we believe that a generalization of the basic concept of spinlock reservation is possible in the systems with a central controller.

The reservation information, also including whether and where to make the reservation, plays a key role in this approach. Currently the information is derived by manually analyzing the application, which requires much tedious work. A tool support for an automatic information analysis will be very helpful to accelerate this design flow.

The current work mainly focuses on a flexible and smart control of spinlocks, instead of considering the traffic overhead caused by polling the locks. It is also interesting to combine our approach with other techniques like spinlock backoff.

REFERENCES

[1] Texas Instruments, "TI OMAP," http://www.ti.com/product/omap3530.

[2] ARM, "ARM MPCore Technology," http://www.arm.com/products/processors/cortex-a/cortex-a9.php.

[3] L. Rudolph and Z. Segall, "Dynamic Decentralized Cache Schemes for MIMD Parallel Processors," in *Proceedings of the 11th Annual International Symposium on Computer Architecture*. ACM, 1984, pp. 340–347.

[4] T. E. Anderson, "The Performance of Spin Lock Alternatives for Shared-Memory Multiprocessors," *IEEE Transactions on Parallel and Distributed Systems*, vol. 1, no. 1, pp. 6–16, Jan. 1990.

[5] J. M. Mellor-Crummey and M. L. Scott, "Algorithms for Scalable Synchronization on Shared-Memory Multiprocessors," *ACM Transactions on Computer Systems*, vol. 9, no. 1, pp. 21–65, Feb. 1991.

[6] C. Yu and P. Petrov, "Distributed and Low-Power Synchronization Architecture for Embedded Multiprocessors," in *CODES+ISSS '08: Proceedings of the 6th IEEE/ACM/IFIP International Conference on Hardware/Software Codesign and System Synthesis*. ACM, pp. 73–78.

[7] X. Chen, Z. Lu, A. Jantsch, and S. Chen, "Handling Shared Variable Synchronization in Multi-core Network-on-Chips with Distributed Memory," in *SOCC '10: Proceedings of the 23rd International SoC Conference*, Sept. 2010, pp. 467–472.

[8] T. Craig, "Building FIFO and Priority-Queuing Spin Locks from Atomic Swap," Dept. of Computer Science, Univ. of Washington, Tech. Rep., Feb. 1993.

[9] T. Johnson and K. Harathi, "A Prioritized Multiprocessor Spin Lock," *IEEE Transactions on Parallel and Distributed Systems*, vol. 8, no. 9, pp. 926–933, Sept. 1997.

[10] M. Monchiero, G. Palermo, C. Silvano, and O. Villa, "Efficient Synchronization for Embedded On-Chip Multiprocessors," *IEEE Transactions on Very Large Scale Integration (VLSI) Systems*, vol. 14, no. 10, pp. 1049–1062, Oct. 2006.

[11] J. Castrillon, D. Zhang, T. Kempf, B. Vanthournout, R. Leupers, and G. Ascheid, "Task Management in MPSoCs: An ASIP Approach," in *ICCAD '09: Proceedings of the 2009 International Conference on Computer-Aided Design*. ACM, 2009, pp. 587–594.

[12] P. Ienne and R. Leupers, *Customizable Embedded Processors: Design Technologies and Applications (Systems on Silicon)*. Morgan Kaufmann Publishers Inc., 2006.

[13] T. Limberg, B. Ristau, and G. Fettweis, "A Real-Time Programming Model for Heterogeneous MPSoCs," in *SAMOS '08: Proceedings of the 8th international workshop on Embedded Computer Systems*, pp. 75–84.

[14] S. Park, D.-s. Hong, and S.-I. Chae, "A Hardware Operating System Kernel for Multi-Processor Systems," *IEICE Electronics Express*, vol. 5, no. 9, pp. 296–302, 2008.

[15] D. Zhang, H. Zhang, J. Castrillon, T. Kempf, B. Vanthournout, G. Ascheid, and R. Leupers, "Optimized Communication Architecture of MPSoCs with a Hardware Scheduler: A System-Level Analysis," *IJERTCS*, vol. 2, no. 3, pp. 1–20, 2011.

[16] C. Meenderinck, A. Azevedo, M. Alvarez, B. Juurlink, and A. Ramirez, "Parallel Scalability of H.264," In MULTIPROG Workshop, Jan 2008.

A Multi-banked Shared-L1 Cache Architecture for Tightly Coupled Processor Clusters

Mohammad Reza Kakoee
DEIS, University of Bologna
m.kakoee@unibo.it

Vladimir Petrovic
Elsys Eastern Europe
vladimir.petrovic@elsys-eastern.com

Luca Benini
DEIS, University of Bologna
luca.benini@unibo.it

Abstract—**A shared-L1 cache architecture is proposed for tightly coupled processor clusters. Sharing an L1 tightly coupled data memory (TCDM) among a significant (up to 16) number of processors is challenging in terms of speed. Sharing L1 cache is even more challenging, since operation is more complex, as it eases programming. The feasibility in terms of performance of shared TCDM was shown in ST Microelectronics platform 2012, but the performance cost of supporting shared L1 cache remains to be proven.**

In this paper we show that replacing TCDM with a multi-banked shared-L1 cache imposes limited speed overhead. Of course, it comes at the cost of area and power. We explore the shared L1 cache architecture in terms of number of processing elements (PEs) and cache banks. Experimental results show that our multi-banked shared-L1 cache can operate with almost the same frequency as that of related TCDM architecture if the cache controller uses a cache line of 4 words. Results also show that, the area overhead with respect to TCDM is less than 18% for a cluster containing 16 Leon3 processors and 32 cache banks. We also show that the overhead on $MIPS/Watt$ and $MIPS/mm^2$ is from 5% to 30% depending on the size of processor in the cluster for a 16x32 configuration (16 cores and 32 cache/memory banks).

I. INTRODUCTION

Main SoC and microprocessor manufacturers are migrating to Multi-processor System on Chips (MPSoCs) and Chip Multi Processors (CMPs) for their latest products. In these devices many processing elements and cores are put together in the same chip and, as Moore's law continues to apply in the multi-core era, we can expect to see a geometrically increasing number of processing elements and memories [2]. Intel has announced "Knights Corner" as the first product based on Intel's Many Integrated Core (MIC) architecture in 22nm which will scale to more than 50 processing cores on a single chip [3]. Previously, Intel also developed a chip prototype [4] that included 80 cores (known as TeraFlops Research chip).

When considering a chip with multiple cores, we should decide whether the caches should be shared or local to each core. Implementing shared cache imposes latency overhead, and more wiring and complexity. However, having one cache per chip, rather than core, greatly reduces the amount of space needed, and thus a larger cache can be included on the chip. This trade-off leads to a multi-level cache hierarchy which generally operates by checking the smallest level one (L1) cache first; if it hits, the processor proceeds at high speed. If the smaller cache misses, the next larger cache (L2) is checked, and so on, before external memory is checked. Most of today's CMPs/MPSoCs (including research designs)

assume private L1 caches and a shared L2. In this architecture providing a consistent view of memory with various cache hierarchies is a key problem. This *cache coherence* problem is a critical correctness and performance-sensitive design point for supporting the shared-memory model. Several works in the literature target cache coherency at both hardware and software levels including interconnect snooping, directory coherence, token coherence, and etc [9], [10]. All these techniques impose performance, power and area overhead. In addition, they need a special interconnect which supports cache coherency protocols.

On the other hand, although sharing L1 cache removes the need for coherency, it is undesirable if the hit latency requires each core runs much slower than a single-core chip. This may happen if the number of cores which share an L1 cache is high and the interconnect between cores and L1 has large latency. Thus, going for a shared-L1 cache is reasonable if the number of cores are limited and the interconnect has very low latency.

Recently, several many-core architectures have been proposed that leverage tightly-coupled clusters as a building block. Examples include the HyperCore Architecture Line (HAL) processors from Plurality [5], ST Microelectronics Platform 2012 [1], or even GPGPUs like NVIDIA Fermi [6]. In a shared memory paradigm, these designs try to overcome the scalability limitations encountered when increasing the number of processing elements (PEs) that share a unique interconnection and memory system [12] by creating a hierarchical design where PEs are clustered into small-medium sized subsystems. The number of PEs inside each cluster is normally less than 16 which makes it possible to design a low-latency interconnect between processors and L1 (in-cluster) memories, while scaling to larger system sizes is enabled by replicating clusters and interconnecting them with a scalable medium like a NoC. However in these systems the shared memory is a tightly coupled data memory (TCDM) which is non-cachable like a scratch-pad memory. This explicitly managed memory can improve performance and is more predictable than cache. However, it comes at the cost of manually managing data transfers and data consistency which can make parallel application development more complex. Some of the architectures like NVIDIA Fermi [6], [7] support shared-L1 cache as an alternative to shared-memory; however, they have put limitations on either hit latency (more than 2 cycles) or cache bandwidth. Moreover, they have not explained the detail of their shared-L1 cache architecture [7].

978-1-4673-2895-1/12 $31.00 © 2012 IEEE

Contribution. In this work we propose a shared-L1 cache architecture which similarly to the above designs takes advantage of tightly coupled clusters as a building block. However, instead of TCDM we propose a multi-banked shared-L1 cache inside a cluster. Each cluster contains several processing elements communicating with the shared cache through a low latency interconnection network. The main contribution is to show that a shared L1 cache is feasible for large tightly coupled clusters where the number of processing elements is beyond (up-to 16) what can be achieved with basic Symmetric Multiprocessing (SMP) technology (i.e. 4).

We explore this architecture in terms of number of PEs and cache banks. Clearly, L1 processor-to-memory interconnects must provide a huge bandwidth, coupled with ultra-low latency. For the interconnection network we use the architecture proposed by Rahimi et.al in [8]. They propose a fully combinational Mesh-of-Tree (MoT) interconnection network which is suitable for tightly-coupled processor clusters. An enhanced version of this network is implemented in ST Microelectronics Platform 2012 on 28nm technology [1]. This network provides single-cycle transfer from processor to memory and round-robin arbitration for a fair access to memory banks, as well as fine-grained address interleaving to reduce memory bank conflicts. Since this network is fully combinational, it is vulnerable to timing and delay variations. Authors in [14] propose a variation-tolerant mechanism to make this network reliable in presence of delay variations.

Experimental results show that our multi-banked shared-L1 cache can operate with almost the same frequency as that of equally sized TCDM. Synthesis results also show that the area overhead of our architecture compared to that of TCDM is 18% for the 16x32 configuration (16 LEON3 cores and 32 cache/memory banks).

II. ARCHITECTURE

A. Multi-banked shared cache

In this section we describe our multi-banked shared-L1 cache architecture. This architecture is suitable for tightly-coupled processor clusters. Each cluster contains several PEs which share a multi-banked L1 cache. The cluster architecture is fixed and scaling is enabled by replicating clusters. The number of PEs inside each cluster is large enough (more than 4) so that we cannot efficiently develop an L1 cache coherency (for instance based on snooping) mechanism if we go for a private-L1/shared-L2 architecture. It is also small enough (up-to 16) which makes it feasible to design a high performance and low latency interconnection infrastructure for communication between cores and the shared cache. Figure 1 shows an abstract view of this interconnect which connects 4 cores to 8 banks of caches.

The interconnection network supports non-blocking communication between the processing elements (PEs) and cache banks (CBs), within a single clock. As shown in Figure 1, a combinational path is created through a network of primitive building blocks: routing primitives (circles blocks) and arbitration primitives (square blocks). The former are used to create independent routing paths (routing trees) from the

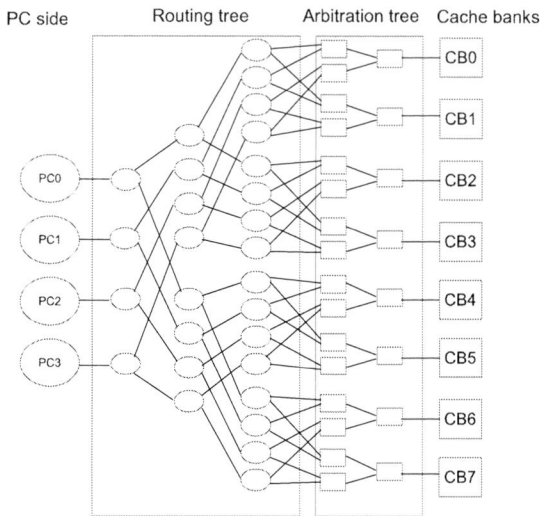

Fig. 1. Log. network 4x8: empty circles represent routing switches and empty squares represent arbitration switches.

PCs to the arbitration tree (and vice-versa). The latter are used to arbitrate concurrent requests (arbitration tree) and to route them up to the CBs ports and vice-versa.

The routing tree consists of simple routing switches which route each packet from processor side to memory side, and vice versa. The packets are routed individually based on packets address field. The switch has two directions: forward (PC ports) which sends out the incoming packet form its input port at processor side to one of its output ports at the cache side; backward which rolls packet back from cache side to processor side (CB ports). The forward packet contains address, data write, and control signal of cache while the backward packet contains the read data, and acknowledgment signal.

During a read/write operation, data and control signals are asserted by PCs. These signals are routed through routing switches, until they reach one of NxM ports of routing tree. In order to reach a cache bank the packet must be arbitrated among the other simultaneous requests for the same cache bank. After passing through all levels of arbitration switches, the request reaches the cache bank, and the read/write operation can be performed.

Request routing and arbitration are performed in a combinational way by using request and acknowledgment signals for arbitration, and address for routing across the switches. Once the request reaches the last level of the arbitration tree and gets the grant, a valid acknowledgement is asserted and propagated back to the related PC through the routing switches (backward). By receiving the acknowledgment signal, PC is able to issue the next read/write operation at the next clock cycle, otherwise it waits for ACK to be received.

In the following section we will briefly describe the internal architecture of a single cache bank.

978-1-4673-2895-1/12 $31.00 © 2012 IEEE

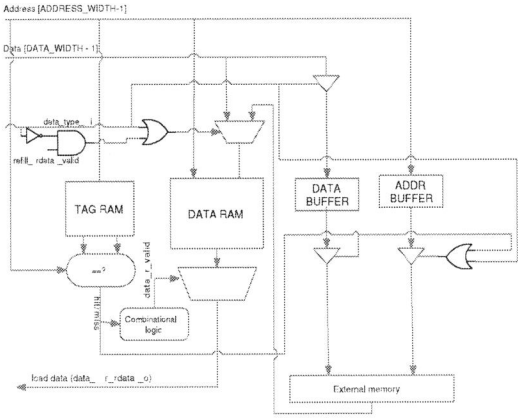

Fig. 2.   Block diagram of a single cache bank.

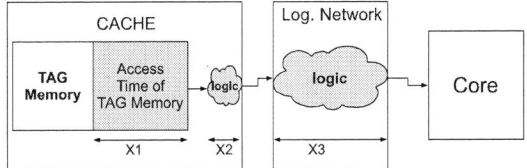

Critical Path of Multi-banked Cache = X1 + X2 + X3

X1: Access time of TAG memory
X2: Comparator and combinational logic for Hit/Miss check
X3: Response network in Logarithmic network

Fig. 3.   Critical path of our multi-banked cache architecture.

B. Cache bank

For this work, we designed and developed a simple write-through cache controller. Figure 3 shows its block diagram. Like any other cache, our simple cache has separated memory blocks for data and tags. It is a direct mapped cache with parameterizable multi-word cache line. The cache also has a write-through buffer allowing us to temporary store data while the data is waiting to be written into the next level memory. We have chosen write-through mechanism as our write strategy because it provides better support for multi-cluster consistency models; write-back strategy is under development for future work.

From the processor side the cache controller accepts requests which are either read or write to the memory. All cache modules are connected to the slave side of logarithmic network. Therefore, every processor which is connected to the master side of this network can send request to every cache. Cache controller gets requests from the network and sends response back to the processor through the response path of the network. If the request is write, cache takes the data from the data input and stores it into the data memory at the address received from the processor and updates tag memory. Also, the same data is stored in the write-through buffer in order to be forwarded and written into the external memory. If the request is read and it hits the cache, related data will be sent to the processor at the next cycle. However, if a cache miss occurs on the read request, we first stall the processor. Then, the cache controller first writes the content of write-through buffer (if it is not empty) into the next level memory and then performs a refill from the external memory. Having completed the refill phase, the related data is sent to the processor and the stall signal goes down.

C. Critical Path

To clearly compare our multi-banked L1-cache with TCDM, we carefully analyzed the timing paths of both architectures. Based on our analysis, the request paths which start from processor, goes through the network and ends at the memory/cache banks are the same in both architectures. Therefore,

we do not have any overhead on the request path compare to that of TCDM. However, the response path related to stalling the processor in case of miss is new in multi-banked cache. Based on our analysis this path is the critical path of our architecture which is shown in Figure 3.

The path shown in Figure 3 does not exist in TCDM, but there is a similar response path in TCDM which starts from the memory bank, goes through the network and ends at the processor. Comparing to the response path in TCDM, the critical path of our multi-banked L1-cache has the overhead of logic path inside the cache controller for checking the tag's data against the request address to determine miss/hit (X2 in Figure 3). However, since the data memory is usually larger than tag memory, the delay related to access time of data memory in TCDM is more than that of tag memory in the cache module (i.e. X1 in Figure 3). Therefore, if we select a small and high speed memory for tags, the delay of the critical path in our architecture becomes close to that of TCDM. We should note that by increasing the cache-line we can reduce the size of tag memory resulting in a lower access time. We will explore this in more details in the experimental results section.

III. EXPERIMENTAL RESULTS

In this section, we discuss the experimental results for the multi-banked shared-L1 cache architecture in terms of delay, power, and area. We quantify the cost of replacing memory banks in TCDM with cache banks for several network cardinalities. To get these results, we synthesized the network and cache modules with the Farady 65nm technology library using Synopsys Design Compiler. In TCDM we considered 8KB memory for each memory module. For multi-banked cache, we put the same data memory (8KB) for each cache bank and we explored the tag memory from 512x20 bits (cache line of 4 words) to 2048x20 bits (cache line of 1 word).

A. Timing Analysis

As described before, the main overhead of replacing TCDM with cache banks on critical path is the response path which starts from tag memory and ends at the processor side. This is the critical path of our architecture. This path contains the access time of tag memory plus the logic inside cache controller to determine hit/miss plus the network's response path. tag's access time depends on the type and size of the

978-1-4673-2895-1/12 $31.00 © 2012 IEEE

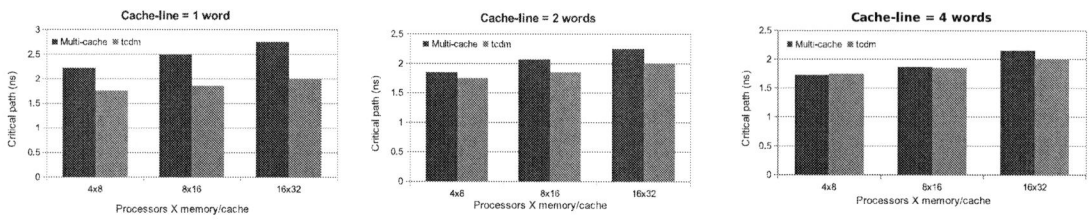

Fig. 4. Timing (ns) analysis of multi-banked cache Vs. TCDM.

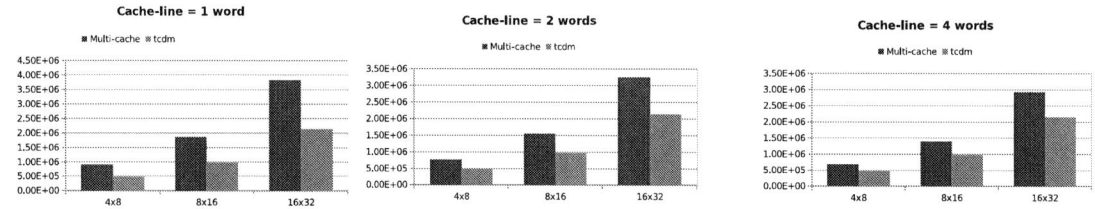

Fig. 5. Area (um^2)of multi-banked cache Vs. TCDM.

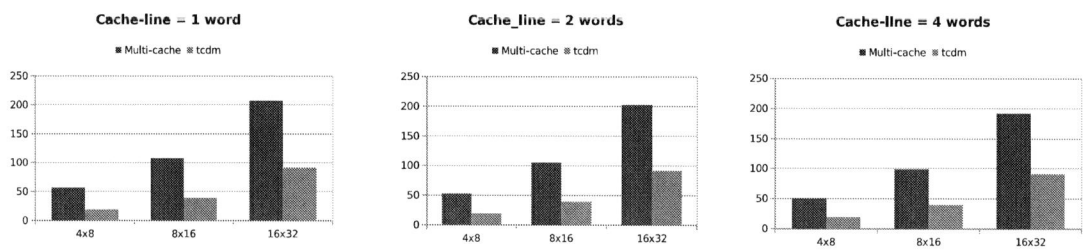

Fig. 6. Power (mW) consumption of multi-banked cache Vs. TCDM.

memory. The wider the cache line the smaller the tag memory. We performed experiments on the architecture with different cache lines and compared the critical path of our design with that of TCDM for 3 different network cardinalities including 4x8, 8x16 and 16x32. Figure 4 shows the related charts for timing analysis.

As it is shown in Figure 4, for all network cardinalities, when the cache line increases the difference between critical path of TCDM and that of our architecture decreases due to the smaller access time of the tag memory. It is clearly seen that for the cache line of 4 words the delay of the critical path in multi-banked cache is almost the same as that of TCDM. For the configuration of 4x8 and cache line of 4 words, the critical path of our architecture is even better than that of TCDM. This is due to the fact that in this configuration delay saving in access time of tag memory compare to the access time of data memory is more than the additional delay imposes by hit/miss logic in cache controller.

B. Area and Power Analysis

We also performed synthesis experiments to compare the power and area of our multi-banked cache architecture with those of TCDM. Figure 5 shows the charts related to area com-

parison. It is clear that the area of multi-banked cache is more than that of TCDM (2× for cache line of 1 word and 1.5× for cache line of 4 words) due to extra memories for tags, write-through buffer and cache controller itself. However, the area overhead decreases when the cardinality of the network and the cache line increase. The area overhead for the cardinality of 16x32 and a cache line of 4 words is 36%. We should note that, in these experiments we considered only the area of the network and the cache/memory, but in a real system this is a portion of the whole design. In other words, if we consider the area of a whole cluster including processors then the percentage overhead significantly reduces. For instance, considering Leon3 which is a small-medium sized processor [13] with an area of $0.127mm^2$, the area overhead of multi-banked L1-cache with respect to the TCDM is 18% for the 16x32 configuration. The percentage overhead decreases even more if we consider the area of other components inside a cluster like the peripheral network, network interface, cluster controller and etc.

Figure 6 shows the charts related to the power consumption comparison. As can be seen, the trend for the power consumption is different from that of area. This is due to the power overhead in the write-through buffer of our cache module. In

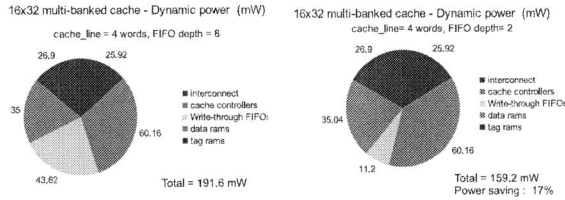

Fig. 7. Break-down of dynamic power consumption in multi-banked cache.

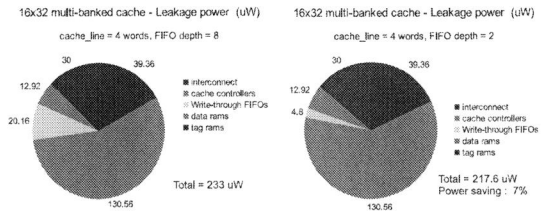

Fig. 8. Break-down of leakage power consumption in multi-banked cache.

these charts the write-through buffer for our cache has a width of 8x64 bits. However, if we reduce the depth of this buffer to 4 or 2 the power overhead decreases.

Figures 7 and 8 show the dynamic and leakage power breakdown of multi-banked cache with a cardinality of 16x32. It is shown that around 44% of dynamic power and 21% of static power is consumed in write-through buffers when their depth is 8. However, it can be seen in the right-side of the figures that if we reduce the depth of write-through FIFO to 2, we can save up-to 17% dynamic power and 7% static power consumption for the 16x32 configuration.

We also performed experiments to see the overhead of shared-L1 cache on $MIPS/mm^2$ and $MIPS/watt$ with respect to those of TCDM when we consider processor's area and power. When considering the processor, the power and area overhead will be heavily reduced in percentage terms, while speed remains the same; therefore, the $MIPS/Watt$ and $MIPS/mm^2$ of L1-cache with respect to those of TCDM complete cluster are not so low. We performed this experiment for various processor areas including that of LEON3. Figure 9 shows the $MIPS/watt$ and $MIPS/mm^2$ of L1-cache normalized to those of TCDM for the 16x32 configuration and for different processor areas. As can be seen, as the area of the processors attached to the cluster increases, the overhead of cache with respect to TCDM on $MIPS/watt$ and $MIPS/mm^2$ decreases. This overhead is between 5 and 30 percent depending on the processor size.

IV. CONCLUSIONS

We proposed a shared-L1 cache architecture for tightly coupled processor clusters. Inside each cluster a number of processing elements communicate with the shared multi-banked cache through a low latency interconnection network. We explored this architecture in terms of number of processing elements (PEs) and cache banks and analyzed overheads in terms of timing, power and area. Our experiments demonstrate

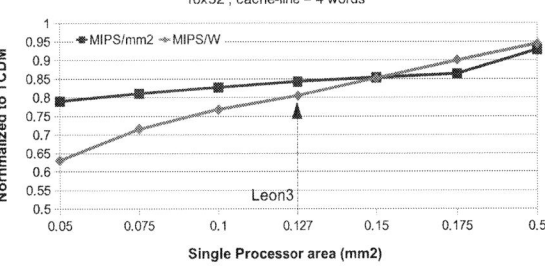

Fig. 9. $MIPS/mm^2$ and $MIPS/Watt$ in 16x32 shared-L1 cache architecture with different processor area attached to the cluster.

that the multi-banked shared-L1 cache can operate with almost the same frequency as an equally sized TCDM architecture if the cache controller uses a cache line of 4 words. We also showed that when considering a complete cluster, the overhead of shared-L1 cache on $MIPS/Watt$ and $MIPS/mm^2$ with respect to TCDM is relatively small for a configuration of 16 medium-sized processors and 32 cache banks. However, we should note that although shared-L1 cache has almost no overhead on the speed with respect to TCDM, architects should also account for the loss in $MIPS/mm^2$ and $MIPS/Watt$ (5%-30% depending on the processor size) coming with the cache.

ACKNOWLEDGMENT

This work is supported by projects: FP7 Virtical (CA 288574) and ERC Multitherman (CA 20110209).

REFERENCES

[1] L. Benini, E. Flamand, D. Fuin, D. Melpignano, "P2012: Building an ecosystem for a scalable, modular and high-efficiency embedded computing accelerator," Design. Automation & Test in Europe Conference & Exhibition (DATE), 2012 , pp.983-987, 2012.
[2] ITRS Home-2010, 2010 International Technology Roadmap for Semiconductors. [Online]. Available: http://www.itrs.net/home.html
[3] George Chrysos, "Knights Corner, Intel's first Many Integrated Core (MIC) Architecture Product," Hotchips 24, Cupertino, CA, August 2012.
[4] http://techresearch.intel.com/articles/Tera-Scale/1449.htm.
[5] Plurality Ltd. The HyperCore Processor. www.plurality.com/hypercore.html
[6] C.M. Wittenbrink, E. Kilgariff, A. Prabhu, "Fermi GF100 GPU Architecture," IEEE Micro, vol.31, no.2, pp.50-59, March-April 2011.
[7] J. Nickolls, W.J. Dally, "The GPU Computing Era," Micro, IEEE , vol.30, no.2, pp.56-69, March-April 2010
[8] A. Rahimi, I. Loi, M.R. Kakoee, L. Benini, "A Fully-Synthesizable Single-Cycle Interconnection Network for Shared-L1 Processor Clusters ," in Proc. of the ACM/IEEE DATE, 2011.
[9] M.R.Marty, "Cache coherence techniques for multicore processors," PhD Dissertation,University of Wisconsin - Madison, 2008.
[10] Yuang Zhang; et.al , "Towards hierarchical cluster based cache coherence for large-scale network-on-chip," in DTIS , pp.119-122, 2009
[11] D. Hackenberg, D. Molka, W.E. Nagel, "Comparing cache architectures and coherency protocols on x86-64 multicore SMP systems," Microarchitecture, 2009. MICRO-42. 42nd Annual IEEE/ACM International Symposium on , pp. 413-422, 2009
[12] Tilera Corp. Product Brief. Tilepro64 processor. 2008.
[13] http://www.gaisler.com
[14] M.R. Kakoee, I. Loi, L. Benini, "A resilient architecture for low latency communication in shared-L1 processor clusters," Design, Automation & Test in Europe (DATE), 2012 , pp.887-892, March 2012.
[15] S. Mitra, K. Brelsford, Kim Young Moon, Lee Hsiao-Heng Kelin, Li Yanjing, "Robust System Design to Overcome CMOS Reliability Challenges," Emerging and Selected Topics in Circuits and Systems, IEEE Journal on , vol.1, no.1, pp. 30-41, March 2011

978-1-4673-2895-1/12 $31.00 © 2012 IEEE

An Automated Framework for the Simulation of Mapping Solutions on Heterogeneous MPSoCs

Antonio Miele, Christian Pilato, Donatella Sciuto

Politecnico di Milano - Dip. Elettronica e Informazione

P.zza L. da Vinci, 32 - I20133 Milano - Italy

{miele|pilato|sciuto}@elet.polimi.it

Abstract—The efficient design of mapping solutions for heterogeneous Multi-Processor Systems-on-Chip (MPSoCs) is usually a challenging task in system-level design, in particular when the architecture integrates hardware cores. This paper proposes a SystemC simulation framework for fulfilling this task featuring (i) an automated flow for the generation of SystemC timing models for the hardware cores starting from the application source code, and, in a second step, (ii) the possibility to simulate different mapping solutions simply by changing an XML descriptor and without any interaction of the designer. The proposed framework has been then applied to a case study considering an image processing application to present the possibility to integrate it as an evaluator in design space exploration environments.

I. Introduction

Nowadays, heterogeneous Multi-Processor Systems-on-Chip (MPSoCs) are the de-facto standard for embedded system design [1] where general purpose processors are connected also with digital signal processors and hardware accelerators to speed up the applications. Platform-based design [2] has been established to tackle the design complexity, but, even when MPSoCs have already been designed, different issues are still open to meet the design requirements, especially when the customization with hardware accelerators needs to be investigated. In such a scenario, the Y-chart design methodology [3] (shown in Fig. 1) is one of the most popular approaches for the design space exploration devoted to the system-level synthesis of MPSoCs: starting from an application and an architectural platform, the *tasks* are assigned to the architectural resources (*mapping*); then, a *performance analysis* of the identified solution is carried out and, finally, the obtained *metrics' values* (e.g., performance, power consumption, area) are analyzed to give a feedback for improving the solution.

In this work, we focus on the evaluation and comparison of different mapping solutions. Besides analytical approaches, based on worst case execution time estimation and static scheduling [4], the designer may require additional information by means of simulations, at different abstraction levels, especially when the actual platform is not implemented yet. These simulators can consider either SystemC performance models (e.g., [5]) or slower but more accurate ISS-based Instruction Set Simulators (ISSs) (e.g., [6], [7], [8], [9]). However, all these approaches present several issues, especially in the case of hardware accelerators. Indeed, in most of the cases, the generation of the models for the hardware accelerators and

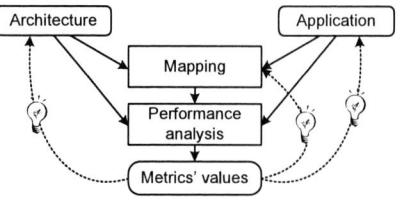

Figure 1. Different phases of the design space exploration.

their interfacing with the application is not automated; thus, these approaches require a manual interaction of the designer.

This paper presents an automated framework for the evaluation of application's mapping solutions on heterogeneous MPSoCs. The framework is composed of an enhanced version of a state-of-the-art simulator called ReSP [7], where MPSoC's components are modeled with SystemC and Transaction Level Modeling (TLM, both available on [10]), and a tool for high-level synthesis (HLS) called Bambu [11]. Starting from the application source code, the synthesis tool allows to automatically generate SystemC models with an accurate timing for the hardware accelerators and integrate them in the ReSP component library; this step is performed only once at the beginning. Then, ReSP can be used for evaluating different mapping solutions simply by changing the list of tasks mapped on hardware. Indeed, our approach does not require any modification in the application source code for interfacing with the hardware components, thanks to the exploitation of the OS *system-call emulation* facilities provided by ReSP. The framework can be seen as a generic environment where any algorithms for an automated HW/SW partitioning and mapping can be integrated. The novel contributions of the presented framework are: (i) a novel SystemC hardware module within ReSP for the simulation of parallel hardware threads; (ii) an extension to Bambu for the generation of the high-level TLM descriptions starting from the Register-Transfer Level (RTL) implementations obtained by the HLS; (iii) an automated infrastructure to make the two tools cooperate.

The rest of the paper is organized as follows. Section II overviews the past work on the MPSoC simulation. Section III presents the proposed framework that is then detailed in Section IV and Section V. In Section VI, the framework has been validated on a case study considering an image processing application and, finally, Section VII concludes the paper and outlines the future directions.

978-1-4673-2895-1/12 $31.00 © 2012 IEEE

II. RELATED WORK

The literature presents a large set of simulation environments for MPSoCs. Most of these simulators are based on SystemC and TLM, which are the current standard for the modeling and simulation of embedded systems.

StepNP [6] is an example of SystemC simulator, even if TLM interfaces are not supported. Synopsys Platform Architect [8] offers the possibility to automatically generate the Instruction Set Simulator (ISS) from the specification of the Instruction Set Architecture (ISA). In both these solutions, a considerable effort is required for the integration of new custom ones and the implementation of the hardware drivers. Furthermore, none of these platforms provides any emulation mechanism for the operating system (OS). This requires to load on the ISS models a real OS supporting software concurrency management to execute a parallel software, dramatically worsening the simulation time. SoCLib [12], differently from previous approaches, is integrated with Gaut [13], a tool for HLS of custom hardware accelerators. Again, SoCLib does not support OS emulation and the generated hardware accelerator has to be manually interfaced with the executed software. To avoid to execute a real OS on the simulated processors, several ISSs (e.g., Sim-It ARM [9] and ArchC [14]) feature a *system-call emulation* mechanism. This monitors the ISS instructions to intercept the calls to OS routines that are then executed in a specific module, directly on the host machine. This allows to execute the application without any modification of its source code, allowing preliminary analysis during the simulations in the early stages of the design process. However, such ISSs can be hardly extended with new hardware models and very few of them support the execution of parallel applications.

ReSP [7] is another simulator featuring an easily integration of new components based on TLM and a tool for the generation of ISS starting from the ISA. Moreover, it provides an advanced system-call emulation mechanism supporting also the execution of parallel applications, based on POSIX threads (or *pthread*), and calls to hardware accelerators specified by the designer. However, the emulator features only the blocking call paradigm for the execution of hardware functions that cannot be thus executed in parallel. Finally, data transfers between the hardware accelerators and the memories are directly managed by the emulator, without modeling the actual traffic on the communication infrastructure and thus implying a loss in timing accuracy. Thus, it cannot fully support the execution of parallel applications mapped both in hardware and in software. As a final note, other simulation environments (e.g. gem5 [15]) include a system-call emulator, however not completely supporting all the features considered in ReSP.

In conclusions, there is no past solution that integrates all the capabilities to provide an effective design mapping evaluation. Therefore, to implement the proposed evaluation framework, we enhance ReSP with a new emulation facility overcoming the discussed limitations. Moreover, as in SoCLib, we also integrate a HLS tool for an automated generation of the hardware modules from the application under analysis.

III. PROPOSED EVALUATION FRAMEWORK

The proposed framework for evaluating mapping solutions onto a heterogeneous MPSoC is shown in Fig. 2. It receives the following inputs: (i) the source code of the application, in C annotated with OpenMP [16] pragmas to describe the task partitioning; (ii) an XML file containing the list of functions that can be potentially mapped in hardware; (iii) the ReSP descriptor of the target architectural platform (in *python* language); (iv) the XML descriptors of the mapping solutions to be simulated. Then the output of the framework is the metrics' values (e.g., performance or power consumption) and the statistics computed during the simulations. The framework is composed of two different parts: the generation of components necessary for the simulation and the actual evaluation of the mapping solutions. The resulting framework is particularly interesting to be adopted as an evaluator of a design space exploration process (as shown in Fig. 1). In fact, by automatically generating all the simulation models in a single initial step, we can reduce the set-up time without requiring any effort to the designer, and thus perform a rapid evaluation of alternative solutions by reusing a single simulation model.

During the first step, a HW/SW co-design framework called PandA [11] is used to extract the task graph of the application from its C source code annotated with the OpenMP pragmas. We assume that the mapping is performed at function level; thus, the input source code is organized in functions, where each of them can potentially represent a task. Then, the application source code is regenerated by adopting the *pthread* API and it is cross-compiled for the target processors. Finally, each hardware functions in the XML list is synthesized and, consequently, a SystemC model is generated.

The second part of the framework consists of the ReSP simulation environment [7]. ReSP instantiates the SystemC TLM architecture specified in the *python* descriptor received in input, and configures the hardware accelerators according to the XML descriptor of the mapping solution. Finally, the simulation is executed, and the required metrics' values and statistics are returned.

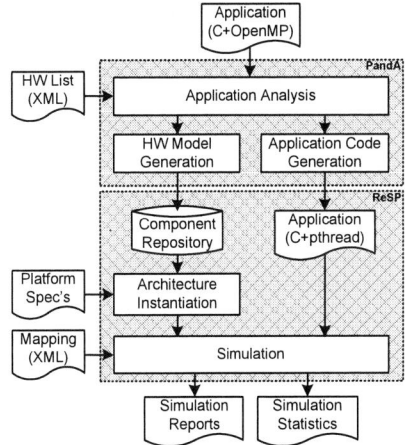

Figure 2. Overview of the proposed evaluation framework.

978-1-4673-2895-1/12 $31.00 © 2012 IEEE

IV. THE ADOPTED SIMULATION ENVIRONMENT: ReSP

ReSP ([7], downloadable from [17]) has been enhanced to be adopted in the proposed evaluation framework. The template architecture simulated by ReSP (shown in Fig. 3) is composed of a set of master processing units, usually ISS models, connected through a communication channel to a set of slave components, such as data/instruction memories, reconfigurable modules or specific accelerators; optionally, master processing units can be provided with local caches featuring a mechanism for global data coherency.

In order to simulate hardware/software parallel applications, we enhanced the basic architecture and the related emulation facilities with a new SystemC model of reconfigurable hardware module acting as a master processing unit and integrating the emulation of hardware threads directly in the *concurrency manager* within the system-call emulator, as shown in Fig. 3. The defined hardware module acts on the communication infrastructure as a master processing unit directly accessing the bus infrastructure to exchange data with the memory; moreover, it contains the mechanisms to simulate reconfigurable capabilities (if required), in particular according to the model of the Xilinx FPGAs [18]. This model allows to register a set of executable functions at design time (i.e., the XML list specified as input for the overall framework). At design time, the source code of the hardware function is generated according to a specific template (further details are provided in Section V-B) and compile in a specific library. Then, when instantiating the system to be evaluated, the designer only provides an XML file containing the list of functions mapped in hardware, without any change to the application code.

The execution of a hardware function can be invoked by the application in two different ways: a blocking call and a non-blocking one. The *blocking call*, inherited by the previous version and improved, can be used for simulating sequential tasks. The behavior is shown in Fig. 4: when a call of a function mapped in hardware is trapped by the system-call emulator, the request, together with the related parameters, is forwarded to a hardware module capable to execute the task. If the module is busy, the request will wait the end of the on-going computations to be executed. Thus, the processor is blocked until the end of the hardware function; when the required computation ends, the result are sent back to the processor which resumes the execution.

A *non-blocking call* has been introduced to simulate parallel threads mapped in hardware; the concurrency manager has been consequently enhanced to support their execution. As shown in Fig. 5a, when a thread creation is trapped, the request is forwarded to the concurrency manager, while the processor resumes its execution. Thus, the concurrency manager creates the new hardware or software thread according to the mapping solution, and schedules it on the required resource by using the selected scheduling policy (e.g., first-in-first-out or round-robin). Note that differently from software threads, the hardware counterparts cannot be preempted in case the round-robin policy is used. Then, when the software thread executes the join function (Fig. 5b), the call is trapped and forwarded to the concurrency manager. Thus, the software thread is blocked and, if other threads are ready, the concurrency manager will schedule one of them on the processor. Finally, when the hardware threads return, the parent software thread is unblocked and thus resumed on a free processor.

When considering a complex multi-processor system, it is necessary to arbitrate conflicts in the execution of hardware functions, such as the blocking calls, since the non-blocking calls are already managed by the concurrency manager. Thus, we defined the following policies. First, in case different modules exposes the same required function, the request will be forwarded to the first available module. Then, first-in-first-out queues are used for managing waiting blocking calls on each hardware module in order to be executed in the arrival order. Moreover, in case all hardware modules are busy, the blocking call is forwarded to the first hardware module with the shortest queue of waiting calls. Finally, blocking calls have a higher priority than non-blocking ones. It is worth noting that we defined very generic policies, aiming at performing the proof-of-concept experimental sessions discussed next; they can be changed according to the designer needs.

V. GENERATION OF THE SIMULATION MODELS

The analysis of the input application and the generation of the simulation models for ReSP have been automated by means of a set of tools developed in a HW/SW co-design framework called PandA [11]. The proposed approach starts from the input application, in C language with OpenMP pragmas, where each function can potentially represent a task to be executed by any of the platform resources. It is also required the XML list of functions that can be potentially executed in hardware, as shown in Fig. 6. Then, it generates the SystemC models for the functions in the XML list to be integrated in the ReSP's component repository and the application code with *pthread* API instead of OpenMP pragmas for the simulation.

Note that, even if ReSP can simulate applications with OpenMP pragmas, this is not suitable for simulating mapping solutions in the proposed environment. In fact, in the common implementations of `libgomp` (i.e., the library that translates OpenMP pragmas into *pthread* API), it is not possible to maintain a direct 1:1 relationship among OpenMP sections and threads during the simulation. For this reason, it is necessary to explicitly translate the OpenMP pragmas into calls to the *pthread* API function, as further described in Section V-A.

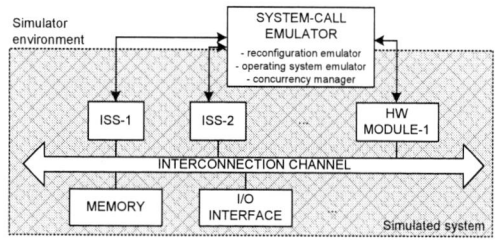

Figure 3.   Architecture of the multi-processor system simulated by ReSP.

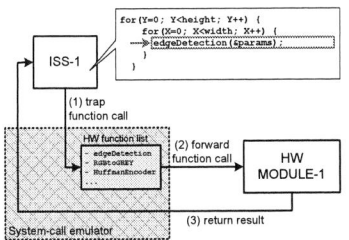

Figure 4.   Behavior of the blocking call of a HW function.

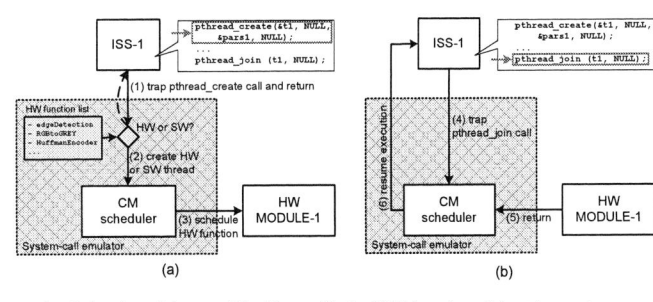

Figure 5.   Behavior of the non-blocking call of a HW function: (a) `pthread_create()` and (b) `pthread_join()`.

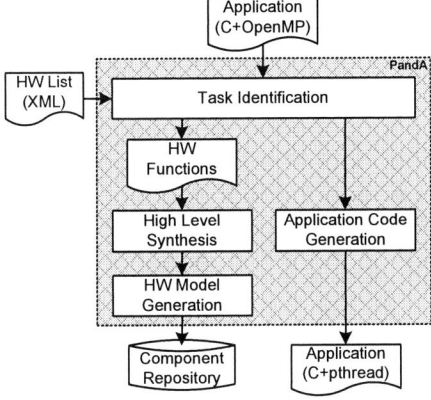

Figure 6.   Details of the steps to generate the simulation models.

For the sake of simplicity, we restricted the set of supported pragmas only to the `#pragma omp parallel sections` (starting a set of parallel tasks) and the `#pragma omp section` (starting one of these parallel task). Other pragmas, such as the `#pragma omp for`, should be converted into these supported pragmas to have the number of parallel sessions to be started known at compile time. Moreover, it is assumed that the body of each parallel `section` is only composed of the call to the function representing a task. In this way, function calls within parallel `sections` represent the execution of parallel tasks (implemented through the non-blocking paradigm), while stand-alone function calls represent sequential tasks (implemented through the blocking paradigm).

Concerning the hardware accelerators, we adopted a high-level synthesis tool called Bambu [11] to generate the RTL implementation starting from the functions' code. The tool allows to generate the accurately-timed models of the hardware functions and also to identify for each of them several implementations with different area/time trade-offs.

A. Application Code Generation

A simple back-end tool has been implemented into the PandA framework for the rewriting of the application source code. As described above, each task is represented with a function; moreover, the application contains parallel tasks, which invocation is not blocking up to the barrier, and sequential ones, where the execution cannot continue until the task accomplishment. For the parallel ones, as described in Section IV, we adopt the `pthread` API to specify that the

function has to be considered as non-blocking, and, for the sequential ones, we can just invoke the function that will be thus considered as blocking by the system-call emulator (if mapped in hardware). It is important to remember that we aim at exploring different function mapping (either hardware or software); moreover, the model of the hardware core has to have the same interface for blocking and non-blocking calls. For this reason, coherently to the `pthread` API, we defined a common C prototype for the function that can be potentially accelerated, that is constituted by a pointer to a data structure (in memory) where all the actual parameters have been stored. This allows it to be compliant with `pthread` API for passing the parameters to the threads and it has been adopted as a common interface for all the cores that can be potentially executed in hardware. Therefore, the defined back-end tool rewrites the source code in order to explicit the parameters passing for each function potentially accelerated in hardware.

Fig. 7 shows an example of how the OpenMP pragmas (on the left) are translated into calls to the `pthread` API (on the right). Given the function prototype (box **1**), the data structure corresponding to the parameters is generated (box **A**). Then, we update the function interface to represent the parameters of the function just as a `void` pointer (box **B**), to be properly managed inside the function body; we also update the corresponding implementation to retrieve the values from this data structure and use them as like as the original parameters. Then, the function calls are updated to fill the data structure with the actual parameters and to invoke the function itself (blocking calls) or the pthread's functions to dispatch the threads (non-blocking calls). In the former case, considering the following call in box **2**, the corresponding blocking call is shown in box **C**. In the latter case, in case of OpenMP pragmas (box **3** and box **4**), the calls have to be reproduced as non-blocking calls, where the application only stops at the barrier (represented in box **5**). Box **D** and box **E** create and fill the data structures for the two calls, invoking the `pthread_create` functions to dispatch the two tasks to the concurrency manager. Indeed, the two threads are executed up to the `pthread_join` functions (box **F**) and, then, after their completion, the return values are retrieved.

B. Hardware Model Generation

The second step of the generation of the simulation models is the implementation of the hardware functions to be registered in the hardware module presented in Section IV. For

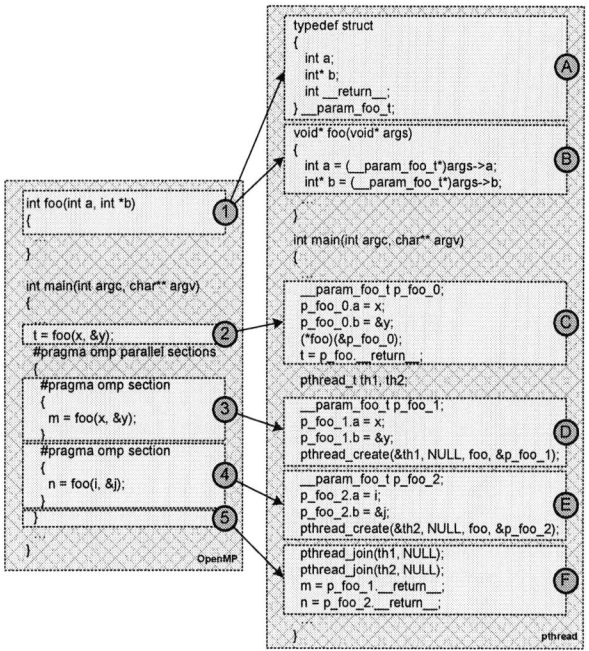

Figure 7. Example of source code transformation from OpenMP pragmas (on the left) to pthread API (on the right).

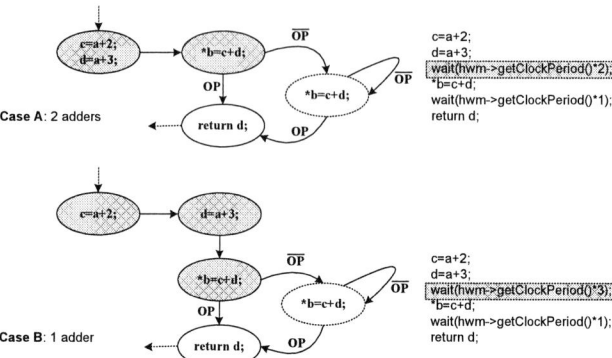

Figure 8. Different hardware implementations will correspond to different STGs and then different hardware models.

this purpose we adopted an enhanced version of Bambu, a high-level synthesis framework that allows to obtain a hardware specification fully equivalent to the software counterpart. Bambu generates an internal representation of the synthesized RTL circuit based on a State Transition Graph (STG); then, a new back-end of the synthesis tool is used for the regeneration of the SystemC TLM implementation of the synthesized hardware functions. For this purpose, the timing information of the synthesized function are directly annotated in its algorithmic description by computing the delays according to the clock period and invoking the SystemC wait() function. Then, the defined hardware module provides two functions for read/write accesses on the bus, that are the read_word() requiring the desired memory location, e.g.:

```
int a = (int) hw_module->read_word(args + 0);
```

and the write_word() requiring the target address and the value to be written, e.g.:

```
hw_module->write_word(args + 0, value);
```

When generating the hardware models by means of Bambu, we have to take into account two different aspects. First, it is possible to generate different area/performance trade-off (e.g., by varying the number of available resources) and timing characteristics have to be adequately considered. Second, the SystemC model can be implemented at different abstraction levels, i.e., with different timing accuracies, from pure functional to cycle-accurate ones, very accurate with respect to the internal RTL implementation. We adopted the loosely-timed modeling approach defined in TLM for the internal timing characterization of the hardware function, but the extension to different models is straightforward since ReSP is able to simulate also at different abstraction levels. It only requires to specify the proper models for the architectural components

(processors, memories, etc.) and, then, generate the hardware functions with the proper timing accuracy.

We defined an approach similar to [19] for abstracting the synthesized circuit to such a loosely-timed model. In particular, this requires to represent the delay of the memory operations and the number of clock cycles between the memory accesses. Thus, we based our model on the STG resulting from the synthesis, with timing annotations at each branch point, since it effectively represents the implementation behavior at each clock cycle. We can thus obtain the code after the synthesizer's compiler optimizations and, then, write back the exact implemented behavior. Moreover, the STG contains information about the operations executed in each clock cycle, the number of clock cycles between read/write operations and different STGs will represent different trade-off implementations between area and timing. For example, Fig. 8 represents two different STGs for the same specification, along with the generated timed models for the simulation with ReSP. This figure shows that, varying the number of resources, it is possible to obtain different implementations, with different timing behaviors, that are reflected in the generated models. In fact, in the former case (case **A**), only 2 cycles have to be spent before the memory operation, while, in the latter one (case **B**), it is necessary to spend 3 cycles due to the resource limitations. Note that these delays, along with the one related to the return operation, have to be specified *before* the clock cycle they refer to in order to have a correct timing of the system. Moreover, the dashed states represent the states where the controller waits for the completion of the corresponding memory operation, which delays are included in the bus model instead. Finally, all the generated hardware models are compiled and stored in ReSP component repository.

VI. CASE STUDY

We employed the proposed framework in a case study considering an image processing application. The aim has been to analyze different mapping solutions for the considered algorithm on an architectural platform with a variable number of processors and hardware accelerators. Fig. 9 shows the task graph extracted from the application's C code; it is composed of the following stages, each one described as a function called in sequence in the *main()*: (i) loading of the image from an I/O interface (*imageLoading*); (ii) conversion of the RGB colors in gray scale (*rgb2gray*), (iii) edge detection (*edgeDetection*), (iv) overlapping of the edges on the original image (*edgeOverlapping*), and (v) transmission of the image to an I/O interface (*imageStoring*). We decided to map always in hardware the *imageLoading* and *imageStoring* functions since they are actually implemented as a transmission of data between the memory and an I/O hardware component, while the *edgeOverlapping* is always executed in software. Finally, the most computational-intensive stages (*rgb2gray* and *edgeDetection*) have been split in four parallel sections acting on different parts of the image. The architecture template (shown in Fig. 3) is composed of a variable number of ARM9 processors and a Xilinx Virtex5 FPGA device hosting a configurable number of hardware modules. They are connected through a system bus to a shared memory; moreover, each master is provided with local caches featuring a global coherency system. Finally, we set $333Mhz$ as operating frequency for the processors, $100Mhz$ for the hardware modules, and $10ns/word$ as transmission latency for the memory and the bus. In a first step, Bambu was used for generating the hardware implementations of the *rgb2gray* and *edgeDetection* functions (targeting a clock period of $10ns$); the resulting implementations occupy 2,501 and 7,413 slices, respectively. Finally, PandA was used to rewrite the application's source code by means of *pthread* API.

We simulated various system configurations obtained by varying the number of processors and hardware cores, and the functions' mapping. Table I reports the execution times of the considered configurations, along with the device resources required for implementing the different instances of the hardware modules, as specified in the "Mapping" column. As expected, the best performance ($7,751\mu s$) is obtained by accelerating both the functions in hardware and creating 4 different hardware modules, each of them containing both the function implementations; however, this requires a very large FPGA with at least 40,000 slices. If not available, the designer can choose a different solutions fulfilling the area constraints while optimizing the execution time. For example, with a Virtex5 LX110T FPGA (having 17,280 slices) the best solution features only one accelerator implementing both the functions that requires only 9,914 slices and takes only $9,849\mu s$. In conclusion, the results show how the proposed framework supports a detailed analysis on the implementation of the application, but also of its architectural parameters.

Table I
PERFORMANCE OF THE VARIOUS MAPPING SOLUTIONS.

# Procs	Mapping		Execution time (μs)	HW area (FPGA slices)
	rgb2gray	*edgeDetection*		
1	SW	SW	165,541	0
1	SW	1 HW	17,932	7,413
1	1 HW	1 HW	9,849	9,914
1	2 HWs	2 HWs	8,434	19,828
1	4 HWs	4 HWs	7,751	39,656
2	SW	SW	124,245	0
2	SW	1 HW	14,914	7,413
2	1 HW	SW	119,179	2,501
4	SW	SW	122,755	0

VII. CONCLUSIONS AND FUTURE WORK

This paper presented an automated framework for the evaluation of mapping solutions onto heterogeneous MPSoCs. The framework separates the synthesis of the SystemC models for the hardware cores (carried out only once) from the actual simulation (performed for each design solution), obtaining a significant reduction of the effort and time required by the designer. Our future work will be devoted in the automation of the design space exploration for system-level synthesis.

REFERENCES

[1] W. Wolf, A. A. Jerraya, and G. Martin, "Multiprocessor System-on-Chip (MPSoC) technology," *IEEE Trans. on CAD of Integrated Circuits and Systems*, vol. 27, no. 10, pp. 1701–1713, 2008.

[2] A. Sangiovanni-Vincentelli and G. Martin, "Platform-based design and software design methodology for embedded systems," *IEEE Design & Test of Computers*, vol. 18, no. 6, pp. 23–33, nov. 2001.

[3] A. Pimentel, C. Erbas, and S. Polstra, "A systematic approach to exploring embedded system architectures at multiple abstraction levels," *IEEE Trans. on Computers*, vol. 55, no. 2, pp. 99–112, feb. 2006.

[4] F. Ferrandi, P. Lanzi, C. Pilato, D. Sciuto, and A. Tumeo, "Ant colony heuristic for mapping and scheduling tasks and communications on heterogeneous embedded systems," *IEEE Trans. on CAD of Integrated Circuits and Systems*, vol. 29, no. 6, pp. 911–924, june 2010.

[5] T. Kempf *et al.*, "A modular simulation framework for spatial and temporal task mapping onto multi-processor soc platforms," in *Proceedings of DATE*, 2005, pp. 876–881.

[6] G. Beltrame, D. Sciuto, C. Silvano, D. Lyonnard, and C. Pilkington, "Exploiting TLM and object introspection for system-level simulation," in *Proceedings of DATE*, 2006, pp. 100–105.

[7] G. Beltrame, L. Fossati, and D. Sciuto, "ReSP: A Nonintrusive Transaction-Level Reflective MPSoC Simulation Platform for Design Space Exploration," *IEEE Trans. on CAD of Integrated Circuits and Systems*, vol. 28, no. 12, pp. 1857–1869, 2009.

[8] "Synopsys Platform Architect." [Online]. Available: http://www.synopsys.com/Systems/ArchitectureDesign

[9] W. Qin and S. Malik, "Flexible and Formal Modeling of Microprocessors with Application to Retargetable Simulation," in *Proceedings of DATE*, 2003, pp. 556–561.

[10] Accelera Systems Initiative, "http://www.accellera.org."

[11] Politecnico di Milano, "Panda web site." [Online]. Available: http://panda.dei.polimi.it

[12] "SoCLib." [Online]. Available: http://www.soclib.fr/

[13] "GAUT - High-Level Synthesis tool from C to RTL." [Online]. Available: http://www-labsticc.univ-ubs.fr/www-gaut/

[14] S. Rigo, G. Araujo, M. Bartholomeu, and R. Azevedo, "ArchC: a SystemC-based architecture description language," in *Proceedings of SBAC-PAD*, 2004, pp. 66–73.

[15] "gem5 simulator system." [Online]. Available: http://www.m5sim.org/

[16] M. Sato, "OpenMP: parallel programming API for shared memory multiprocessors and on-chip multiprocessors," in *Proceedings of ISSS*, 2002, pp. 109–111.

[17] Politecnico di Milano, "ReSP web site." [Online]. Available: http://code.google.com/p/resp-sim/

[18] Xilinx Inc, http://www.xilinx.com.

[19] N. Bombieri, F. Fummi, and G. Pravadelli, "Automatic Abstraction of RTL IPs into Equivalent TLM Descriptions," *IEEE Trans. on Computers*, vol. 60, no. 12, pp. 1730–1743, dec. 2011.

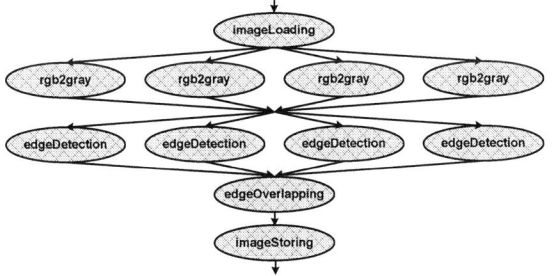

Figure 9.   Task graph of the considered image processing application.

Instrumentation-Driven Model Detection for Dataflow Graphs

Ilya Chukhman, William Plishker, Shuvra S. Bhattacharyya
Department of Electrical & Computer Engineering, and
Institute for Advanced Computer Studies
University of Maryland, College Park, MD, USA
{ilya, plishker, ssb}@umd.edu

Abstract—Dataflow modeling offers a myriad of tools to improve optimization and analysis of signal processing applications, and is often used by designers to help design, implement, and maintain systems on chip for signal processing. However, maintaining and upgrading legacy systems that were not originally designed using dataflow modeling can be challenging. To facilitate maintenance, designers often convert legacy code to dataflow graphs, a process that can be difficult and time consuming. We propose a method to facilitate this conversion process by automatically detecting the dataflow models of the core functions. The contribution of this work is twofold. First, we introduce a generic method for instrumenting dataflow graphs that can be used to measure various statistics and extract run-time information. Second, we use this instrumentation technique to demonstrate a method that facilitates the conversion of legacy code to dataflow-based implementations. This method operates by automatically detecting the dataflow model of the core functions being converted.

Keywords-Dataflow graphs, models of computation, signal processing systems.

I. INTRODUCTION

Modern digital signal processing (DSP) systems run sophisticated algorithms on high-performance systems based on FPGAs, programmable digital signal processors (PDSPs), and multiprocessor system-on-chip (MPSoC) devices. As a result, designing these systems is a complex process prone to inefficiencies and mistakes. Design tools, including dataflow modeling, are often used to help with the design process. Modeling DSP applications through coarse-grain dataflow graphs is widespread in the DSP design community, and a variety of dataflow models have been developed for dataflow-based design (DBD). DBD allows a designer to decompose a complex system into simpler sub-functions (*actors*) that are connected to form a graph. A variety of dataflow modeling tools can then be used to verify correctness of the graph and optimize the entire system (e.g., see [1], [2], [3]).

When employing DBD techniques, it is useful for a designer to find a match between their actors and one of the well-studied models, such as homogeneous synchronous dataflow (HSDF), synchronous dataflow (SDF) [4], cyclo-static dataflow [5], or boolean dataflow (BDF) [1]. When such a match is found, one can systematically exploit specialized characteristics of actors that conform to the

models, and take advantage of more effective, model-specific methods for analysis and optimization. For example, if a dataflow model match cannot be found, a less efficient, generic scheduler and more conservative memory allocation may need to be employed.

Economic factors necessitate reuse of existing designs with periodic upgrades to keep up with technological advances while saving on the non-recurring engineering costs associated with new designs. For example, the Large Hadron Collider (LHC) used for high energy physics experiments is planned to undergo a periodic series of large technology upgrades to allow for new experiments and the expansion of existing experiments [6]. Having a dataflow representation of such a system can alleviate this upgrade process by facilitating correctness verification, and in some cases enabling the use of an automatically generated implementation for the new hardware [7], [8]. DSP systems that are not designed using DBD, including legacy systems, are more difficult to upgrade, since implementation details can lead to errors that are hard to detect. For this reason, deriving dataflow graphs for these systems is beneficial and is increasingly done even though converting existing DSP code to dataflow graphs can be difficult and time consuming (e.g., see [9]).

In this paper, we introduce a method to facilitate this conversion process by automatically detecting specialized dataflow models that can be used to represent key functions being converted. By taking more generally described actors and automatically identifying them as instances of specific dataflow models, we enable the application of model-specific analysis and optimization techniques, which are often much more powerful than general purpose techniques [10].

To implement and experiment with our proposed model detection methodology, we have employed the DSPCAD Integrative Command Line Environment (DICE) [11], which is a framework for facilitating efficient management of design and software projects. DICE defines platform- and language-agnostic conventions for describing and organizing tests, and uses shell scripts and programs written in high-level languages to run and analyze these tests.

To create a generic method for instrumenting dataflow graphs, we used a DBD framework called the Lightweight Dataflow Environment (LIDE) [12], which is supported by DICE. This framework supports dynamic dataflow applications with a semantic model called core functional dataflow (CFDF) [3]. Through its foundation in CFDF semantics, LIDE enables dynamic behavior through structured application descriptions, making it an effective platform to instrument dataflow graphs, and prototype techniques for automated dataflow model detection.

As with unit testing, we rely on significant *coverage* of the behaviors of an actor instead of requiring a formal solution to analyze the code of the actor itself. While this requires good tests to exercise all of the behaviors that would occur during running the application in a real world environment, our approach need not "understand" the language or build process of the target-specific actor, which makes it applicable to a wide variety of DSP design scenarios. Designers are free to focus on the correct, efficient implementation of the actor, while the proposed model-detection design instruments the dataflow graph to generate trace information during each test. The trace information is then automatically processed to infer the dataflow model. Although our current implementation detects only static dataflow models — in particular, CSDF, HSDF, and SDF — the design is extensible to detect dynamic models as well.

We demonstrate that correct dataflow models can be extracted with minimum overhead for different components of the triggering system in the LHC. The performance of the model detection algorithm is not related to the complexity of the actor, but rather is a function of the trace file length. However, in order to achieve appropriate coverage, analysis of complex actors may result in longer trace files. Our initial approach for model detection has a run time complexity of $O(kn \log n)$, where n is the length (number of actor firings) of the trace file, and k is the number of actor ports. Careful selection of inputs for maximum coverage helps to maximize the accuracy of dataflow model detection while minimizing the run time cost.

II. RELATED WORK

The complexity of modern designs has caused the use of unit testing to become common practice in signal processing system design and software engineering [13], [14]. Most design and verification tools are language-specific and are often tied to fundamental constructs of the target language.

For example, [15] presents a method to classify general dataflow actors into known models of computation. The approach uses formal analysis of the SystemC finite state machine (FSM) describing the actor to identify the actor as SDF or CSDF. While this formal analysis based approach can definitively identify the dataflow model, it requires the actor to be represented by an FSM, which is not always possible, and furthermore, this approach is language specific. Such language-specific approaches provide the designer convenient methods to test individual functions, but lack the ability to provide arbitrary insight into the state of a system as a whole. Our proposed method is complementary

to this approach — for example, our method can be used to provide post processing for instrumentation-driven detection of model properties that are not detected using the formal methods applied in [15].

Like DICE, the Test Anything Protocol (TAP) [16] achieves language independence by defining the protocol that manages the communication between unit tests and a test harness. Individual tests (TAP producers) communicate test results to the testing harness (TAP consumers) in a language-agnostic way. TAP enables multi-platform and multi-language design, but only at the communication boundary. Unit tests need only adhere to the communication design, leaving test writers with no specific language independent mechanism for writing the tests themselves.

In contrast to these efforts, we propose a generic instrumentation method that utilizes the LIDE framework to enable the developer to analyze dataflow graphs by extracting run-time information and measuring performance. Our approach provides the ability to extract actor-specific as well as system-wide state, which can then be used to debug graph components and extract various statistics. We build on earlier work in [10], where dataflow models were detected manually through human analysis, and introduce in this paper a novel method to automatically extract dataflow models. By utilizing the proposed instrumentation framework and DICE unit-testing environment, we are able to automatically determine specialized dataflow models that can be associated to functions in legacy code, thereby alleviating the difficulty in converting legacy code to dataflow graphs.

Our work is also useful when working with system descriptions in general dataflow programming environments, such as CAL [17], by allowing tools to automatically detect specialized dataflow models that can be used to help streamline later stages of the design flow. However, unlike automated tools that have been developed for CAL and related frameworks (e.g., see [18]), the methods that we propose in this paper are trace-driven, and hence do not depend on any specific DBD language or intermediate representation.

III. DATAFLOW MODELS

Many dataflow models of computation have been developed for both actors and graphs to enable realization of a wide variety of applications and design techniques in DBD environments [9]. Some of the associated actor models are highly restrictive and can be used to infer powerful system-wide properties, which help, for example, to optimize scheduling and memory management. Other actor models offer flexibility that enables their use in a variety of applications not conducive to restrictive models; however, the same flexibility makes it more difficult to reason about these models and analyzing them becomes much more complex (e.g., see [1]).

The different dataflow models that have been created over the years are not all easily related to one another. As a result, it is not always possible to compare dataflow models in order to, for example, determine whether or not one model A is more restrictive than a different model B. Here, by "more restrictive", we mean intuitively that the class of computations that can be represented by A is a proper subset of the class that can represented by B. However, there are useful groups of models that can be compared in such a way — i.e., that can be compared in terms of this restrictivity notion. We refer to such a group (subset) of dataflow models as a *comparable* group. In this paper, we limit our discussion to comparable groups, and demonstrate our methods on a comparable group that is supported in LIDE. Extending these methods to groups that are not comparable is a useful direction for future investigation.

Fig. 1 shows three classes of dataflow models within the universe of dataflow models that are currently supported in LIDE: data independent, data dependent, and mode based. As illustrated in Fig. 1, data independent models are most restrictive and mode based models are least restrictive. Outer classes in Fig. 1 generalize the inner classes, so that for example, data-dependent models also contain data independent actors, and mode based models can also contain data dependent actors.

Before we define these classes of models, it is useful to define the token transfer rates of dataflow actors — i.e., consumption rates and production rates. These kinds of rates are commonly used in characterizing dataflow actors. Consumption and production rates are associated with individual input and output ports, respectively, and are measured as the numbers of tokens that are consumed and produced when the enclosing actor fires (i.e., when it executes a single unit of execution).

In homogeneous synchronous dataflow (HSDF), all consumption and production rates are restricted to be equal to unity [4]. Thus, an actor is an HSDF actor if every input port consumes exactly one token per firing and every output port produces exactly one token. A more general model is synchronous dataflow (SDF), where the consumption and production rates of actor ports must be constant (positive integer) valued [4] (i.e., they cannot vary as a function of data or state). The cyclo-static dataflow (CSDF) model introduces the concept of actor *phases*, where in each phase the actor conforms to an "extended SDF" model (extended in the sense that zero-valued production and consumption rates are also allowed) [5]. In addition, the phases cycle through a periodic sequence, so that on each actor port, one can observe a periodic pattern of token consumption or production. If the consumption and production rates are the same in each phase, and they are all positive valued, then the CSDF actor conforms to the more restrictive SDF model. Similarly, if the consumption and production rates of an SDF actor are all equal to one, then the actor conforms

to the more restrictive HSDF model. The HSDF, SDF and CSDF models have the property of being data independent, where their consumption and production rates are not related to values of the data inputs.

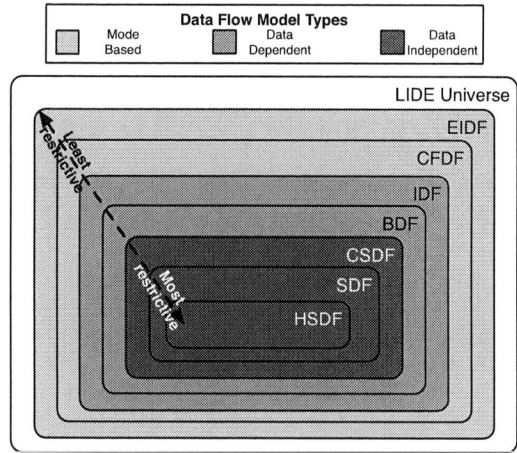

Figure 1. A classification of dataflow models supported in the LIDE framework

The next class of models depicted in Fig. 1 encapsulates what we refer to as the data dependent models. In a data dependent model, a dataflow graph can contain one or more data dependent dataflow (DDD) actors. A DDD actor is one in which one or more inputs (data values consumed) or the actor state (or both inputs and state) determine(s) how much data is consumed and produced by a given actor firing. Boolean dataflow (BDF) is an example of a data dependent model. In BDF, Boolean-valued input tokens on designated ports are used to determine the production and consumption rates for ports where the rates are data dependent [1]. For example, when a control input is TRUE, the actor could consume tokens from one data port, while it consumes tokens from a different data port when the control input is FALSE. It is more difficult to reason about and analyze BDF graphs; however, useful quasi-static scheduling and analysis techniques have been developed that exploit the control-token-based dynamic dataflow structure of BDF (e.g., see [1]).

Integer-controlled dataflow (IDF) is a natural generalization of BDF where the inputs used to control data dependent actors are integer valued [19], [7], and correspondingly, the variations in consumption and production rates for individual ports can span integer numbers of different values. For example, IDF can be used to represent a generic multiplexer, where a control input selects data from a specific graph instance among N possible instances, each of which can be structured based on a different dataflow model.

The most general dataflow model that we consider in this work is the enable-invoke dataflow (EIDF) model [3]. In EIDF, an actor is specified in terms of a set of modes, such that in each mode the production and consumption rates must be constant, non-negative integer values. Intuitively then, in each mode, the actor can be viewed as an extended SDF actor. However, different modes of an actor can have different production and consumption rates, and dynamic dataflow behavior can be achieved in this way. This form of dynamic dataflow is distinguished from the data dependent class introduced previously by the decomposition of actor operation into distinct modes. For dataflow models that employ decomposition into distinct modes, where production and consumption rates need not be constant or periodic across distinct modes, we introduce the third (most general) class of models, the mode based models, represented in Fig. 1.

In EIDF, when a given mode m is invoked, the invoking function returns a set $N(m)$ of possible *next modes* of execution. The provision in EIDF for multi-element sets of valid next modes allows for non-determinism, as in its next firing, an actor can be invoked in any mode within the next mode set. A more restrictive, deterministic form of EIDF is the core functional dataflow (CFDF) model [3]. CFDF enforces that the set of next modes must always have exactly one element — i.e., $|N(m)| = 1$. Note that CFDF modes are different from CSDF phases in that the selection of the next mode in CFDF can be data dependent.

In summary, a group of dataflow models in a given DBD environment can cover a wide spectrum of trade-offs between expressive power and formal analysis potential. This is demonstrated by the spectrum of comparable models illustrated in Fig. 1. Model detection helps designers identify the most restrictive dataflow model a given actor conforms too, thereby helping to identify the most powerful sets of analysis and optimization methods that can be applied to subsystems that contain the actor.

IV. DATAFLOW GRAPH INSTRUMENTATION

Dataflow graph instrumentation provides a modular and flexible approach for extracting run-time information, which can be used to help debug incorrect behavior, measure performance, or see how different forms of execution state evolve as a graph executes. The enable-invoke interface provided by the LIDE framework is a convenient mechanism to build upon for instrumenting dataflow graphs. As described in [3], the `enable` and `invoke` functions correspond to testing for sufficient input data, and executing a single firing (invocation) for a given actor, respectively.

An example of an EIDF graph along with a simple form of scheduler, called the *canonical scheduler*, for EIDF is shown in Fig. 2(a). The canonical scheduler is usually not efficient for implementation purposes, but for simulation and testing processes, such as those relevant to model detection, it is useful as a simple, generally-applicable scheduling method that can easily be applied for instrumentation purposes. The instrumentation approach described below can easily be extended to work with other scheduling mechanisms.

When an actor is executed by the canonical scheduler, the `enable` function for the actor is called. This function, which is a basic actor primitive in the EIDF model, returns TRUE if and only if the actor has a sufficient number of tokens on each input edge to allow for a complete firing of the actor in its next mode. If the `enable` function returns TRUE, then the actor is executed using the `invoke` function; otherwise, the `enable` function of the next scheduled actor is called. The canonical scheduler applies this two-phase process (a call to `enable` followed by a conditional call to `invoke`) on a given sequence of graph actors.

To support model detection and related applications of dataflow graph instrumentation in a structured way, we propose "instrumentation extensions" just prior to and just after execution of the `invoke` function, as illustrated in Fig. 2(b). By inserting appropriate forms of instrumentation before and after an actor fires, developers can expose powerful insight into the actor's state, and patterns or useful statistics that can be derived to characterize the progression of actor execution state over time. Such an approach enables the developer to precisely capture relevant changes in a graph state caused by the firing of an actor, which is a powerful technique that can be used for debugging purposes, as well as for understanding characteristics of actor and subsystem operation.

Based on the returned value of the `enable` function, we can precisely determine if a given actor will fire, and we can insert *pre-invoke instrumentation* (`pre_ins`), as shown in Fig. 2(b). After the `invoke` function finishes, the *post-invoke instrumentation* (`post_ins`) can execute to complete instrumentation of the actor associated with its most recent firing. For example, by observing the populations of the input FIFOs before and after an actor fires, one is able to compute the consumption rate. Similarly, the actor execution time can be obtained by recording the clock during `pre_ins` and comparing it to the clock obtained in `post_ins`, after the actor finishes executing.

For relatively simple forms of instrumentation and coarse-grain actors, the proposed instrumentation approach adds minimal overhead to the scheduler. Furthermore, in the simulation/testing time context where we use instrumentation

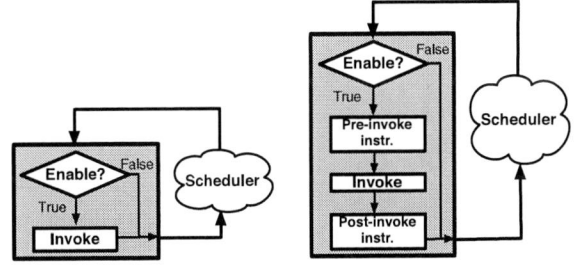

(a) An illustration of an EIDF graph (b) An instrumented dataflow graph. that is executed by the canonical EIDF scheduler.

Figure 2. Using the enable-invoke interface to instrument dataflow graphs.

for model detection, even significant overheads can often be tolerated (compared to actual run time overhead in an implementation). In addition to executing the `enable` and `invoke` functions for the given actor ordering, our *instrumentation augmented scheduler* (*IAS*) executes `pre_ins` and `post_ins` functions associated with the actor of interest. The designer determines the exact behavior of the `pre_ins` and `post_ins` functions such that only the state and statistics of interest are examined. By having full control of instrumentation functions, and allowing definition and use of arbitrary state within these functions, the designer can instrument the actor a specific number of times (e.g., the first time the actor fires, every time the actor fires, every other time the actor fires, or every time the actor is in a specific mode).

V. MODEL DETECTION

Before defining our algorithm for model detection, we first review some basic dataflow graph notation, and provide a formal definition of dataflow model detection. A dataflow graph G is an ordered pair (V, E), where V is a set of vertices (*actors*), and E is a set of directed edges. A directed edge $e = (v1, v2) \subset E$ is an ordered pair of a source vertex $v_1 \subset V$ and a sink vertex $v_2 \subset V$. Actors represent computations while edges represent communication links between them.

We define:

$$\mathcal{A} = \{a_1, a_2, \ldots, a_n\} \qquad (1)$$

as the set of all actors of interest in a given DBD scenario (e.g., a DSP system design project or group of related projects). For example, these could be the set of all actors that are available across all of the actor libraries accessible to the design team.

In the same design scenario, suppose that

$$\mathcal{M} = \{m_1, m_2, \ldots, m_z\} \qquad (2)$$

is a group of comparable models that make up the "universe" of available models (analogous to how the LIDE Universe is depicted in Fig. 1). Furthermore, assume that the m_is are ordered in increasing generality (m_a is more restrictive compared to m_b whenever $a < b$). Intuitively, $\mathcal{M}$ is the set of available dataflow models of computation in the given design scenario, and we assume that the models in $\mathcal{M}$ are comparable, as discussed in Section III. In conjunction with the notion that $\mathcal{M}$ is the model universe, we assume that each actor in $\mathcal{A}$ conforms to at least one of the models in $\mathcal{M}$.

We define the *actor set* $\mathcal{A}_k \subset \mathcal{A}$ of each model m_k as the set of all actors in $\mathcal{A}$ that conform to model m_k. It is important to note that some actors can be represented by multiple models, which means that $\mathcal{A}_k \cap \mathcal{A}_l$ can be nonempty for $k \neq l$. In fact, since $\mathcal{M}$ is assumed to be ordered in terms of increasing generality, we will have $\mathcal{A}_k \subset \mathcal{A}_l$ for $k < l$.

We define $\mathcal{R}(a)$ as the set of all models in $\mathcal{M}$ that actor a conforms to, and we define the most specialized model (MSM) for an actor a as:

$$\mathrm{MSM}(a) = \min\{i \mid m_i \in \mathcal{R}(a)\}. \qquad (3)$$

That is, $\mathrm{MSM}(a)$ is the most specialized model in the model universe to which a conforms.

For a given actor, the model detection problem can then be defined as: given an actor $a \in \mathcal{A}$, determine $\mathrm{MSM}(a)$.

For example, consider the LIDE universe of Fig. 1. We can represent this model universe as

$$\mathcal{M} = \{\mathrm{HSDF}, \mathrm{SDF}, \mathrm{CSDF}, \mathrm{BDF}, \mathrm{IDF}, \mathrm{CFDF}, \mathrm{EIDF}\}, \qquad (4)$$

and given an actor a, the model detection problem amounts to determining which of these seven models is the most specialized model that a conforms to.

Note that we have assumed that each actor conforms to at least one m_i only for simplicity and conciseness. The formulation in this section can easily be adapted to handle actors in $\mathcal{A}$ that do not belong to any of the models in the universe (e.g., because of bugs in the implementation or documentation of the "misfit" actors). In such cases, the model detection problem formulation can be extended to allow for the additional "output" value of $\bot$, which represents that the given actor does not conform to any of the models in the universe.

VI. MODEL DETECTION IMPLEMENTATION

We have implemented our model detection design using a two stage process. The legacy code is converted to a generic LIDE-compatible dataflow format in the first stage. The dataflow instrumentation methodology discussed in Section IV is then used to analyze the LIDE-compatible component and determine a specific dataflow model. Both of these stages are performed in conjunction with DICE features for unit testing.

Fig. 3 shows the steps in converting original legacy code to an LIDE-compatible format. The initial transformation step entails adding an LIDE-supported FIFO for each input and output port. In the next step, the `invoke` and `enable` functions required for LIDE compatibility are created. In the example of Fig. 3, the `invoke` function is set to `fnc`, such that when the `invoke` function for this block is called, `fnc` would execute. The `enable` function is created to return `TRUE` when the input FIFOs are nonempty and `FALSE` otherwise. Our approach to validating correctness of this transformation relies on the availability of a unit test that can be used to populate the input buffers with an appropriate quantity of tokens, such that all of the input tokens are used after some number of invocations of the `invoke` function.

The LIDE-compatible block created with this transformation conforms to the generic EIDF model. Doing further analysis to determine whether the LIDE-compatible block conforms to a more restrictive model, such as SDF or CSDF,

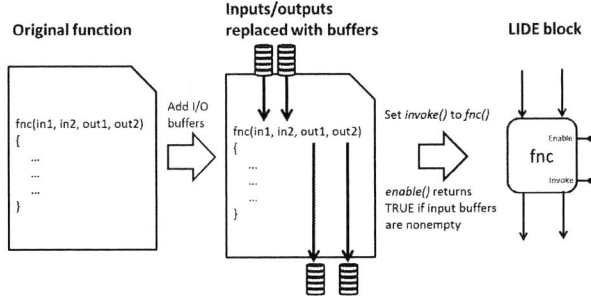

Figure 3. Converting legacy code to an LIDE-compatible dataflow block involves adding input/output buffers and creating `enable` and `invoke` functions.

can enable the use of stronger analysis and optimization techniques than those available for EIDF models. This further analysis step is done utilizing a unit test framework.

A typical unit test is depicted in Fig. 4(a), where test inputs are fed to the module under test (MUT), which in our context is the intermediate dataflow actor being tested, and the outputs of the MUT are saved in the output file. After all the inputs have been processed, the outputs file is compared to the expected outputs. The unit test is considered `PASSED` if the expected outputs match the generated outputs, otherwise, the test is considered `FAILED`.

By enhancing an actor's unit test with dataflow graph instrumentation, as introduced in Section IV, the designer can glean key properties needed to determine the MSM for the actor. The general design of such an *enhanced unit test* is shown in Fig. 4(b), where the first step is to provide the actor's interface information to the `pre_ins` and `post_ins` functions, denoted by the *model detector* block. As the unit test executes by feeding inputs to the actor (shown in Step 2a), the instrumentation functions monitor the state of the actor and extract the consumption and production information (denoted by C&P rates in the figure), as well as the coverage information, as shown in Step 2b. Using the knowledge about inputs, outputs, and the consumption and productions rates, the model detector can test this data for certain dataflow properties, which in turn can be used to determine the MSM. The result of the

enhanced unit test is no longer a PASS/FAIL criterion, but is instead the hypothesized MSM (*detected MSM*) of the actor. The accuracy of this hypothesis is generally as good as the coverage of the associated test suite, and can be improved as the test suite evolves, just as the designer's confidence in functional correctness can be improved.

The model-detector block is depicted in Fig. 5. The input to the block is interface information of the MUT, and the output is the detected MSM of the MUT. The model detector instruments the MUT and extracts runtime information, including the consumption and production values after each firing, as well as the coverage information. As discussed in Section I, coverage knowledge indicates how much of the MUT's typical behavior has been covered, and in some instances, enables generating inputs to exercise new code paths. The hypothesis-generator block cycles through the supported dataflow models and provides the expected pattern for a given dataflow model to the hypothesis-tester block. The hypothesis-tester block uses pattern matching functions to test whether the actor outputs conform to the expected pattern of the MSM hypothesis. If the hypothesis-tester finds a given hypothesis to be `TRUE`, then model detection is complete and that dataflow model is the detected MSM for the MUT. However, low coverage values may lead to false findings of the MSM. For those cases, it is recommended to generate new inputs to more fully exercise the MUT and rerun the model detection test.

A pseudo-code specification of the `hypothesis_tester` function is shown in Fig. 6. This function tests whether the inputted consumption/production information conforms to the specific model being tested. Our initial implementation can detect HSDF, SDF, and CSDF models. The loops testing for HSDF and SDF each take $O(n)$ time, where n is the length of `rates`.

A CSDF model is defined by consumption/production rates that repeat in a consistent pattern. We utilize the `findreps` algorithm introduced in [20] to find the positions of all repetitions in `rates` in $O(n \log n)$ time, where n is the length of `rates`. Next, `dist_reps` is computed by taking the difference between consecutive elements of the

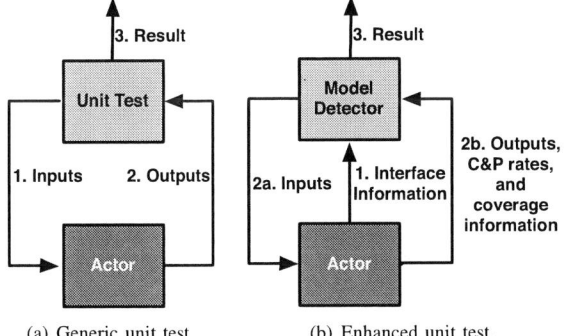

(a) Generic unit test. (b) Enhanced unit test.

Figure 4. The generic unit test can be enhanced to capture the actor's state information used by our model detection algorithm.

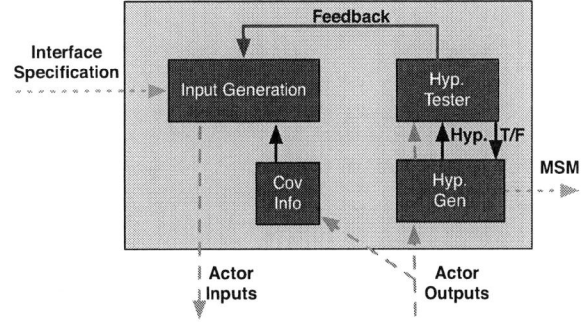

Figure 5. Our model detector uses the actor interface information to generate inputs to exercise the actor. The actor outputs are used by the hypothesis generator and tester components to determine the MSM.

repetitions array in $O(n)$ time. Finally, we determine that the test for CSDF is TRUE if separation between all the patterns is the same, and the patterns span the length of rates. The entire CSDF test takes $O(n \log n)$ time.

The pseudo-code specification of the hypothesis_generator function is shown in Fig. 7. This function inputs the consumption/production rates of all the ports and outputs the detected MSM of the actor, as defined by Equation 3. The models array contains all the models being tested, sorted from most restrictive to least restrictive. A dataflow model is detected for each port by testing whether the observed consumption/production rates for that port are consistent with the dataflow model being analyzed. After the model for each port is identified, the detected MSM for the actor can be found by selecting the *least restrictive* model of all the ports. The total time to check all of the ports is $O(kn \log n)$, where n is the length of rates and k is the number of ports.

VII. MODEL DETECTION RESULTS

We used the DICE testing framework to implement dataflow model detection of high energy physics actors that are part of the Trigger system of the Compact Muon Solenoid (CMS) Detector of the Large Hadron Collider (LHC) at CERN [21]. Following recommended design practices, unit tests have been created for the core components of the CMS Level-1 Trigger [6] and added to our DICE-based test suite and LIDE-based design framework for these high energy physics design components. We modified these existing unit tests by augmenting the scheduler to instrument the MUT, as discussed in Section VI.

The *jet reconstruction* component attempts to identify a group of particles using a sensor grid. The actor has one input port, which consumes an 8x8 grid, where each entry represents a sensor. This results in the consumption of 64

```
function hypothesis_tester(rates, model)
    if model == HSDF
        //all values have to be 1
        for i=1:length(rates)
            if rates[i] != 1
                return FALSE
        return TRUE
    if model == SDF
        k = rates[1]
        //all values have to be k
        for i=2:length(rates)
            if rates[i] != k
                return FALSE
        return TRUE
    if model == CSDF
        //findreps returns positions of all repetitions
        //in rates
        repetitions = findreps(rates)
        dist_reps = diff(repetitions);
        //distance between reps has to be the same
        //pattern needs to span the entire space
        if (length(unique(dist_reps)) == 1)
        && (repetitions(end)+dist_reps(1) > length(rates))
            return TRUE
        return FALSE
```

Figure 6. Pseudocode for the hypothesis tester block

```
function hypothesis_generator(all_rates)
    //models ranked from most to least restrictive
    models = {HSDF, SDF, CSDF, NONE}
    for p in ports
        detected_models[p] = NONE
        rates = all_rates[p]
        //test models from most to least restrictive
        for m in models
            if (hypothesis_tester(rates, m)
                detected_models[p] = m
                break
    //select the least restrictive model of all ports
    msm = max(detected_models);
    return msm
```

Figure 7. Pseudocode for the hypothesis generator block

tokens each time the actor executes. The jet reconstruction actor also has two output ports, which correspond to the total energy detected, and a Boolean value indicating if a jet has been identified. The actor produces one token on each of the two output ports.

An example of the trace file containing state information produced by the jet reconstruction actor is shown in Table I. Notice that consumption values are negative while the production values are positive. The actor has one mode and two output ports conforming to the HSDF model, and one input port conforming to an SDF model. Since SDF is less restrictive than HSDF, the jet reconstruction actor was determined to conform to the SDF model.

The dataflow model detection results for the CMS actors are summarized in Table II. Having confidence that all of the actors in the system have static dataflow models, the designer can utilize an aggressive scheduler. Combining this dataflow model information with inter-dependency information, which can also be obtained from the dataflow graph, the developer can strategically partition the system to maximize parallelism. Finally, an efficient buffer size for each dataflow graph edge can be derived from the consumption and production rates extracted by the instrumentation code, as well as from knowing that each actor conforms to a static dataflow model (e.g., see [9]).

The trace information for a *block adder* conforming to a CSDF model is shown in Table III. The block adder has two input ports and one output port. Unlike the previous examples where all the actors had static behavior and only one phase, the block adder cycles through three phases. The

Table I
INSTRUMENTATION RESULTS FOR A JET RECONSTRUCTION ACTOR.

Mode	In[0]	Out[0]	Out[1]
1	-64	1	1
1	-64	1	1
1	-64	1	1
1	-64	1	1
Model Detected	SDF	HSDF	HSDF

Table II
DETECTED DATAFLOW MODELS OF VARIOUS CMS ACTORS

Actor	#Inputs	#Outputs	Cons./Prod. Rates	Model Detected
Jet Reconstruction	1	2	64/1	SDF
Cluster Threshold	12	12	1/1	HSDF
Cluster Compute	12	6	1/1	HSDF
Cluster Isolation	1	2	64/8	SDF

actor consumes one token from input port *In[0]* when in phase 2; it consumes one token from input port *In[1]* when in phase 3; and it produces one token to output port *Out[0]* when in phase 1. This results in all of the ports having repeating firing patterns: {-1,0,0} for *In[0]*, {0,-1,0} for *In[1]* and {0,0,1} for *Out[0]*, indicative of a CSDF dataflow model, which is what the model detector determined. Again, negative values indicate consumption, while positive values indicate production.

Our results on the CMS Detector demonstrate the utility of our model detection approaches on a complex and important application. The low run time complexity of our model detection techniques (see Section VI) enhances this utility, and facilitates high confidence detection from large traces associated with coverage-intensive test suites.

VIII. CONCLUSION

A common problem of modern, high-performance system on chip designs is the need for frequent upgrades to keep up with current technology or evolving application requirements. Designers can convert legacy code to dataflow based implementations to help alleviate this upgrade process, though the conversion can be laborious and time consuming. In this work, we have proposed a method to facilitate this conversion process by automatically detecting the dataflow models of the core functions. We have also developed a generic instrumentation approach, and demonstrated the approach using the lightweight dataflow environment (LIDE) framework and the DSPCAD integrative command line environment (DICE). In addition to supporting our proposed model detection features, this instrumentation-driven approach can be useful in debugging dataflow graphs and measuring performance. Useful directions for future work include extending our instrumentation and model detection approaches to detect data dependent and mode based dataflow models, and integrating run-time model verification to detect and act on deviations from the assumed dataflow models. Such run-time support enables graceful termination of an application or a fall-back to a more general, dataflow-model-agnostic configuration.

IX. ACKNOWLEDGMENTS

This material is based on research that was sponsored in part by the U.S. National Science Foundation (ECCS0823989).

REFERENCES

[1] J. T. Buck, "Scheduling dynamic dataflow graphs with bounded memory using the token flow model," Ph.D. dissertation, Department of Electrical Engineering and Computer Sciences, University of California at Berkeley, September 1993.

[2] F. Siyoum, M. Geilen, O. Moreira, R. Nas, and H. Corporaal, "Analyzing synchronous dataflow scenarios for dynamic software-defined radio applications," in *Proceedings of the International Symposium on System-on-Chip*, 2011, pp. 14–21.

[3] W. Plishker, N. Sane, M. Kiemb, K. Anand, and S. S. Bhattacharyya, "Functional DIF for rapid prototyping," in *Proceedings of the International Symposium on Rapid System Prototyping*, Monterey, California, June 2008, pp. 17–23.

Table III
INSTRUMENTATION RESULTS FOR A BLOCK ADDER.

Mode	In[0]	In[1]	Out[0]
2	-1	0	0
3	0	-1	0
1	0	0	1
2	-1	0	0
3	0	-1	0
1	0	0	1
2	-1	0	0
3	0	-1	0
1	0	0	1
2	-1	0	0
3	0	-1	0
1	0	0	1
Model Detected	CSDF	CSDF	CSDF

[4] E. A. Lee and D. G. Messerschmitt, "Synchronous dataflow," *Proceedings of the IEEE*, vol. 75, no. 9, pp. 1235–1245, September 1987.

[5] G. Bilsen, M. Engels, R. Lauwereins, and J. A. Peperstraete, "Cyclo-static dataflow," *IEEE Transactions on Signal Processing*, vol. 44, no. 2, pp. 397–408, February 1996.

[6] A. Gregerson, M. J. Schulte, and K. Compton, "High-energy physics," in *Handbook of Signal Processing Systems*, S. S. Bhattacharyya, E. F. Deprettere, R. Leupers, and J. Takala, Eds. Springer, 2010.

[7] T. Miyazaka and E. A. Lee, "Code generation by using integer-controlled dataflow graph," in *Proceedings of the International Conference on Acoustics, Speech, and Signal Processing*, 1997.

[8] H. Oh and S. Ha, "Efficient code synthesis from extended dataflow graphs for multimedia applications," in *Proceedings of the Design Automation Conference*, 2002, pp. 275–280.

[9] S. S. Bhattacharyya, E. Deprettere, R. Leupers, and J. Takala, Eds., *Handbook of Signal Processing Systems*. Springer, 2010.

[10] W. Plishker, C. Shen, S. S. Bhattacharyya, G. Zaki, S. Kedilaya, N. Sane, K. Sudusinghe, T. Gregerson, J. Liu, and M. Schulte, "Model-based DSP implementation on FPGAs," in *Proceedings of the International Symposium on Rapid System Prototyping*, Fairfax, Virginia, June 2010, invited paper, DOI 10.1109/RSP_2010.SS4, 7 pages.

[11] S. S. Bhattacharyya, W. Plishker, C. Shen, N. Sane, and G. Zaki, "The DSPCAD integrative command line environment: Introduction to DICE version 1.1," Institute for Advanced Computer Studies, University of Maryland at College Park, Tech. Rep. UMIACS-TR-2011-10, 2011, http://drum.lib.umd.edu/handle/1903/11422.

[12] C. Shen, L. Wang, I. Cho, S. Kim, S. Won, W. Plishker, and S. S. Bhattacharyya, "The DSPCAD lightweight dataflow environment: Introduction to LIDE version 0.1," Institute for Advanced Computer Studies, University of Maryland at College Park, Tech. Rep. UMIACS-TR-2011-17, 2011, http://hdl.handle.net/1903/12147.

[13] P. Hamill, *Unit Test Frameworks*. O'Reilly & Associates, Inc., 2004.

[14] H. G. T. Dohmke, "Test-driven development of a PID controller," *IEEE Software*, vol. 24, no. 3, pp. 44–50, 2007.

[15] C. Zebelein, J. Falk, C. Haubelt, and J. Teich, "Classification of general data flow actors into known models of computation," in *Proceedings of the International Conference on Formal Methods and Models for Codesign*, 2008, pp. 119–128.

[16] S. Cozens, *Advanced Perl programming*, 2nd ed. O'Reilly & Associates, Inc., 2005.

[17] J. Eker and J. W. Janneck, "CAL language report, language version 1.0 — document edition 1," Electronics Research Laboratory, University of California at Berkeley, Tech. Rep. UCB/ERL M03/48, December 2003.

[18] R. Gu, J. Janneck, M. Raulet, and S. S. Bhattacharyya, "Exploiting statically schedulable regions in dataflow programs," in *Proceedings of the International Conference on Acoustics, Speech, and Signal Processing*, Taipei, Taiwan, April 2009, pp. 565–568.

[19] J. T. Buck, "Static scheduling and code generation from dynamic dataflow graphs with integer-valued control systems," in *Proceedings of the IEEE Asilomar Conference on Signals, Systems, and Computers*, October 1994, pp. 508–513.

[20] M. G. Main and R. J. Lorentz, "An O(n log n) algorithm for finding all repetitions in a string," *Journal of Algorithms*, vol. 5, no. 3, pp. 422–432, September 1984.

[21] CMS Collaboration, "CMS TriDAS project : Technical design report; 1, the trigger systems," CERN. Geneva. LHC Experiments Committee, Tech. Rep. CERN-LHCC-2000-038, 2000.

Thermal/Performance Trade-off in Network-on-Chip Architectures

Davide Zoni, Simone Corbetta and William Fornaciari

Politecnico di Milano – Dipartimento di Elettronica e Informazione
Via Ponzio 34/5, 20133 Milano, Italy
Email: {zoni,scorbetta,fornacia}@elet.polimi.it

Abstract—**Multi-core architectures are a promising paradigm to exploit the huge integration density reached by high-performance systems. Indeed, integration density and technology scaling are causing undesirable operating temperatures, having net impact on reduced reliability and increased cooling costs. Dynamic Thermal Management (DTM) approaches have been proposed in literature to control temperature profile at run-time, while design-time approaches generally provide floorplan-driven solutions to cope with temperature constraints. Nevertheless, a suitable approach to collect performance, thermal and reliability metrics has not been proposed, yet. This work presents a novel methodology to jointly optimize temperature/performance trade-off in reliable high-performance parallel architectures with security constraints achieved by workload physical isolation on each core. The proposed methodology is based on a linear formal model relating temperature and duty-cycle on one side, and performance and duty-cycle on the other side. Extensive experimental results on real-world use-case scenarios show the goodness of the proposed model, suitable for design-time system-wide optimization to be used in conjunction with DTM techniques.**

I. INTRODUCTION

Aggressive technology scaling has lead continuous miniaturization of transistors, making modern processors experiencing an exponential increase of performance in terms of clock rate, however with power consumption going as faster as clock rate [18]. Higher power consumption density in lower area regions makes operating temperatures increase up to the point reliability is mainly affected by thermal hot-spots: it has been shown that 50% of failures in CMOS integrated circuits are due to thermal issues [16], [22]. The transition to multi-core architectures introduced an opportunity for performance to grow faster than power consumption [5], allowing for a fine grain control on power densities and operating temperatures. Nevertheless, the increasing performance attained by multi-core and many-core processors are again raising the issue of integration capability and inter-core communication for future Multi-Processors System-on-Chip (MPSoC) design [12]. Network-on-Chip (NoC) architectures [3] have been proposed to cope with increasing performance requirements in massively parallel systems, but routers and link drivers consume a non-negligible amount of chip power [14], with a net impact on the chip temperature. In particular a few commercial designs show that the NoC can contribute up to 28% of total chip power [10]. Thermal Design Power (TDP) is the most challenging design constraint that accounts for and, sometimes, determines

the feasibility of the final system. In this perspective, thermal issues must be accounted at each design step, both at early design stages and at run-time. However, one of the main challenges in this perspective is to find a set of appropriate metrics that allows to manage and optimize such sensible chip design aspects, i.e. performance, thermal profile, power.

This work addresses thermal performance trade-off in a particular scenario, where multicore architectures are used to ensure security in critical web-service transactions and each workload/application must be mapped on a single core with no overlap. Traditionally, virtualization techniques are used to provide a logical separation between workloads on the same system, mainly for security reasons. However, such isolation is mild since it relies on virtual machine software components, that are usually designed for performance and can be violated quite easily [21]. Unlike existing software isolation techniques, the new trend on security seems to address isolation problem by a proper set of hardware modules, where workloads are physically isolated on a single core [2].

A. Novel contributions

The novel contributions of the work presented in this paper are many-fold. This research work focuses on the joint optimization of performance and temperature profile in multi-core architectures with NoC interconnect. The objective of our work is to analyse the thermal/performance trade-off in parallel architectures employed in high-performance server systems. To this extent, the following contributions are discussed in this paper:

- *Thermal/performance optimization* - an optimization methodology to jointly deal with performance and thermal profile trade-off is proposed as design-time optimization framework. The proposed work is general enough to be employed to constraint chip temperature, while maximizing core performance allowing for minimum core-to-core performance differences. In particular, we want to obtain a per-core maximum performance level with two conditions: chip temperature is maintained below a certain threshold and performance differences between cores are minimized;

- *Performance and duty-cycle* - we use clock-gating to control the performance of the cores; a valuable relationship between the applied clock-gating level (i.e., duty-cycle specification) and the performance degradation is then proposed and validated against extensive experimental results;

- *Temperature and duty-cycle* - a valuable relationship between the applied clock-gating level and the operating temperature is proposed and validated against a rich set of experimental results;
- *System-wide optimization* - by employing the chip topology, we introduce the concept of topological rings to deal with thermal and performance trade-off. We propose a novel system-wide optimization methodology for multi-core architectures underpinned by the new concept of topological ring;
- *Real use case scenario* - to demonstrate the validity of the proposed approach, we cast our methodology on a specific available multi-core architecture [20]. After an in-depth use case analysis, we have exploited the multi-core architecture specificities providing strengthen results on our methodology.

To attain the contributions of the proposed research work, we propose two different tools. First, we developed a linear optimization model, to deal with thermal/performance trade-off. Moreover, we have cast both the performance and temperature empirical relations, extracted from data, as linear equations. Second, experimental results have been collected through an ad-hoc simulation framework, capable of cycle-accurate and thermal simulation of multi-core architectures with standard NoC interconnects.

B. Paper structure

This paper is organized as follows. Section II will give a brief overview of the state-of-the-art thermal optimization techniques, both at design-time and run-time. Section III introduces the proposed formal model for joint thermal/performance optimization under either absolute temperature constraints. Experimental results are discussed in Section IV, and conclusions will be drawn in Section V.

II. RELATED WORKS

The reliability dependence on increasing operating temperatures of microelectronics systems makes the control of the temperature profile of utmost importance in multi-core processors. Thermal management refers to a set of techniques and design choices that leads to the optimization of the temperature profile of a chip: hard-faults mechanisms such as electromigration and stress-migration are known to be exponentially related to operating temperature [27]. Optimization techniques can be employed either at design-time or at run-time. The former approaches have the advantage of finer-grain control (e.g., circuit-level techniques or microarchitecture-level techniques) at the cost of reduced flexibility and increased silicon area. The latter approaches, on the other hand, has greater flexibility, but generally require additional software complexity (e.g., additional data structures to hold temperature information on a per-core granularity) and might have non-negligible effects on performance, without the opportunity to trade performance and thermal off in an easy way.

A. Design-time thermal optimization

Design-time thermal management techniques can be conveniently organized in two broad classes [15]: microarchitecture-level techniques and floorplanning optimizations. At the microarchitecture level we can find several works for general purpose applications, ranging from techniques targeting processor cores only, or techniques for on-chip memory caches. In the first case the processors can be restructured according to a cluster-based architecture, or by duplicating portions of the processor that are known to be thermal hot-spots. Functional units are duplicated in [9], with increased hardware area and cost: these units are used alternatively to reduce the stress on each single unit (e.g., an ALU or register files). Similar work has been done in [25] in which the only register file has been duplicated, and activity migration is directed toward the spare unit under dynamic thermal constraints. Functional units can also be resized to accommodate a lower power density [23], but with a reduction of the clock frequency and negative impact on processor performance. Floorplan can also be conveniently designed to accommodate thermal hot-spots as done in [19].

Design-time tools are generally required to perform predictions on the benefits of the thermal management solution under investigation, such that to modify where appropriate the entire design. A few works have tried to integrate performance, power and thermal analysis in a single framework. The `Polaris` framework [26] allows to estimate power and area of NoC-based designs, but does not allow to provide detailed power consumption profile for the processors and memory hierarchy. The work in [11] proposes an integrated framework for power, area and thermal modeling for large-scale computing systems. In this work, application traces are emulated rather than collected from cycle-accurate simulation, thus without considering the real behavior of reference use-case scenarios. The authors in [4] propose an integrated approach based on Virtutech Simics functional simulator, employing power and thermal models from real hardware characterization. The advantage of this approach relies on the possibility to develop, analyze and tune different control algorithms for thermal and power management, based on high-level Matlab descriptions. However, the power and thermal models are bound to a particular architecture and floorplan (an Intel©Xeon X7350 system), and also the simulation is not cycle-accurate. These aspects make the approach in [4] unsuitable for accurate thermal evaluation of MPSoC architectures with NoC communication channel running different core configurations and floorplans.

B. Dynamic thermal management

In the context of Dynamic Thermal Management (DTM), several approaches have been presented in literature for the run-time optimization of thermal profile in single-chip multiprocessor architectures. The major concern in this kind of works is the lack of an appropriate metric specifying the impact of temperature-related decisions on the performance degradation of the system (e.g., on the impact of CPI). Indeed, several authors provide a methodology based on simple temperature predictive control to avoid exceeding a predefined threshold value [28]. History-based approaches in this sense have been proposed in [30]. The performance impact of many DTM techniques for high-performance microprocessors has been extensively discussed in [6].

III. PROPOSED METHODOLOGY

This section details the four main aspects of the methodology proposed in this paper. Before presenting in details the formal model, it is worth giving some basic definitions that will be used throughout the entire section. The reference architecture is multi-core and composed of *tiles* placed in a 2D-mesh topology. Each tile is composed of a processor core, a router and a L2 cache bank; the router is used to interface to the distributed (shared) L2 cache. The 2D-mesh topology is logically composed of a set $R := \{1, 2, ..., n_R\}$ of n_R rows and a set $C := \{1, 2, ..., n_C\}$ of n_C columns. We also consider a set $D := \{1, 2, ..., n_D\}$ of duty-cycle islands. A duty-cycle island is composed by a set of tiles with a common clock rate. Each tile belongs to one and only one duty-cycle island.

The remainder of this section is organized as follows: at first, an optimization linear model to deal with the thermal/performance trade-off is sketched in Section III-A; such model is underpinned by two formal analytical relations on temperature and performance. Temperature and performance linear relations are discussed in details in Section III-B and Section III-C, respectively. Last, Section III-D details how the 2D-mesh topology has been exploited to support design time thermal performance analysis.

A. Thermal/performance linear model

We consider three sets of variables for the optimization linear model. For each tile $(i, j) \in R \times C$, the integer variable $p_{i,j} \in \{0\%, 1\%, ..., 100\%\}$ represents the performance degradation level of the tile with respect to the base-case where performance is 100%, and $t_{i,j} \geq 0$ defines its temperature. It is worth to notice that we measure core performance degradation level with respect to the maximum performance of the same core. We employ clock-gating to tune performance of each core such that the maximum performance is intended as duty-cycle equal to 1, without any clock-gating action. For each island $d \in D$, the integer variable r_d specifies its duty-cycle, i.e. the fraction of time the core in the tile is active, with respect to the time clock-gating stops its execution. We aim at maximizing the minimum performance for each tile, as specified by the following objective function, where the max-min formulation is satisfied by Equation 2.

$$max \; q \qquad\qquad\qquad\qquad\qquad (1)$$
$$q \leq p_{i,j} \qquad\qquad \forall \; (i, j) \in R \times C \qquad (2)$$

The first constraint to bind the frequency of each tile to its own duty-cycle island is as follows:

$$p_{i,j} = 1 - r_{f(i,j)} \qquad\qquad \forall \; (i, j) \in R \times C, \qquad (3)$$
$$f := (i, j) \to D, \qquad (4)$$

where f represents a mapping function between the Cartesian coordinates (i, j) of the tile in the 2D-mesh topology, and the duty-cycle island. The proposed methodology is biased toward this function, and further details will be given in Section III-D. Temperature-aware designs constraint the maximum operating temperature to a predefined threshold temperature T_{max}, determining the reliability of the processor chip. This constraint can be defined as a simple relation, as follows:

$$t_{i,j} \leq T_{max} \qquad\quad \forall \; (i, j) \in C \times R \qquad (5)$$

The optimization model presented so far sets a threshold temperature to the chip (Equation 5), meanwhile maximizing the performance of the worst-case task (Equation 1 and Equation 2). The result of this joint optimization lies in fairness of performance degradation across tiles belonging to different duty-cycle islands.

B. Thermal linear model

The linear model presented in Section III-A allows to maximize performance, under a maximum operating temperature constraint. However, the intrinsic simplicity of the maximum temperature requirement lacks of a suitable formulation to be employed in the linear optimization model. This section details a derived linear thermal equation that is meant to be employed in the proposed linear optimization model; this model is derived from extensive and accurate simulation measurements using a cycle-accurate simulation of homogeneous architectures (refer to Table I for more details on this). The thermal model of each tile is defined as:

$$t_{i,j} := \sum_{d \in D} (\alpha_d \cdot r_d) \qquad \forall \; (i, j) \in R \times C, \qquad (6)$$

where the temperature $t_{i,j}$ of tile (i, j) is linearly dependent on the duty-cycle r_d of island $d \in D$, and weighted by an unknown coefficient α_d, to be determined. In order to characterize Equation 6, i.e. quantifying α_d coefficients, we use a least square approach, since regressors are supposed to be independent. This means that duty-cycle islands are decoupled each other, with the advantage of finer grain control, but at increased hardware cost (associated to the control circuitry). To characterize the model, we have extracted a rich set of per-tile temperature measurements, using different duty-cycle combinations, using the cycle-accurate simulation framework presented in Section IV. Experimental results have shown a strong linear relation between regressors and temperature, strengthen by an analysis of the R^2 fitting coefficient, that is very close to 1. Moreover, experimental data generate a very well conditioned matrix A with $cond(A) \leq 10$ in all of conducted experiments on both 16 and 36 cores.

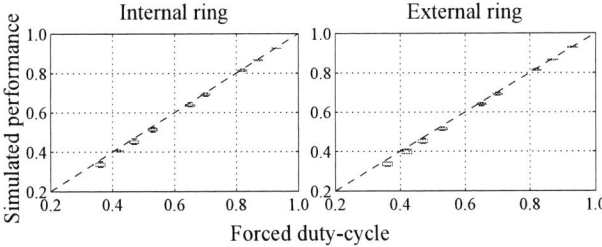

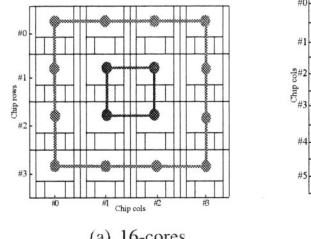

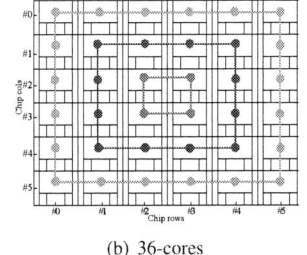

(a) 16-cores
(b) 36-cores

Fig. 1. Normalized simulated performance as a function of the applied clock-gating, for cores belonging to internal and external ring in 16-cores 2D-mesh.

Fig. 2. Rings in 16-cores and 36-cores 2D-mesh: a topological ring groups tiles that have similar thermal-related characteristics.

C. Performance linear model

The optimization model in Section III-A uses performance and temperature measurements to exploit the thermal/performance trade-off. This section details the linear relation that binds processor performance to the duty-cycle it belongs to. Notice that the validity of the proposed linear model is underpinned by the fact that the reference processor is an in-order core. Although the validity of the model is coupled with a specific and simple architecture, it is worth noticing that such in-order processors are still used in high-performance systems, such as Web servers or Data centers [20]. In addition, each core in the multi-core processor can be assumed to be isolated from the rest of the chip, because each core is assumed to serve a single request, to maximize response throughput and to ensure logical and physical security [2].

Processors run at a fixed clock frequency, and the performance is related to the number of committed instructions, bound to the level of duty-cycle specified by the island the processor belongs to. Moreover, for simple and only in-order cores without multi-thread capabilities, the clock rate is tightly coupled to all the executed instructions.

We have experimentally validated such relation by an extensive set of experiments on our cycle-accurate simulation framework using benchmarks from different test suites, finding a strong linear correlation between committed instructions and duty-cycle. Figure 1 shows the linear relationship between the number of committed instructions (*Simulated performance* on vertical axis) and the applied duty-cycle (*Forced clock-gating* on horizontal axis) for a 16-cores architecture in both internal and external topological rings. The dotted line represents the theoretical linear relation between committed instructions and clock-gating level, while the box-and-whiskers plots represent the simulated performance: for each clock-gating level, the maximum, minimum and median simulated performance are reported. The height of each box plot is tied to the variability of the simulated measurements: 50% of the simulated values fall in this interval. The width of the box plot, on the other hand, has no statistical meaning, but for graphical intent. As already stated, Figure 1 presents a strong linear relation, with very low variance at almost every clock-gating level. However, the variance increases, i.e. greater box height, with performance decreasing (higher clock-gating levels).

D. Ring-based view in 2D-meshes

The optimization model presented in Section III-A employs the mapping function f to bind tile performance to duty-cycle island it belongs to; however, a suitable analytical formulation of such function has not been provided, yet. This section details the f mapping function formulation to exploit the 2D-mesh topology for thermal/performance trade-off.

The rationale of our mapping function proposal is based on a simple yet effective observation: thermal hotspots are generally located in the centre of a 2D-mesh architecture, independently of the size of the mesh (refer to Section IV for additional details). This fact is tied to the thermal coupling phenomenon: cores surrounded by other cores (as it happens for those located in the centre of the chip) are under the direct influence of core-to-core heat exchange (e.g., through conduction), such that their operating temperature increases up to a point where the thermal management solution is able to dissipate the total system heat. Moreover, the hotspot trend is independent of the mesh size, with the maximum temperature reached by the centre of the chip, and gradually decreasing toward the edges. The only impact of the mesh size is on the maximum operating temperature, with increasing absolute values for aggressive integration made possible by continuous technology scaling. Another key observation relies on the symmetry property of a 2D-mesh thermal map, with respect to all dimensions. Starting from these two observations, the proposed methodology constructs a concentric ring-based set of duty-cycle islands. The concept of *topological ring* is shown in Figure 2 for both 16-cores and 36-cores architectures, with 2 and 3 topological rings respectively. A topological ring is associated with a set of cores in the architecture, and rings are placed concentric each other. Each core belongs to one and only one ring, and cores belonging to the same ring share similar temperature dissipation properties: for instance, all the cores belonging to the outermost topological ring are placed against the chip edge, with direct impact on the way heat is dissipated and temperature is exchanged with package and ambient [24].

IV. EXPERIMENTAL RESULTS

The methodology proposed in Section III is general, while its validity is hereby shown for a reference architecture. In this perspective, we have focused on a real environment scenario

978-1-4673-2895-1/12 $31.00 © 2012 IEEE

to validate the goodness of the methodology in a real-world context, to demonstrate the practical solution found. Section IV-A details simulation setup and experimental settings, and the steps to assess the proposed methodology. Section IV-B reports and discusses a preliminary analysis on the role of thermal coupling in setting the operating temperature of a multi-core architecture: we will show that the high density of cores in a multi-core architecture makes the central region of the silicon die more spotted to reliability concerns. Section IV-C reports strengthening results obtained on the selected reference architecture. Section IV-D shows how the proposed model can constraint the operating temperature, given a tunable threshold: in reliable designs, this is of utmost relevance in determining the lifetime of the device. Last, temperature/performance trade-off is shown in Section IV-E.

A. Experimental setup and methodology evaluation

We conducted several experiments using a modified version of GEM5 as an appropriate cycle-accurate simulator (http://gem5.org), a modified version of McPAT [17] and Orion [13] detailed models for cores and routers power consumption estimates, and the widely used HotSpot thermal model [24] to generate chip temperature map. The reference architecture we target is an Alpha21364 network architecture [20], that is used in real Web-servers and Data-centre contexts; commercial examples exist for this kind of architecture, based on the Alpha21264 processor core. We selected and simulated two different architecture configurations, with 16 and 36 cores based on the Alpha21364 architecture. We conducted the experiments with the architecture configuration presented in Table I for typical 45nm technology node. Each tile in the network architecture is composed of a single Alpha21264 core, 1.75MB local (shared) L2 cache memory and a router to interface to the NoC; its logical architecture is reported in Figure 3 for reference.

We assess the soundness of the proposed methodology within four main steps. First, a set of $500 + 500$ experiments are conducted on both 16-cores and 36-cores architectures to collect representative samples for both temperature and performance related to different duty-cycle levels. Each experiment runs for 2×10^7 instructions per core with a different benchmark mix randomly selected from our representative pool of benchmark suites. We used WCET benchmarks from Mälardalen University [7], SPLASH2 [29] from the University of Delaware, and MiBench [8] to cover a broad range of applications, with a mix of integer, floating-point and memory instructions. For each experiment different duty-cycle levels are set for each topological ring in the architecture. Starting from such raw data, we have estimated both the thermal and performance model, described in Section III-B and III-C respectively, using a least squares approach. Then, for a selected set of $10 + 11$ temperature levels, we run the optimization model to obtain duty-cycle levels for each ring to achieve the desired chip temperature.

We run 20 different simulation for each optimized temperature duty-cycle, for a total of 10×20 simulation on 16-cores

TABLE I
TILE, CORE ARCHITECTURE AND TECHNOLOGY PARAMETERS.

Processor core	3GHz, in-order based on Alpha21264 core
Int-ALU	4 integer ALU functional units
Int-Mult/Div	4 integer multiply/divide functional units
FP-Mult/Div	4 floating-point multiply/divide functional units
L1 cache	64kB 2-way set assoc. split I/D, 2 cycles latency
L2 cache	1.75MB per bank, 8-way associative
Router	2-stage wormhole switched (Garnet network [1])
Topology	2D-mesh based on Alpha21364 network processor
Technology	45nm at 1.1V

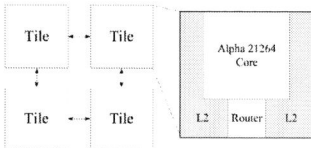

Fig. 3. Alpha 21364 tile architecture, adapted from [20].

and 11×20 simulation on 36cores. Last, we compared the maximum simulated chip temperature against the predefined threshold, under the performance level found by the optimization model, as reported in Section IV-D and IV-E respectively.

B. Preliminary analysis on topological rings

The methodology presented in this paper is driven by a ring-based view of the target multi-core chip: the processor floorplan is divided into concentric rings, each ring being composed of a predefined set of tiles. The optimization linear model presented in this paper allocates clock-gating levels to cores, according to their placement (i.e., according to the ring they belong to) and according to the desired optimization (e.g., maximum absolute temperature). The rationale of the ring-based methodology has been sketched in Section III, and it is hereby further detailed with experimental results. Figure 4 shows the temperature profile of the 16-cores processor running different applications. The temperature map has been generated after executing 2×10^7 instructions, and after having collected microarchitecture-level statistics to be passed to the power and thermal models. Two aspects are clear from this scenario: the centre of the die has an higher operating temperature with respect to the edges of the silicon die, even though the power consumption of each single core is comparable. This phenomenon is related to the thermal coupling between adjacent cores, causing the centre of chip to increase the heat dissipation density, increasing the operating temperature. This phenomenon has been shown to get worse with technology scaling [12], but for two to four-cores architectures only. With more cores integrated in the same silicon die, the problem is exacerbated. From a reliability view-point, higher operating temperatures introduce several problems. The thermal profile from Figure 4 presents some variability while crossing horizontally adjacent cores, and this is due to the L2 caches that are known to be cold spots. Routers, on the other hand, contribute to the higher temperature value between cores that are vertically adjacent in the matrix.

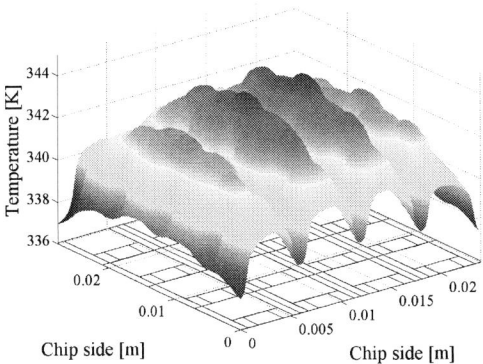

Fig. 4. Temperature profile of a 16-cores processor based on Alpha21364 tiles.

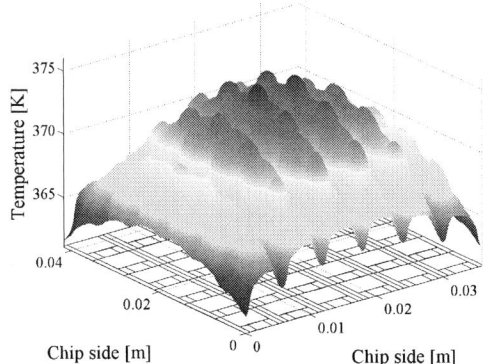

Fig. 5. Temperature profile of a 36-cores processor based on Alpha21364 tiles.

TABLE II
INSTRUCTIONS BREAKDOWN AND POWER CONSUMPTION PROFILE FOR
EACH CORE IN THE 16-CORES PROCESSOR, WHOSE TEMPERATURE
PROFILE IS GIVEN IN FIGURE 4.

| Core # | Placement | | Instructions | | | Power |
	Row	Col	Int	FP	Mem	[W]
1	0	0	12.8%	6.6%	80.6%	7.132
2	0	1	34.6%	36.5%	28.9%	6.810
3	0	2	64.4%	30.4%	5.3%	7.185
4	0	3	40.8%	29.6%	29.6%	6.717
5	1	0	40.8%	29.6%	29.6%	6.956
6	1	1	64.4%	30.4%	5.3%	6.630
7	1	2	40.8%	29.6%	29.6%	6.849
8	1	3	67.7%	32.3%	0.0%	6.632
9	2	0	66.1%	33.9%	0.0%	6.909
10	2	1	80.7%	19.2%	0.1%	6.790
11	2	2	99.1%	0.9%	0.0%	6.717
12	2	3	68.0%	31.9%	0.0%	6.848
13	3	0	69.4%	30.6%	0.0%	6.626
14	3	1	66.1%	33.9%	0.0%	7.131
15	3	2	34.6%	36.5%	28.9%	7.042
16	3	3	12.8%	6.6%	80.6%	6.790

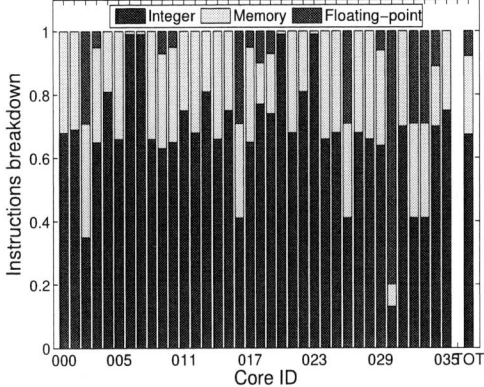

Fig. 6. Instructions breakdown per core, and total mix for the 36-cores architecture.

C. Preliminary analysis on applications

The validation of the proposed methodology on a real architecture allows to demonstrate the goodness of our solution, giving us the possibility to exploit the architecture itself to strengthen and generalize our results. Experiments show that considering in-order cores organized in a 2D-mesh, allows to provide an optimal solution that is roughly application-independent. In particular, a detailed view of the applications shows different power consumptions, as sketched in Table II; however, such power differences do not greatly impact thermal map, since this is overwhelmed by the thermal coupling effects. Simply put, we can say that for this specific architecture the effect of different workload is negligible compared to the thermal coupling effects. This result, that is extensively supported by experimental data, allows us to cast a single design-time optimization solution in terms of constrained chip temperature and performance level. Such solution is valid for each application mix that is mapped on the multi-core, providing a great design-time optimization result.

The same situation is seen in 36-cores architectures, where the high number of cores pushes temperature toward further

high values. Figure 5 reports the temperature surface of a 36-cores processor, and the relative floorplan. In this case, the workload assignment and instructions breakdown is given in Figure 6, for each core and the total mix.

D. Constraining absolute operating temperature

Reliable designs focus on minimizing operating temperature, to increase the MTTF and reduce the probability of faults. In this work we address hard-faults and not transient ones, and consider two main mechanisms that are known to cause several problems to high-performance processors in scaled technologies [27]: electromigration and stress-migration. We compute the MTTF for these two mechanisms, through the expressions given in Equation 7, taken from [27]: E_{EM} and E_{SM} are the energy activation for electromigration and stress-migration respectively, k is the Boltzmann's constant, T the operating temperature and T_0 the reference temperature for stress-migration (melting temperature), and n is a technology-dependent parameter. We use the values for these parameters as given in [27]. Notice that we consider only the exponential contribution from electromigration, instead of considering the current density since we are assuming to compare results at

TABLE III
MODEL OUTPUT ACCURACY FOR 16-CORES PROCESSOR.

Target threshold temperature [K]	Max. simulated temperature	
	Average [K]	Variance
332.0	332.06	0.005
333.0	333.11	0.002
334.0	333.92	0.007
335.0	334.96	0.007
337.0	337.13	0.018
338.0	337.97	0.013
339.0	339.01	0.021
340.0	340.07	0.022
341.0	340.86	0.033
342.0	341.92	0.047

TABLE IV
MODEL OUTPUT ACCURACY FOR 36-CORES PROCESSOR.

Target threshold temperature [K]	Max. simulated temperature	
	Average [K]	Variance
350.0	350.13	0.034
352.0	352.48	0.126
354.0	354.32	0.018
356.0	356.15	0.002
358.0	358.20	0.027
360.0	360.16	0.137
362.0	361.95	0.143
364.0	363.93	0.099
366.0	365.82	0.193
368.0	367.74	0.211
370.0	369.67	0.379

different operating temperatures but equal operating conditions (e.g., supply voltage and frequency).

$$MTTF_{EM} \propto exp\left\{\frac{E_{EM}}{k\,T}\right\}$$

$$MTTF_{SM} \propto |T_0 - T|^{-n} \cdot exp\left\{\frac{E_{SM}}{k\,T}\right\} \quad (7)$$

Figure 7 shows the reliability projection of the system for 16-cores and 36-cores processor, while constraining absolute operating temperature. The dotted line shows the theoretical trend of MTTF values with changing temperatures, while circular and diamond markers show the projections ensured by our model: the reliability values are those obtained while employing the optimization model presented in Section III averaged across different runs. The horizontal axis reports the target reliability improvement relative to base case when reliability equals 1. The vertical axis reports the operating temperature required to accommodate such improvement: for example to increase by 40% MTTF caused by electromigration in 16-cores processor, temperature should be diminished to 337K from the 342K base case. Our model ensures that the maximum operating temperature is 337.13K, achieving the expected reliability with an error of less than 1%.

Extensive experimentation has shown a good match between the temperature ensured by the proposed optimization model, and the maximum temperature requirements. Table III and Table IV report the results for the 16-cores and 36-cores processors, respectively. Data is given as an average and variance. Results show a very good match of the computed temperature against the target one, with an average (absolute) error of less than 0.1K and variance in the order of 0.02.

E. Temperature/performance trade-off

The temperature/performance trade-off is depicted in Figure 8 and Figure 9 for 16-cores and 36-cores processor respectively. The plots show a linear relation between the operating temperature and the desired performance. Performance is reported as a percentage over the base-case, when no clock-gating is applied and performance is at 100%. Trade-off linearity is experienced in both architectures, meanwhile presenting a linear relation throughout the entire performance degradation interval from 93% down to 36%. It is worth noticing that

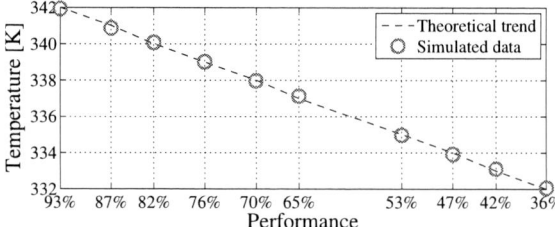

Fig. 8. Temperature/performance trade-off for 16-cores architecture: theoretical trend and simulated data.

Figure 8 and Figure 9 show results for one single generic tile, since the proposed optimization model flattens performance degradation equally on each tile, for each experiment. It is belief of the authors this is a relevant result, since it gives suitable control over operating temperature through a simple relation with respect to core performance.

V. CONCLUSIONS

A joint thermal/performance optimization model has been proposed for design-time optimization of multi-core architectures, as opposed to state-of-the-art Dynamic Thermal Management solutions. Performance and clock-gating have been shown to be linearly related, such that it is possible to use clock-gating as control-knob to seize performance and temperature. Indeed, temperature and performance have been demonstrated to follow a linear relation. The proposed sound and formal model has been used to provide a system-wide optimization framework, focusing on real-world 16-cores and 36-cores processors for high-performance servers and

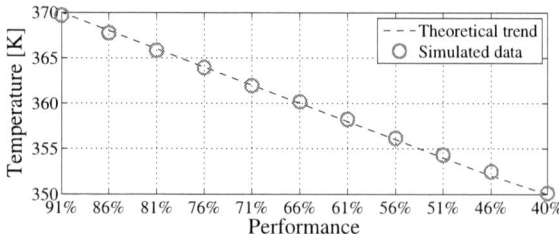

Fig. 9. Temperature/performance trade-off for 36-cores architecture: theoretical trend and simulated data.

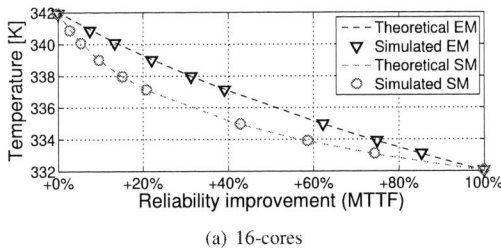

(a) 16-cores

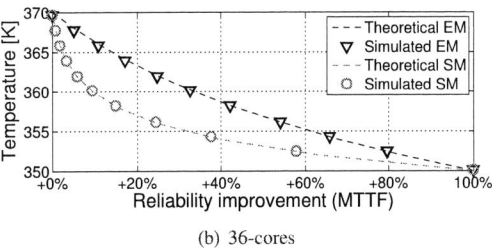

(b) 36-cores

Fig. 7. Target reliability improvement against base-case sceanrio (MTTF = 1), and required operating temperatures: theoretical and simulated data.

data-centres. Extensive experimental results have shown the goodness of the proposed optimization model, with respect to 16-cores and 36-cores processors running a predefined set of representative benchmarks. The linear relations have been shown to cover a broad range of temperature and performance situations, such that the proposed methodology is suitable to be employed in real-case scenarios.

REFERENCES

[1] N. Agarwal, T. Krishna, L.-S. Peh, and N. Jha. Garnet: A detailed on-chip network model inside a full-system simulator. In *Performance Analysis of Systems and Software, IEEE International Symposium on*, pages 33 –42, april 2009.

[2] A. M. Azab, P. Ning, and X. Zhang. Sice: a hardware-level strongly isolated computing environment for x86 multi-core platforms. In *ACM Conference on Computer and Communications Security'11*, pages 375–388, 2011.

[3] A. Banerjee, R. Mullins, and S. Moore. A Power and Energy Exploration of Network-on-Chip Architectures. In *NOCS '07*, pages 163–172. IEEE Computer Society, 2007.

[4] A. Bartolini, M. Cacciari, A. Tilli, L. Benini, and M. Gries. A virtual platform environment for exploring power, thermal and reliability management control strategies in high-performance multicores. In *GLSVLSI'10*, pages 311–316, New York, NY, USA, 2010. ACM.

[5] S. Borkar. Thousand core chips: a technology perspective. In *Annual ACM IEEE Design Automation Conference*, 2007.

[6] D. Brooks and M. Martonosi. Dynamic Thermal Management for High-Performance Microprocessors. In *High Performance Computer Architecture. International Symposium on*, 2001.

[7] J. Gustafsson, A. Betts, A. Ermedahl, and B. Lisper. The mälardalen wcet benchmarks - past, present and future. In *Proceedings of the 10th International Workshop on Worst-Case Execution Time Analysis*, July 2010.

[8] M. R. Guthaus, J. S. Ringenberg, D. Ernst, T. M. Austin, T. Mudge, and R. B. Brown. Mibench: A free, commercially representative embedded benchmark suite. In *Proceedings of the Workload Characterization. IEEE International Workshop*, pages 3–14, Washington, DC, USA, 2001.

[9] S. Heo, K. Barr, and K. Asanović. Reducing power density through activity migration. In *Proceedings of the 2003 international symposium on Low power electronics and design*, ISLPED '03, pages 217–222, New York, NY, USA, 2003. ACM.

[10] Y. Hoskote, S. Vangal, A. Singh, N. Borkar, and S. Borkar. A 5-ghz mesh interconnect for a teraflops processor. *Micro, IEEE*, 27(5):51 –61, sept.-oct. 2007.

[11] M.-y. Hsieh, A. Rodrigues, R. Riesen, K. Thompson, and W. Song. A framework for architecture-level power, area, and thermal simulation and its application to network-on-chip design exploration. *SIGMETRICS Perform. Eval. Rev.*, 38:63–68, March 2011.

[12] M. Janicki, J. H. Collet, A. Louri, and A. Napieralski. Hot spots and core-to-core thermal coupling in future multi-core architectures. In *Annual IEEE SEMI-THERM Symposium*, pages 205–210. IEEE, Feb. 2010.

[13] A. Kahng, B. Li, L.-S. Peh, and K. Samadi. Orion 2.0: A fast and accurate noc power and area model for early-stage design space exploration. In *DATE.*, pages 423 –428, april 2009.

[14] J. S. Kim, M. B. Taylor, J. Miller, and D. Wentzlaff. Energy characterization of a tiled architecture processor with on-chip networks. In *Proceedings of the International Symposium on Low Power Electronics and Design*, pages 424–427, New York, NY, USA, 2003. ACM.

[15] J. Kong, S. Ching, and K. Skadron. Recent thermal management techniques for microprocessors. *ACM Computing Surveys. To appear*, 2012.

[16] C. J. M. Lasance. Thermally driven reliability issues in microelectronic systems: status-quo and challenges. *Microelectronics Reliability*, 43(12):1969–1974, 2003.

[17] S. Li, J. H. Ahn, R. Strong, J. Brockman, D. Tullsen, and N. Jouppi. Mcpat: An integrated power, area, and timing modeling framework for multicore and manycore architectures. In *Microarchitecture. Annual IEEE/ACM International Symposium on*, pages 469 –480, dec. 2009.

[18] A. Majumdar. Helping chips to keep their cool. *Nature Nanotechnology*, (1):214–215, 2009.

[19] M. Monchiero, R. Canal, and A. González. Design space exploration for multicore architectures: a power/performance/thermal view. In *Proceedings of the 20th annual international conference on Supercomputing*, ICS '06, pages 177–186, New York, NY, USA, 2006. ACM.

[20] S. S. Mukherjee, P. Bannon, S. Lang, A. Spink, and D. Webb. The alpha 21364 network architecture. In *Proceedings of the The Ninth Symposium on High Performance Interconnects*, pages 113–, Washington, DC, USA, 2001. IEEE Computer Society.

[21] T. Ormandy. An Empirical Study into the Security Exposure to Hosts of Hostile Virtualized Environments.

[22] M. Pedram and S. Nazarian. Thermal Modeling, Analysis, and Management in VLSI Circuits: Principles and Methods. *Proceedings of the IEEE*, 94(8):1487–1501, August 2006.

[23] M. D. Powell and T. N. Vijaykumar. Resource area dilation to reduce power density in throughput servers. In *Proceedings of the 2007 international symposium on Low power electronics and design*, ISLPED '07, pages 268–273, New York, NY, USA, 2007. ACM.

[24] K. Skadron, M. Stan, W. Huang, S. Velusamy, K. Sankaranarayanan, and D. Tarjan. Temperature-aware microarchitecture. In *Computer Architecture, 2003. Proceedings. 30th Annual International Symposium on*, pages 2 – 13, june 2003.

[25] K. Skadron, M. R. Stan, K. Sankaranarayanan, W. Huang, S. Velusamy, and D. Tarjan. TSkadronTACO0oemperature-aware microarchitecture: Modeling and implementation. *ACM Transactions on Architecture and Code Optimization (TACO)*, 1(1), 2004.

[26] V. Soteriou, N. Eisley, H. Wang, B. Li, and L.-S. Peh. Polaris: A system-level roadmap for on-chip interconnection networks. In *ICCD 2006.*, pages 134 –141, oct. 2006.

[27] J. Srinivasan, S. Adve, P. Bose, and J. Rivers. The case for lifetime reliability-aware microprocessors. In *Computer Architecture. Annual International Symposium on*, pages 276 – 287, 2004.

[28] J. Srinivasan and S. V. Adve. Predictive dynamic thermal management for multimedia applications. In *ICS '03: Proceedings of the 17th annual international conference on Supercomputing*, 2003.

[29] The modified SPLASH-2 benchmark suite from University of Delaware http://www.capsl.udel.edu/splash/.

[30] I. Yeo, C. C. Liu, and E. J. Kim. Predictive dynamic thermal management for multicore systems. In *45th Annual Design Automation Conference*, 2008.

978-1-4673-2895-1/12 $31.00 © 2012 IEEE

A Double Data Rate 8T-Cell SRAM Architecture for Systems-on-Chip

Saleh M. Abdel-Hafeez, Mohammad Shatnawi
Department of Computer Engineering
Jordan University of Science and Technology
22110 Irbid, Jordan, P.O.BOX 3030
sabdel@just.edu.jo

Ann Gordon-Ross
Department of Electrical and Computer Engineering
University of Florida, Gainesville, FL 32611, USA
ann@ece.ufl.edu

Abstract—**The substantial increase in market demand for hand-held devices drives the need for low-power high-speed data access SRAM for systems-on-chip (SoCs). In this paper, we present a novel low-power SRAM architectural design that provides high-noise margin double data rate (DDR) read/write accesses using a conventional 8T-Cell and a partitioned architectural structure consisting of even and odd modules (corresponding to even and odd addresses), which are accessed alternatingly. Write accesses occur at both clock edges such that the even modules are accessed at the rising edge and the odd modules are accessed at the falling edge. Similarly, the read accesses occur at both clock edges such that the even modules are assumed to be evaluated at the rising clock edge, while concurrently the odd modules are pre-charged, and vice versa. We implement a 128-bit X 64-bit SRAM with DDR accesses and an 8T-Cell structure using a standard 0.09μm/1V CMOS TSMC process. Simulation results reveal that our architecture operates with a 1GHz read/write cycle, a data throughput of 2GHz/64-bit, and an average power consumption of 23.4mW.**

Keywords—*Double-Data-Rate (DDR) Memory, 8T-Cell, SRAM, System-on-Chip (SoC).*

I. INTRODUCTION

The demand for high data throughput and low power operation SRAM designs for systems-on-chip (SoCs) is ever increasing due to the market demand for hand-held devices and applications. Since many of these applications process a tremendous amount of data, both high throughput and low power consumption are critical for the advancement and success of these applications.

Data processing throughput can be directly increased by accessing data on both the rising and falling clock edges—double data rate (DDR) throughput. Commercial products leverage this increased throughput using a DDR system input/output (I/O) bus that interfaces with off-chip DDR SDRAM. However, this bus is typically only available for accessing off-chip main memory and not for transfers between on-chip memory modules, which are used for temporary storage and manipulation of data between arithmetic units. Since these on-chip SRAM circuits typically leverage different geometric variations, the SRAM design can be based on a wide topology of memory cell structures (e.g., [1][3][4][6][8][9]). Whereas these structures are appropriate for the recent features required by memory designs for SoC products, since the on-chip system I/O bus does not support DDR access, on-chip DDR SRAMs cannot be used. In order to support on-chip DDR memory communication, we propose a DDR SRAM architecture using a typical 8T-Cell structure [1][3][4], which is amenable to continued technology scaling [3], in Sections II and III. Section IV presents HSPICE simulation results and a comparison with related work. Section V summarizes and concludes our work.

II. DDR SRAM DESIGN SPECIFICATIONS

In this section, we present our proposed DDR SRAM design specifications for a 128-bit (height) X 64-bit (width) SRAM using the 8T-Cell and the I/O signals denoted in TABLE 1 and illustrated in Figure 1. The design's timing constraints depend on a self-timing design methodology [2], where modifying the transistors' sizes and inserting gates are the key elements for reducing the skew timing and ensuring proper operation. For reference, TABLE 2 lists all of the abbreviated timing constraints.

Figure 2 illustrates the write timing constraints for the write address and data buses with respect to the required setup and hold time constraints for the address and data buses, and in addition, to the rising and falling edge of the write clock. The read timing constraints are similar and are omitted for brevity. Without loss of generality, even data addresses are written/read to even modules after the rising edge of *WCLK*/*RCLK* and odd data addresses are written/read to odd modules after the falling edge of *WCLK*/*RCLK*. This method allows for DDR throughput to be fully utilized during read or write operations by toggling between consecutive even and odd addresses, which imposes

TABLE 1: SYMBOL DEFINITIONS FOR THE ARCHITECTURAL SIGNALS

SYMBOL	DEFINITION	Maximum input capacitance (Cin) in pf
FFI (63:0)	Input data	0.01
WCLK	Write clock	0.01
RCLK	Read clock	0.01
RDA (6:0)	Read address	0.01
WRA (6:0)	Write address	0.01
FFO (63:0)	Output data	------------

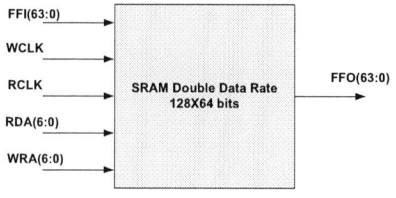

Figure 1: 128-bit X 64-bit DDR SRAM and relevant signals

TABLE 2: SYMBOL DEFINITIONS FOR THE TIMING SIGNALS

SYMBOL	DEFINITION
T_{Eas}/T_{Eah}	Even module address setup/hold time
T_{Eds}/T_{Edh}	Even module data setup/hold time
T_{Oas}/T_{Oah}	Odd module address setup/hold time
T_{Ods}/T_{Odh}	Odd module data setup/hold time
$T_{Wacc} = T_{Eah} = T_{Oah}$	Write access
$T_{Eracc} = T_{Oracc}$	Read access

essentially no throughput degradation on the design's operation since data, in many applications, are usually processed in blocks of consecutive even-odd addresses.

Our proposed DDR SRAM design's core cells are implemented using the 8T-Cell depicted in Figure 3 [1][3][4]. The 8T-Cell structure provides a read mechanism that does not disturb the internal node of the cell with a high read-write noise margin, thus, the 8T-Cell is amenable to continued technology scaling with low supply voltage [4]. In addition, the 8T-Cell realizes a low power consumption sensing I/O circuit that is considered among the least power of all counterpart memory cells [4]. For brevity, we refer the reader to [1][3]**Error! Reference source not found.** for additional details on the 8T-Cell's advantages. The proposed sizes under 90nm technology are depicted in Figure 3 as dedicated sizes derived by several foundries [11][12] for general SRAM SoC products, which we leverage in our design's simulation.

III. DDR SRAM ARCHITECTURE

A. Modules and Decoder Structure

The SRAM DDR architecture is partitioned into two main modules of sizes 64-bits X 64-bits, which are depicted as the even and odd modules in Figure 4 and are constructed using arrays of 8T-Cells. This partitioned approach provides a regular structure with sufficient driving capabilities and reduces the skew timing variations between the cells, and thus minimizes the design iterations necessary for modifying the sizes of the buffers and logic gates.

The least most significant address bus bits (RDA_0 and WRA_0) distinguish between the even and odd modules, while the remainder of the address bus bits (RDA_6 to RDA_1 and WRA_6 to WRA_1) are evaluated in the pre-decoder module. This parallelization minimizes the even and odd decoders' switching activities and results in efficient addressing power

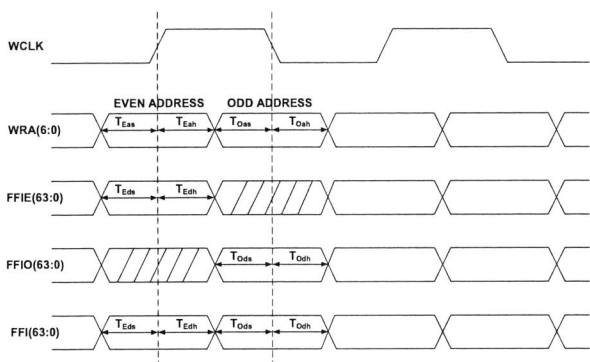

Figure 2: Write timing constraints between the write clock and write address and data buses

consumption. We connect RDA_0 and WRA_0 to the last stage of the even and odd decoders with $WCLK/RCLK$ in order to preserve all timing constraints with the addition of a minimum pre-decoder and decoder gate activity.

The decoder delay structure inhibits balance delay among all of the decoder's selected output pins with respect to all of the input pins in order to maintain a constant setup and hold time with respect to the memory clock systems ($RCLK$, $WCLK$). We refer the reader to [2] for the complete self-timing decoder design.

B. Write Operation

Data is driven through simple CMOS inverters—input buffers—as depicted in Figure 5 (a). Every input buffer is associated with one data bit as input that generates two write bit lines (WBL, $WBLB$) as outputs, which are the complement of each other. The write bit lines for any particular data bit input are associated with a column array of 64 8T-Cells. Consequently, every write bit line (WBL, $WBLB$) is connected to 64 diffusion capacitances as shown by the vertical dashed line in Figure 6. The write bit line delay T_{WBL} can be approximated using [13]:

$$T_{WBL} = 0.35 \times R_{WBL} \times C_{WBL} \times L^2_{WBL} \qquad (1)$$

where R_{WBL} and C_{WBL} are the distributed components of the write bit line including the two overlap diffusion capacitances between every two adjacent cells in the column.

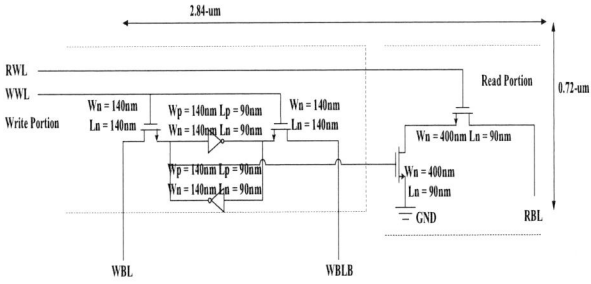

Figure 3: Schematic of the 8T-Cell with size 2.84μm X 0.72μm

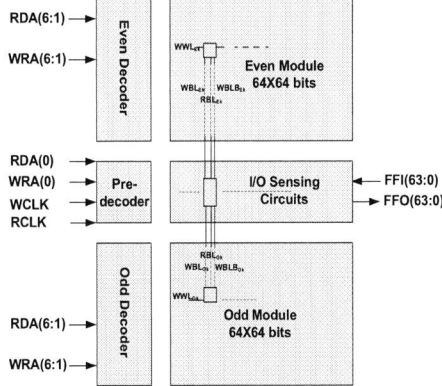

Figure 4: Proposed partitioned DDR SRAM architecture

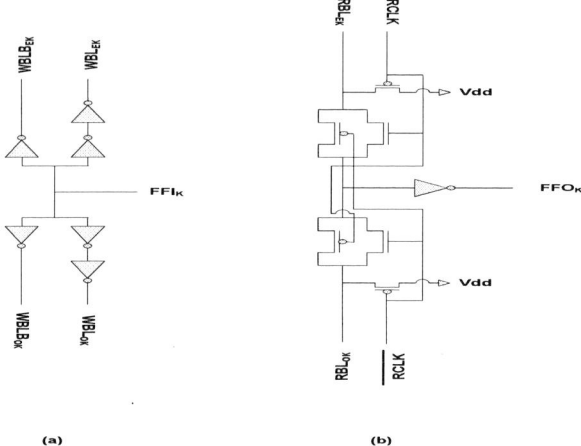

(a)   **(b)**

Figure 5: I/O sensing circuit: (a) input buffer; (b) output buffer

The length of the write bit lines L_{WBL} is the length of the memory column, which equals 64 x 0.72 µm = 46.08 µm.

The write word lines (WWL) are associated with a row array of 64 8T-Cells where each cell comprises of two gates, as shown by the horizontal line in Figure 6, The write word line delay T_{WWL} can be approximated using [13]:

$$T_{WWL} = 0.35 \text{ x } R_{WWL} \text{ x } C_{WWL} \text{ x } L^2_{WWL} \qquad (2)$$

where R_{WWL} and C_{WWL} are the distributed parasitic components of the write word lines including the two gate capacitances per cell in 64 cells of the row. The length of the write word lines L_{WWL} is the length of the memory row, which equals 64 x 2.84 µm = 158.72µm.

Each $WBL_k/WBLb_k$ data bit line must preserve a setup and hold time with respect to WWL_k at each cell based on the timing depicted in Figure 2.

C. Read Operation

The 8T-Cell separates the read and write operational logic and the read bit lines (*RBL*) must be pre-charged before evaluation. A stored value of 1 is considered the critical path read delay, which can be approximated using [13]:

$$T_{RACC} = 0.35 \text{ x } R_{RWL} \text{x } C_{RWL} \text{ x } L^2_{RWL} + 0.35 \text{ x } R_{RBL} \text{x } C_{RBL} \text{ x } L^2_{RBL} \qquad (3)$$

where R_{RWL} and C_{RWL} are the distributed parasitic values including the gate capacitances of the cell for 64 cells of the read word lines (RWL) and length of the read word lines L_{RWL} equals 64 x 2.84 µm =181.76 µm. Alternatively, R_{RBL} and C_{RBL} are the distributed parasitic values of the read bit lines including the diffusion capacitance of the read portion per cell where the length of the read bit lines L_{RBL} equals 64 x 0.72 µm = 46.08 µm.

The read access time depicted in Equation (3) depends on the address setup and hold times and the rising/falling edge of *RCLK*. In order to ensure proper DDR read operation, the read access time T_{RACC} must be completed within a quarter of the read clock cycle T_{RCLK} (i.e., $T_{RACC} \le \frac{1}{4} T_{RCLK}$).

Figure 5 (b) depicts the sensing circuit that multiplexes between the two modules of read bit lines ($RBLO_k$, $RBLE_k$)

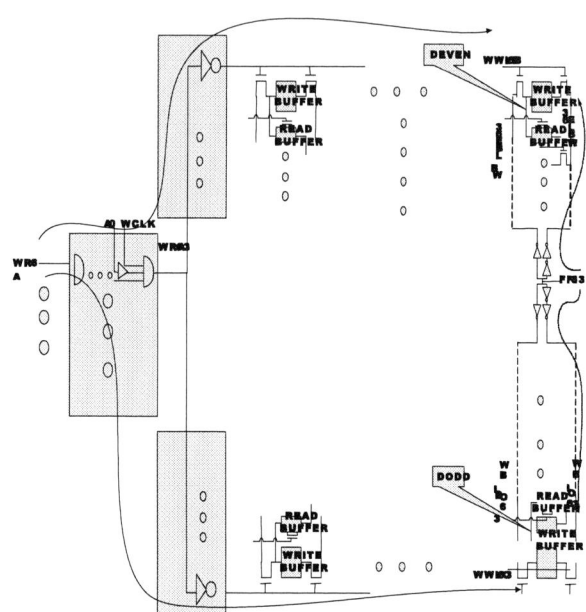

Figure 6: Write critical paths for even and odd corner cells

such that one bit line (i.e., in the even module) is pre-charged while the other bit line (i.e., in the odd module) is evaluated. Each sensing circuit is associated with two columns of the 64 8T-Cell array each in opposite directions. This structure results in a total of 64 sensing circuits where each sensing circuit has two input lines, $RBLE_k$ and $RBLO_k$

The pre-charge time for $RBLE_k$ and $RBLO_k$ depends only on the assertion or de-assertion of *RCLK*, which activates the sense circuits for pre-charging the even or odd module's read bit lines. The pre-charge time occurs within one half cycle of *RCLK* and can be approximated using [13]:

$$T_{pre\text{-}charge} = 0.35 \text{ x } R_{RBL} \text{ x } C_{RBL} \text{ x } L^2_{RBL} \le \frac{1}{2} T_{RCLK} \qquad (4)$$

IV. HSPICE SIMULATIONS AND COMPARISONS

This section shows the HSPICE simulations for the write operation (for brevity, we omit similar read operation results) for our proposed DDR SRAM constructed with 8T-Cells at a size of 128-bit X 64-bit (8,192 bits of total storage) with the following specifications: a supply voltage of 1V, a temperature of 25°C, a write operating frequency of 1GHz, and 90nm TSMC CMOS technology. To fully verify and demonstrate the DDR operation, we simulated writing from the even and odd modules. We show a sample portion of the simulation consisting of the upper right corner cell for an even module and lower right corner cell for an odd module since these corner cells present the worst case delay.

Figure 7 shows the write timing simulation with respect to the toggling write address *WRA63*, gated clock *WCLK*, and input data *FFI63*. The address is propagated through the pre-decoder and is gated with WRA_0 and *WCLK* in the decoder, thereby generating the *WWLE63/WWLO63* signal, which propagates horizontally as shown by the dashed line in Figure 6 and approximated by Equation (2). The data bit line *FFI63* arrives at the upper cell from *WBLE63/WBLO63* with

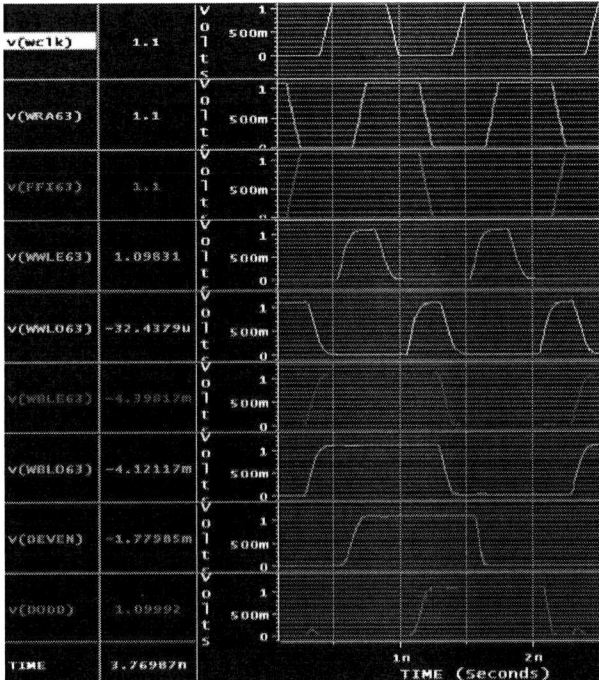

Figure 7: Write timing simulation @ 1 GHz at 90 nm TSMC technology

TABLE 3: COMPARISON WITH SIMILAR FEATURES DESIGNS

Design	Size (K-bit)	Tech	Power (mW)	Clock access	Type
Ours	8	90nm/1V	23.4	1ns	DDR
[10]	64	90nm/1V	12.9	1.2ns	SDR
[7]	8	180nm/1.8	20.5	2ns	SDR
[5]	8	65nm/1V	10.7	2ns	SDR

leverages DDR throughput for read/write access by leveraging a partitioned architecture wherein the memory module is partitioned into two modules—an even and an odd module—which alternately operate on the data at rising and falling edges of the memory clock. Additionally, we architected an I/O low-power sense multiplexed circuit to facilitate the DDR read operation. Simulations verified our design's correctness with a 64-bit I/O bus at a read/write operating frequency of 1GHz with DDR throughput of 2GHz/64-bit.

VI. REFERENCES

[1] S. Abdel-hafeez and S. P. Sribhashyam, "System and Method for Efficiently Implementing a Double Data Rate Memory Architecture", US patent No. 6,356,509 B1; March 12, 2002.

[2] S. M. Abdel-hafeez and A. S. Matalkah, "CMOS Eight-Transistors Memory Cell for Low-Dynamic-Power High-Speed Embedded SRAM," Journal of Circuits, Systems, and Computers, Vol. 17, No. 5, World Scientific Publishing Company, Jan. 22, 2009, pp. 845-863

[3] Anandtech (Intel I7): http://www.anandtech.com/show/2594/10

[4] L. Chang, R. K. Montoye, Y. Nakamura, K. A. Batson, R. J. Eickemeyer, R. H. Dennard, W. Haensch, and D. Jamsek, "An 8T-SRAM for variability Tolerance and Low-Voltage Operation in High-performance CACHES,"IEEE Journal of Solid-State Circuits, Vol. 43, Issue 4, April 2008, pp. 956-963.

[5] A. T. Do, K. S. Yeo, J. Y. S. Low, J. Y. L. Low, and Z. H. Kong, "An 8T SRAM Cell with Column-based Dynamic Supply Voltage for Bit-interleaving," Conference on Circuits and Systems (APCCAS) IEEE Asia Pacific, 2010, pp. 704-707

[6] K. Nii, Y. Masuda, M. Yabuuchi, Y. Tsukamoto, S. Ohbayashi, S. Imaoka, M. Igarashi, K. Tomita, N. Tsuboi, H. Makino, K. Ishibashi, and H. Shinohara, "A 65nm Ultra-High-Density Dual-port SRAM with 0.71μm² 8T-cell for SoC," Symposium on VLSI Circuits Digest of Technical Papers, 2006, pp.130-131

[7] S. Reddy G M and P. C. Reddy, "Design and Implementation of 8K-bits Low power SRAM in 180nm technology," Proceedings of the International Conference of Engineers and Computer Scientists 2009, Vol. III, IMECS 2009, March 18-20, pp. 100-105

[8] T. Suzuki, S. Moriwaki, A. Kawasumi, S. Miyano, and H. Shinohara, "0.5-V, 150-MHz, Bulk-CMOS SRAM with Suspended Bit-Line Read Scheme," Proceedings of the ESSCIRC, 2010, pp. 354-357

[9] T. Suzuki, H. Yamauchi, Y. Yamagami, K. Satomi, and H. Akamatsu, "A table 2-Port SRAM Cell Design Against Simultaneously Read/Write-Distributed Accesses," IEEE Journal of Solid-State Circuits, Vol. 43, No. 9, Sept. 2008, pp.2109-2119

[10] K. Takeda, Y. Hagihara, Y. Aimoto, M. Nomura, Y. Nakazawa, T. Ishii, and H. Kobatake, "A Read-Static-Noise Margin-Free SRAM Cell for Low-VDD and High-Speed Applications," IEEE Journal of Solid-State Circuits, Vol. 41, Issue 1, January 2006, pp. 113-121

[11] TS Taiwan Semiconductor Manufacturing Corp., "0.09 μm CMOS ASIC Process Digests," 2005.

[12] United Microelectronics Corporation (UMC), "0.09 μm CMOS ASIC Process Digests," 2005

[13] J. P. Uyemura, *CMOS Logic Circuit Design*, Kluwer, 1999

enough setup and hold time with respect to *WWLE63/WWLO63*, which ensures a valid write data operation on the even and odd modules' cells. In this case, the hold time is considered the write access time. Finally, the content of the even/odd cell is realized by the *DEVEN/DODD* signal, which shows correct DDR write operations on the rising and falling edges of *WCLK*.

TABLE 3 depicts a comparison with analogues designs [5][7][10], where the previous works' data are reported directly from the literature without any scaling. Since, to the best of our knowledge, there is no reported DDR SRAM for on-chip communication with arithmetic units, the compared designs are single data rate (SDR). Although our design's circuit structure is implemented to support DDR throughput in contrast with the compared designs' SDR throughput, our design uses the same 8T-Cell and the same decoder logic with the same timing constraints, the only difference being the multiplexed I/O sense circuit. Therefore, the majority of the 8T-Cell's advantages with respect to SDR SRAM are applicable to our DDR SRAM design, such as low power consumption, competitive silicon area, fast access, and large noise margin between read and write that support continued technology scaling. Furthermore, the proposed design provides twice the throughput at a competitive memory clock speed (1GHz) and in addition, competitive power consumption against dynamic throughput activities.

V. CONCLUSIONS

In this paper, we presented a double data rate (DDR) SRAM design for communication between the memory modules on a system-on-chip (SoC) that is independent of the DDR input-output (I/O) system bus. Our design

Scalability Analysis of Release and Sequential Consistency Models in NoC based Multicore Systems

Abdul Naeem, Axel Jantsch and Zhonghai Lu

Department of Electronic Systems, KTH-Royal Institute of Technology, Sweden

E-mail: {abduln, axel, zhonghai}@kth.se

Abstract—**We analyze the scalability of the Release Consistency (RC) and Sequential Consistency (SC) models which are realized in the Network-on-Chip (NoC) based distributed shared memory multicore systems. The analysis is performed on the basis of workloads mapped on the different sizes of networks with different data sets. The experiments use a configurable platform based on a 2D mesh NoC using deflection routing algorithm. The results show that under the synthetic workloads using different distributed locks, the performance of the RC model is increased by 17.6% to 54.6% over the SC model in the 64-cores system. For the application workloads, as the network size grows from 1 to 64 cores, the execution time under the RC model decreases relative to the SC model which depends on the application and its match to the architecture. The performance improvement of the RC model over the SC model tends to be higher than 50% observed in the experiments, when the system is further scaled up.**

Keywords- Scalability; Memory consistency; Release consistency; Distributed shared memory; Network-on-Chip

I. INTRODUCTION

Parallelization as a key means to enhance performance and reduce power can be achieved at the computation, communication and memory architectures in the system [1]. The distributed nature of the Network-on-Chip (NoC) based systems can be exploited by using on-chip Distributed Shared Memory (DSM) architectures. Since the shared memory operations may be reordered in the network, the DSM system may show unexpected behavior. Memory consistency defines the execution order of the shared memory operations for the correct behavior of the DSM systems. Different memory consistency models [2] enforce different ordering constraints on the shared memory operations, implying different system performance. The *Sequential Consistency* (SC) model is a strict consistency model [3] and it cannot exploit the system optimizations. Therefore, several *relaxed* consistency models [2][4-7] are proposed to alleviate the ordering constraints on the memory operations to exploit these system optimizations.

Memory consistency and cache coherence are two main issues in the DSM systems. The former issue arises due to the unconstrained shared memory operations, while the later issue is due to the different cached copies of the same shared data in the DSM systems. Different memory consistency models and cache coherence protocols are proposed to handle these issues. In some situations, when a cache is not used like in the hard real time applications or when these two problems have different requirements on the size of the cache block and the consistency object, independent implementation schemes for the two problems are preferred [8][9].

This paper analyzes the scalability of the RC and SC models [8][9] which are realized in the Multicore NoC (McNoC) systems. The key performance metrics like *execution time, performance, speedup, overhead* and *efficiency* are evaluated as a function of the network size. The scaling behavior of both the RC and SC models are analyzed by mapping the workloads on the different sizes of the network with different data sets and also on the basis of application types and the system design perceptions (e.g. distributed locks and DSM architectures).

For the experiments, a configurable McNoC platform is used with distributed locks, DSM and 2D mesh *Nostrum* NoC [10] using a deflection routing policy. The *scalability study* of the RC and SC models is performed in the McNoC systems with 1 to 64-cores. The experimental results show the scaling behavior of the RC and SC models in the McNoC systems.

The rest of the paper is organized as follows. The next section overviews the related work. In section III, the SC and RC models, DSM based McNoC platform and the implementation of the SC and RC models are discussed. The simulation results and scalability analysis of the RC and SC models up to 64-cores systems are presented in section IV. In section V, our contributions are summarized.

II. RELATED WORK

In general multiprocessors DSM systems, several memory consistency models are discussed in the literature [2-7]. Adve et al. [2] discussed the memory consistency models from the system optimizations point of view. The SC model enforces a *total* order on the shared memory operations [3]. The *total store ordering* model relaxes the ordering constraint in the case of a *write followed by a read* operation. The *partial store ordering* model provides an additional relaxation among the *write* operations [2]. The *Weak Consistency* (WC) model [5] classifies the shared memory operations as *data* and *synchronization* operations. The data (read, write) operations issued between the two consecutive synchronization points can be reordered with each other. The RC model [6] further classifies the synchronization operations as *acquire* and *release* operations. The RC model is implemented in the DASH project [7] which depends on the *directory* based cache coherence protocol [11]. However, the directory based coherence protocols have some issues in the larger networks like, i.e., extra coherence traffic, directory overhead, additional latencies and complexities. In NoC based DSM systems, the proposed mechanism in [12] is very restrictive and allows one outstanding transaction of an initiator at a time in the network. The streaming consistency [13] is based on the software cache coherence protocol. However, polling the circular buffer at

978-1-4673-2895-1/12 $31.00 © 2012 IEEE

each request level is not a scalable approach. A protocol stack for on-chip interconnects is proposed [14] at different levels of the SoC design. They briefly outline the mechanisms to implement the RC model at the memory-mapped stack. But, the implementation detail is not discussed. The AXI [16] and OCP [18] protocols enforce the *ordering models* by using transactions IDs and thread IDs, respectively. In [16], transactions of the same master with *different* IDs can be reordered, but transactions with the *same* ID are not allowed to be reordered. In [18], *tagged* transactions of the same master using thread IDs are allowed to be reordered, but *non-tagged* transactions are strictly ordered. In [8], the SC model is realized in the McNoC systems by *stalling* the processor on the issuance of an operation till its completion. In [15], two *Transaction Counters* (*TCs*) based approach is adopted to realize the RC model in the McNoC systems. The *TC1* and *TC2* are used to keep track of the outstanding data operations issued in the non-critical and critical sections, respectively. In [9], a single *TC* based approach is used to realize the RC model in the McNoC systems. In this paper, we further analyze the scalability of the RC model [9] and SC model [8] in the McNoC systems.

III. SC and RC Models in NoC based Systems

The ordering constraints to be enforced on the shared memory operations under the SC and RC models are given in Figure 1(a) and (b). An *arrow* between the two variables indicates an ordering constraint between the operations on these variables. For example, G→H indicates that an operation on variable G is followed by an operation on variable H in the program and these two operations are not allowed to be reordered with each other. The variables to the left side of the assignment operators are written and those to the right are read.

A. SC Model

According to the SC model [3][8] (Figure 1(a)), the shared memory operations are completed in the *program order*. The *sequential order* is maintained by interleaving operations on lock (L) among processors in the system. The SC model enforces the global orders (Figure 1(c)) on the shared memory operations. We refer to these global orders in the later part.

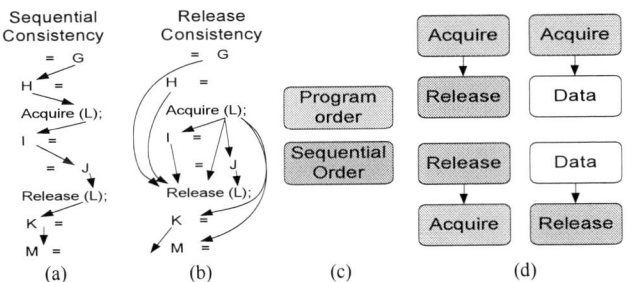

Figure 1. a) SC model b) RC model c) Global orders under SC model
d) Global orders under RC model

B. RC Model

The RC model [6][9] is a refinement of the WC model [10]. It classifies the *synchronization* operations as *acquire*

and *release* operations. The acquire operation delays the following data operations until the lock is obtained and does not wait for the completion of the previously issued data (read, write) operations. The release operation is to inform about the completion of previously issued data operations and does not delay the subsequent data operations. According to the RC model (Figure 1(b)), the independent data (read, write) operations on (G, H) are allowed to be reordered with each other, with the acquire operation on lock (L) and with the data operations on (I, J) in the critical section. They are not permitted to be reordered with respect to the release operation on lock (L). The data operations (I, J) can be reordered and overlapped with respect to each other, but they are not allowed to be reordered with the acquire and release operations on lock (L). The data operations on (K, M) are allowed to be reordered with each other, with the prior outstanding release operation on lock (L) and with the prior outstanding data operations on (I, J). However, they are not permitted to be reordered with respect to the prior acquire operation on lock (L). The global orders to be enforced on the shared memory operations under the RC model are given in Figure 1(d).

C. Platform Architecture

A homogenous McNoC platform is shown in Figure 2(a). As demonstrated in Figure 2(b), each Processor-Memory (PM) node consists of a processor, transaction controller (TCTRL), Synchronization Handler (SH), Network Interface (NI) and the local memory. The platform uses 2D mesh packet switched *Nostrum* NoC [10] with an adaptive routing algorithm. It is a buffer-less network and only buffers are used at the NIs to store the packets before injection into and after ejection from the network. The NI connects a PM node to the network. It performs packetization, de-packetization, queuing, arbitration and communication over the network. The platform uses the DSM in the network. All shared parts in the local memories constitute the DSM in a single global address space. The local memory is connected to the local processor within the node and to the remote processors via the network.

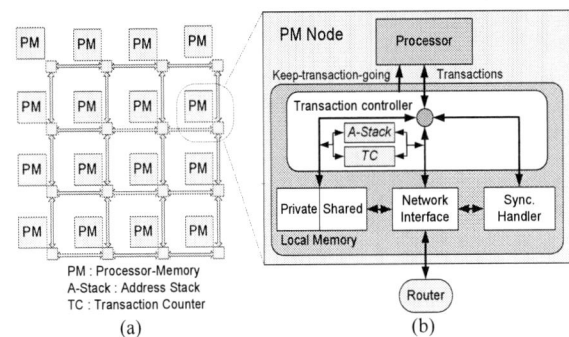

Figure 2. a) Homogeneous McNoC b) PM node

The platform also uses the *distributed* locks in the network. The SH controls N locks maintained in the global address space. Every lock is accessed in a sequential order by multiple processors in the system. The synchronization (acquire, release) requests to the SH either come from the local processor or from the remote processor via the network. If the requested lock is available then it is acquired. Otherwise, a negative acknowledgement is sent back and the source node sends again the same request until the lock is gained. A release request makes the lock available for the next acquire on it. The data (read, write) operations to the local shared memory and synchronization (acquire, release) operations to the local SH are accomplished within the node. For the remote accesses, message passing is carried out to the remote node via the network. The customized interface (TCTRL) like any standard interface [15][17] integrates the processor with the rest of the system. It also implements the *memory consistency protocols* using the hardware structures (*TC, Address-Stack*). The TCTRL is developed specifically for the LEON3 IP-core [17] which is used in each node of the network.

D. Implementation of the SC Model

The SC model is implemented [8] by enforcing the required global orders (Figure 1(c)) on the shared memory operations. **Program Order:** is enforced by *stalling* the processor on the issuance of a shared memory operation till its completion. On the completion of a previously issued memory operation, the next operation is issued in the program.
Sequential Order: The multiple processors mutually agree on a common lock to sequentially access the critical resource.

E. Implementation of the RC Model

The RC model is implemented [9] by enforcing the required global orders (Figure 1(d)) on the shared memory operations. **Data → Release:** To enforce this global order, a *Transaction Counter* (TC) is used in each node of the network to keep track of the outstanding *data* (read, write) operations issued before the release operation. The TC is incremented by the issuance of a data operation. It is decremented by the completion of a data operation. The issuance of a release operation is delayed by stalling the processor till the completion of previously issued outstanding data operations, i.e., (TC=0).
Acquire → Data/Release: To enforce these global orders, the processor is *stalled* on the issuance of an acquire operation till the acquisition of the lock. The lock is gained by a processor before entering to the critical section and before trying to release it.
Release → Acquire: This global order is enforced by sequential ordering on a lock in the multiprocessor system. The lock is released by a processor before the next acquire on it.
Data operations to the same location: are constrained for the purpose of correctness. To that end, an *address stack (A-Stack)* is used in each node of the network to ensure the parallel program correctness [9].

IV. EXPERIMENTS AND RESULTS

A. Experimental Setup

We experimented on a configurable cycle-accurate McNoC simulation platform constructed in VHDL (Figure 2). The LEON3 processor [17] is used in each node of the network. The size of the shared memory in each node is 16 MB. The SH maintains 256 locks in each node of the network. The TC is 32 bits and the *A-Stack* can stack up to 64 addresses each with 24 bits. The size of the *A-Stack* is kept small and it

is utilized efficiently. The addresses are popped from the *A-Stack* continuously by the completion of operations in a pipelined manner. The packet formation in the NI uses 7 fields (96 bits). The buffering capacity at the NI is 64 packets. The *Nostrum* NoC [10] uses 2D mesh regular topology and deflection routing policy.

We have developed some in-house synthetic and application workloads to evaluate the performance of the McNoC systems. Synthetic workloads are small in size and used to test a particular aspect of the system, while application workloads give accurate and deeper evaluations of the system [19][20]. The developed applications are light weight, specific to NoC/embedded architectures and communication centric.

B. Performance Metrics for the Scalability Analysis

To study the scalability of the RC and SC models in the McNoC systems (Figure 2), application workloads are mapped on the different sized networks. The performance metrics like *execution time, performance, speedup, overhead* and *efficiency* are evaluated as the network scales up. The *execution time* (ET) of a workload is the time from the start of the execution on the first processor to the end of the execution on the last processor in the system. The *performance* is the reciprocal of the ET. The *speedup* (SP) is the ratio of the single core execution time (Ts) and the execution time of the multicore (Tm) system. We define the *communication overhead* (OH) of the multicore system as: $Nc*Tm-Ts$, where Nc is the number of cores in the system. The *efficiency* (EF) of the multicore system is defined as the ratio SP/Nc. For the synthetic workloads, the ET of the multicore normalized to the single core is defined as (Nor-ET1C). In the experiments, the effects of the network size on the ET, performance, SP, OH and EF are investigated under the RC and SC models in the McNoC systems.

C. Synthetic Workloads

To demonstrate the benefits of the *distributed locks*, we evaluate the RC and SC models with synthetic workloads manually mapped on the LEON3 processors in the systems (Figure 3(a)). The same sequence of transactions is generated by the processor in each node of the network. The workloads have both data and synchronization operations. For the lock and protected (critical section) data operations, hotspot traffic pattern is generated (Figure 3(b)). We consider an 8x8 network. For *k* locks, the network is divided into *k* equal segments. All nodes within a segment synchronize over a common lock in a node that belongs to the same segment.

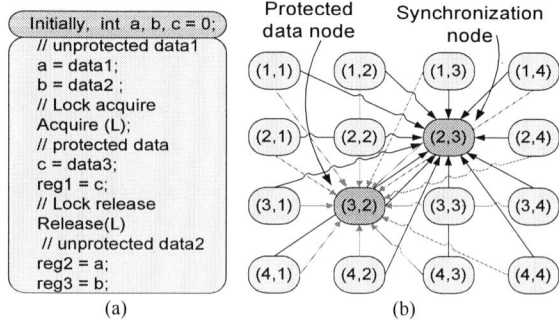

Figure 3. a) Sequences of transactions generated b) Traffic Patterns

The performance of the RC and SC models are compared using different number of segments/locks in the 8x8 network (Figure 4(a)). As the number of segments/locks increase in the network, the performance quickly increases due to the fact that different segments synchronize over different distributed locks in the network. The average lock acquire wait time is reduced as the network traffic/congestion decreases. The performance is higher under the RC model compared to the SC model due to reordering and relaxation in the shared memory operations. The average performance under the RC model for 1 to 32 locks is increased by 17.6% to 54.6% over the SC model. The ET of the 64-cores normalized to the single core (*Nor-ET1C*) is shown in Figure 4(b). The ideal *Nor-ET1C* is 1 by assuming zero communication overhead in the 8x8 system and the ET of the 64-cores system is equal to that in the single core system. Note that, the same sequence of transactions is generated by identical processors in the system. The deviation of the actual *Nor-ET1C* under both the memory models from the ideal case decreases as the number of the segments/locks increases in the network.

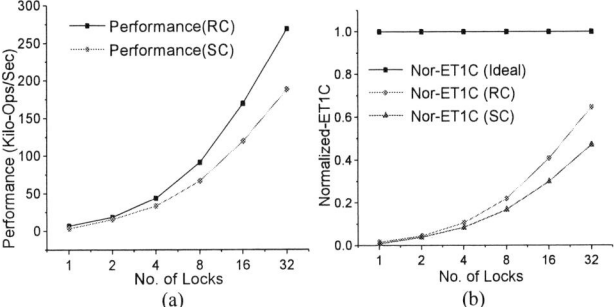

Figure 4.   (a) Performance (b) Normalized ET of 64-cores to single core

D. *Application Workloads*

1) *Bit Count*

Bit count application analyzes a data vector and calculates the number of set bits in each integer data item. After initialization these data items are read, analyzed and the output values are stored in the DSM. Different sizes vectors are used with (16, 32, 64, 128, 256, 512 and 1024) data elements. The network size is increased from 1 to 64 nodes. When a data vector of 16 elements is mapped on the 4x8 and 8x8 networks, only 16 nodes are involved in the computations. Similarly, when a 32 data vector is mapped on the 8x8, only 32 nodes perform the computations. For the rest of the data vectors all nodes perform the computation in the 8x8 network. Each node operates on the data items in the *randomly* selected node and also writes the output results into the same node.

Figures 5~8 illustrate the ET, speedup, communication overhead and efficiency of the RC and SC models under different sizes of networks and different data sets. As illustrated in Figure 5, the Application workload ET (AET) is decreased as the system size is increased from 1 to 64 cores. This is due to the division of computation cost in the network. The RC model further decreases the AET compared to the SC

model by reordering and overlapping the shared memory operations in the network.

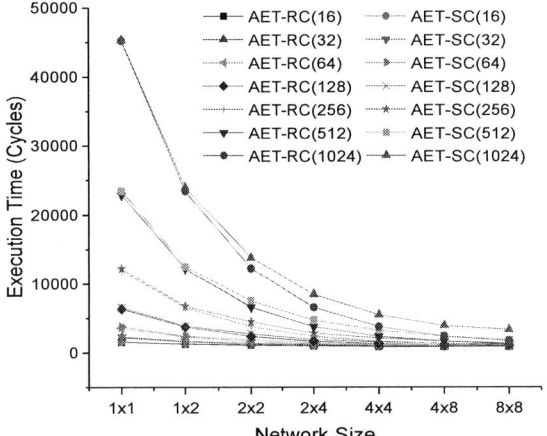

Figure 5.   Execution Time under Bit Count Application

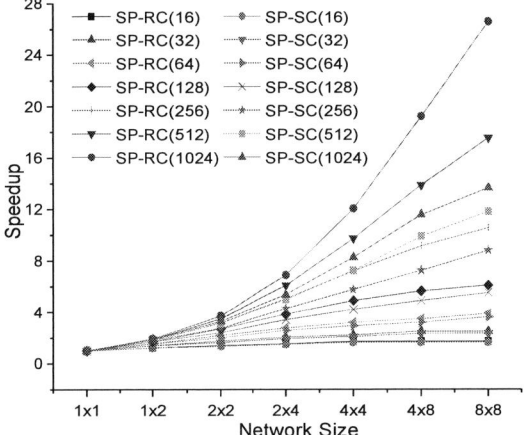

Figure 6.   Speedup under Bit Count Application

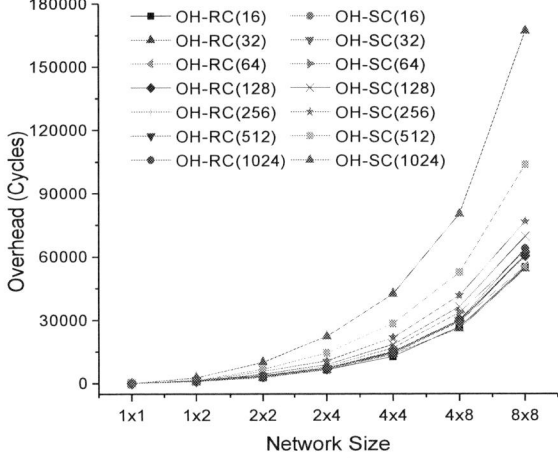

Figure 7.   Communication Overhead under Bit Count Application

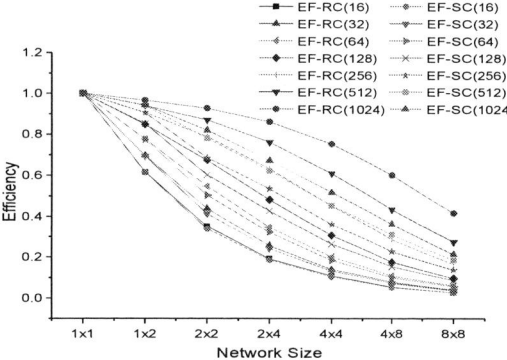

Figure 8. Efficiency under Bit Count Application

The AET reduction also depends on the data set size. It is high under the larger data set due to the parallelization of the significant amount of computation cost in the system. For the 1024 data set, the AETs in the single core system are 26.6 and 13.6 times of that in the 64-cores system under the RC and SC models, respectively. Likewise, the speedup (Figure 6) grows faster under the RC model compared to the SC model as the system size is increased. The speedup under the RC model is even higher under the larger data set. It is because of the efficient handling of the communication overhead (Figure 7) under the RC model by allowing *more* outstanding operations in the network which are overlapped and pipelined with each other. For the 1024 data set, the overhead in the 64-cores system compared to the two-core system is 39.6 and 60.9 times under the RC and SC models, respectively. The efficiency (Figure 8) is maintained high under the RC model compared to the SC model under different data sets when the network size grows up. In general, the RC model demonstrates better scalability and can efficiently utilizes the system resources in the larger networks. The execution time and overhead are lower and the speedup and efficiency are higher compared to the SC model.

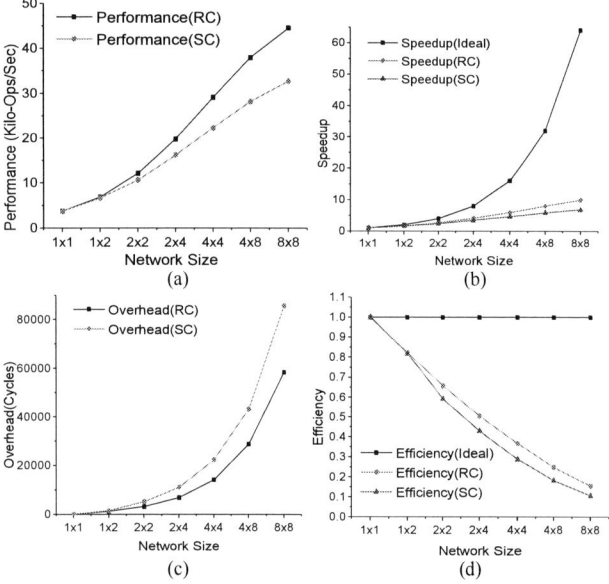

Figure 9. Bit Count: a) Performance b) Speedup c) Overhead d) Efficiency

The *average* performance, speedup, overhead and efficiency under the RC and SC models are given in Figure 9. As shown in Figure 9(a), on average the performance in the 64-cores systems compared to the single core systems are 12 and 8.8 times higher under the RC and SC models, respectively.

2) Pattern Search

The application searches data patterns (P) against the data elements (D), which are initialized in the shared memory across the network. Four different cases are simulated using different combinations of the patterns and data elements. The system size is increased from 1 to 64-cores. *P32-D32:* when 32 patterns and 32 data elements are mapped on the 8x8 network, only 32 nodes participate in the computations. *P32-D64:* For 32 patterns and 64 data elements, one pattern each in the 32 nodes, while one data element each in the 64 nodes is initialized in the 8x8 network. Also, 32 nodes are involved in the computations. *P64-D32:* one pattern each in the 64 nodes, while one data element each in the 32 nodes is initialized in the 8x8 network. All 64 nodes perform the computation. *P64-D64:* one pattern and data element each is mapped in the 8x8 network and each node is involved in the computation. The outputs are the number of times that the patterns appear in the data elements, which are stored in the local node.

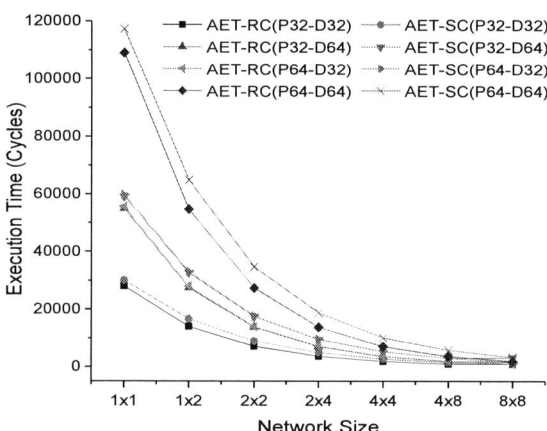

Figure 10. Execution Time under Pattern Search Application

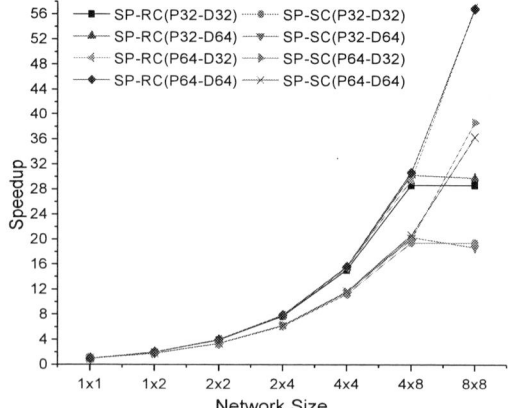

Figure 11. Speedup under Pattern Search Application

978-1-4673-2895-1/12 $31.00 © 2012 IEEE

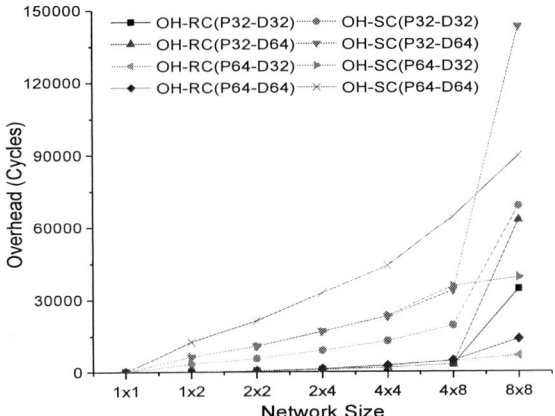

Figure 12. Communication Overhead under Pattern Search Application

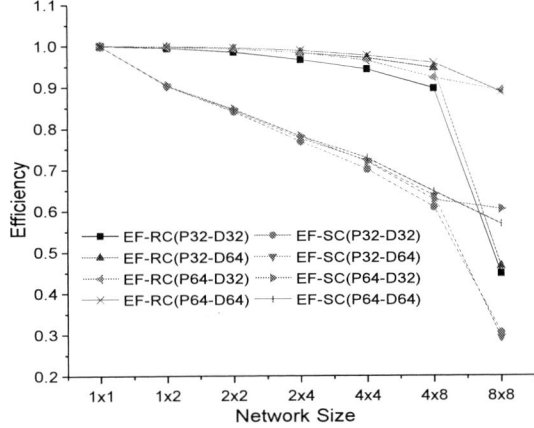

Figure 13. Efficiency under Pattern Search Application

size is increased. It is due to the fact that for each pattern all 64 data elements are searched which are distributed across the network. Also, as the network size is increased, the efficiency (Figure 13) is maintained high under the RC model compared to the SC model. Overall, the RC model again maintains low execution time and high speedup compared to the SC model.

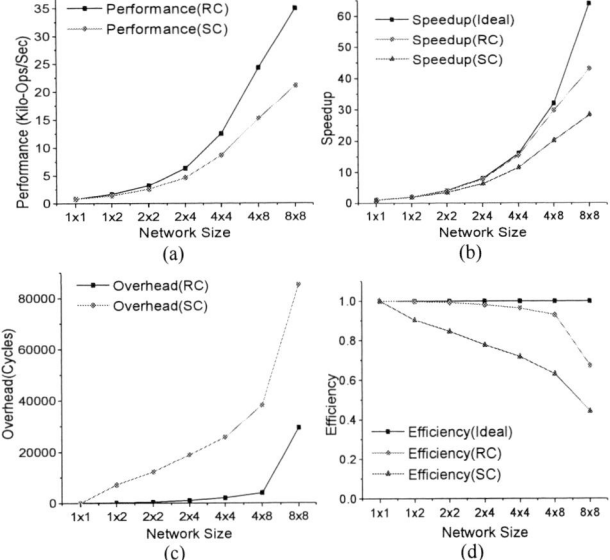

Figure 14. Pattern Search: a) Performance b) Speedup c) Overhead d) Efficiency

For the pattern search application, Figures 10~13 demonstrate the AET, speedup, communication overhead and efficiency of the RC and SC models under different sizes of networks and different data sets. As the system scales up from 1 to 64-cores, the AET reduction is high under both the memory models for the (P64-Dx) cases, because these problem sizes fit well in to the increasing size of network compared to the (P32-Dx) cases (Figure 10). For instance, better scaling behavior can be observed under the (P64-D32) case, where the problem size fits well into the 8x8 network and each node is involved in the computation. Due to parallelization of more computation time among the nodes, the AETs are reduced more as the system size is scaled up. The AETs in the single core systems are 57 and 38.6 time of that in the 64-cores systems under the RC and SC models, respectively. The AET reduction under the RC model over the SC model is high by pipelining and overlapping the shared memory operations. Similarly, the speedup (Figure 11) grows quickly under the RC model compared to the SC model. After 32 nodes the speedup levels off up to 64 nodes under the (P32-Dx) cases, since the same amount of computation is performed in the 4x8 and 8x8 networks. The communication overhead (Figure 12) under both the memory models significantly increases under the (Px-D64) configurations when the network

The average performance, speedup, overhead and efficiency under the pattern search application are given in Figure 14. The increase in the performance and speedup for both the memory models in the 64-cores systems over the single core systems are higher in contrast to the bit count application. This is because of the *low* computation to communication ratio. The computation time per input data is less (9 cycles) under the pattern search application compared to that under the bit count application (21 cycles). In addition, the communication is significant under the pattern search application, because the numbers of input and output data items are more compared to the bit count application. The increase in the average performance (Figure 14(a)) under the RC and SC models is 31.4% and 19.2% higher than that in the bit count application (Figure 9(a)). The average speedup (Figure 14(b)) for the RC model is 43.1 (almost ideal), while for the SC model it is 28.2 in the 64-cores system. The RC model compared to the SC model shows even better and more scalable behavior by allowing *more* outstanding data operations on the network which are reordered and overlapped with each other. The average communication overhead (Figure 14(c)) is controlled efficiently under the RC model with the increasing size of the network. The overhead reduction under the RC model over the SC model is quite high compared to the (Figure 9(c)). As long as the system size increases, the average efficiency (Figure 14(d)) is maintained high (close to the ideal case 1) compared to the (Figure 9(d)). The average efficiency in the 64-cores system for the RC and SC models is 0.67 and 0.44, respectively.

978-1-4673-2895-1/12 $31.00 © 2012 IEEE   43

E. Summary of the Scalability Analysis of RC and SC models

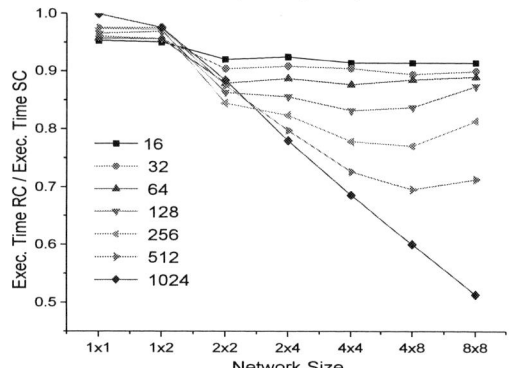

Figure 15. Bit Count: Ratio of AETs (RC/SC)

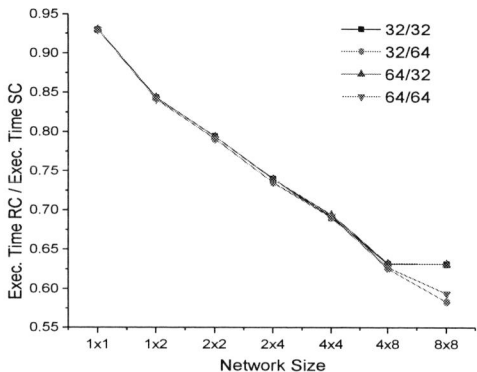

Figure 16. Pattern Search: Ratio of AETs (RC/SC)

In all our experiments, the execution time of RC model has been between 50% and 100% of the SC model. The specific numbers are highly sensitive to the application and depend on how well it matches to the platform. However, the observed trends suggest that the RC model scales inherently better with the network size than the SC model. As shown in Figure 15, the execution time is very similar for the small networks. As the network size grows, the execution time under the RC model decreases relative to the SC model and at some point (network size) the decrease flattens off. It depends sensibly on the application and its match to the architecture, when exactly this leveling off occurs. As long as the speedup increases, the benefits of the RC model over the SC model also increase, but when the nature of the problem makes it harder to utilize the additional parallelism, the benefits of the RC model over the SC model saturate as well. However, problems that scale well, like the 1024-bit count problem, continue to obtain increased benefits from the higher level of parallelism that the RC model offers compared to the SC model. We expect the trend shown in Figure 15 to continue for larger networks, which means that the performance benefits of the RC model continue to increase for well matched problems as long as the network size grows. Exactly, the same trend is visible in Figure 16. Thus, we conclude that the performance increase of the RC model over SC model can be significantly higher than 50% as observed in our experiments.

V. CONCLUSION

The scalability of the RC and SC models is analyzed in the NoC based DSM systems with 1 to 64 nodes based on the workloads mapping on the various sizes of networks with different data sets. The results show that under the synthetic workloads, the performance of the RC model is increased by 17.6% to 54.6% over the SC model using distributed locks in the 8x8 network. Under the application workloads, as long as the system size scales up, the execution time under the RC model decreases relative to the SC model. It depends on the scaling of the problem size and how efficiently the RC model is utilized compared to the SC model. The performance gain for the RC model over the SC model is expected to be higher than 50% as observed in the results, when the network size is further increased.

REFERENCES

[1] Axel Jantsch et al., "Memory architecture and management in an NoC platform," in: Axel Jantsch and D. Soudris, editors, Scalable Multicore Architectures: Design Methodologies and Tools. Springer, 2011.

[2] S. V. Adve et al., "Shared Memory Consistency Models: A Tutorial," Digital Western Research Laboratory, report no. 95/7, USA, 1995.

[3] L. Lamport, "How to Make a Multiprocessors Computer That Correctly Executes Multiprocessor Programs," IEEE Transaction on Computers, Vol. C-28. No. 9, pp. 690-691, September 1979.

[4] David E. Culler et al. "Parallel Computer Architecture: A Hardware/Software Approach," Morgan Kaufmann Publishers,1999.

[5] M. Dubois et al., "Memory access buffering in multiprocessors," in: Proc. of 13th Ann. Inter. Symp. on Comp. Arch. (ISCA'86), 1986.

[6] K. Gharachorloo et al. "Memory consistency and event ordering in scalable shared-memory multiprocessors," Computer Architecture News, 18(2): 15-26, June 1990.

[7] D. Lenoski et al., "The Stanford Dash Multiprocessor," Computer, 87(3), March 1992, pp. 418- 429.

[8] A. Naeem et al., "Realization and Performance Comparison of Sequential and Weak Memory Consistency Models in Network-on-Chip based Multicore Systems," in: Proc. of the 16th (ASP-DAC), 2011.

[9] A. Naeem et al., "Architecture Support and Comparison of Three Memory Consistency Models in NoC based Systems," in: Proc. of Euromicro Conference on Digital Systems Design (DSD), 2012.

[10] A. Jantsch "The Nostrum NoC," in: http://www.ict.kth.se/nostrum.

[11] L. M. Censier et al. A new solution to coherence problems in multicache systems," IEEE Trans. on Computer, c-27(12):1112–1118, 1978.

[12] F. Petrot, A. Greiner, P. Gomez, "On cache coherence and memory consistency issues in NoC based shared memory multiprocessor SoC architectures," in: Proc. of 9th Euromicro (DSD), pp. 53-60, 2006.

[13] J.W. van den Brand and M. Bekooij, "Streaming consistency: a model for efficient MPSoC design," in: Proc. of 10th Euromicro (DSD), 2007.

[14] Andreas Hansson, and Kees Goossens. "An On-Chip Interconnect and Protocol Stack for Multiple Communication Paradigms and Programming Models," In: Proc. of CODES+ISSS'09, France, 2009.

[15] A. Naeem, X. Chen, Z. Lu, and A. Jantsch, "Scalability of Relaxed Consistency Models in NoC based Multicore Architectures," ACM SIGARCH Computer Architecture News, April 2010, 37(5): 8-15.

[16] "AMBA AXI Protocol Specification," in: http://infocenter.arm.com/

[17] http://jorisvr.nl/leon3_insntiming.html

[18] OCP International Partnership. OCP Specification 2.2, 2007.

[19] C. Grecu, A. Ivanov, A. Jantsch, P.P. Pande, E. Salminen, U.Y. Ogras, R. Marculescu, Towards Open Network-on-Chip Benchmarks, First Int. Symposium on Networks-on-Chip (NOCS'07), May 2007.

[20] Zhonghai Lu, A. Jantsch, E. Salminen, and C. Grecu. Network-on-chip benchmarking specification part 2: Micro-benchmark specification. Technical Report Version 1.0, OCP-IP, May 2008.

Resource-shared Custom Instruction Generation under Performance/Area Constraints

Di Wu
TIP In-Memory Platform P*Time
SAP Labs Korea, Seoul, Korea
di.wu@sap.com

Junwhan Ahn, Imyong Lee, Kiyoung Choi
School of Electrical Engineering and Computer Science
Seoul National University, Seoul, Korea
junwhan@snu.ac.kr, bonobono@dal.snu.ac.kr, kchoi@snu.ac.kr

Abstract—**Adding custom instructions is an efficient mechanism for accelerating the performance of an application-specific processor. In general, however, area cost for custom instructions grows rapidly as the performance goal grows. There have been many researches targeting this issue. However, their approaches try to optimize area cost only after the custom instructions have been generated by a separate identification algorithm, thus can hardly generate optimal area-efficient custom instructions. This paper proposes a novel approach to custom instruction generation considering area cost and performance at the same time. Experiments with benchmark examples show that our approach efficiently generates optimal results under various constraints.**

Keywords-custom instruction, resource sharing, extensible processor, area constraint, performance constraint

I. INTRODUCTION

Application-specific instruction-set processors (ASIP) allow designers to customize the instruction-set architecture (ISA) of processors for efficient execution of specific applications. However, designing, optimizing and verifying a fully customized ISA can be time consuming and difficult. As a special type of ASIP, extensible processors are designed for reducing the design time and efforts. A typical extensible processor provides a well-made base processor and an interface for custom instructions (CIs). Performance of the whole system can be improved by adding user-defined CIs. As this approach has its own advantage, many commercial products have emerged on the market, such as Tensilica Xtensa [1], Altera Nios II [2], Xilinx MicroBlaze [3], Synopsys ARC600/700 Family [4], and MIPS Pro Series [5].

Generally speaking, there are two approaches to automatic CI identification. The first one focuses on recurrence of each CI. Related studies include [11-14,22]. Usually it includes two steps. In the first step, relatively small patterns of subgraphs are identified from the given kernels while satisfying the architectural constraints such as maximum number of input ports and output ports provided by the CI interface. Then, in the second step, a set of patterns to be implemented as CIs are selected based on the product of performance merit of each CI and its frequency of execution.

On the other hand, the second approach considers only performance merit of each CI. Related studies include [15-21]. Basic idea of this methodology is subgraph enumeration. Given the graph representations of kernels, all subgraphs satisfying the architectural constraints are enumerated (explicitly or implicitly). By comparing the performance merit of each valid subgraph, the best CI can be identified. Multiple

CI identification is realized by iteratively running the single CI identification algorithm. In each run, the CI with the best performance merit is selected.

Regardless of which approach to take, in order to achieve higher performance improvement, it is inevitable to add more complex CIs to the base processor, which results in significant area overhead to the whole system. As measured in [6,7], area cost for CIs can be similar to or even larger than the base processor of Xtensa. In this case, resource sharing becomes a critical issue in custom instruction design. There are several researches on resource-shared CI generation [8-10]. Brisk et al. [10] proposed an efficient and accurate heuristic algorithm that aggressively transforms a set of CIs into a single data-path on which they are executed. Instead of considering only maximum area saving as was done in [10], Zuluaga et al. [9] introduced latency constraints in their transformation process. A parametric approach is used for exploring the trade-offs that can be achieved between instruction latency and implementation complexity. In addition, Pothineni et al. [8] applied High-Level Synthesis (HLS) techniques to the resource sharing problem for implementing CIs. Lee et al. [24] proposed an HLS-based CI identification algorithm to support memory operations and I/O access serialization in CIs. Our approach is different from theirs in that we consider both area cost and performance at the same time, whereas they focus only on performance merit. The studies in [25,26] also consider area constraints but only for functional units (FUs). In contrast, this paper considers area overhead resulting from HLS by applying a complete HLS process. In addition, it also considers performance constrained problem which is critical in real-time system design.

II. MOTIVATION AND PROBLEM STATEMENT

Performance of an extensible processor generally increases as more custom instructions are added. However, in modern embedded system design, performance is not the only metric to be taken into consideration, but die area should also be considered to reduce the form factor as well as the manufacturing cost. Resource sharing of CIs can reduce die area of an extensible processor significantly.

Generally, traditional flows for resource-shared CI generation include two phases, namely CI identification and resource sharing. Figure 1 shows an example of the traditional flows. The latest research [8] also includes I/O access serialization phase between the aforementioned two phases. In the first phase, CIs that satisfy the architectural constraints are identified from the given target source code. Notice that the metric for the CI identification in this phase is usually only the

---

This work was supported by the National Research Foundation of Korea (NRF) grant funded by the Korea government (MEST) (No. 2012-0006272) and Ministry of Knowledge Economy (MKE) and IDEC Platform center (IPC) at Hanyang Univ.

978-1-4673-2895-1/12 $31.00 © 2012 IEEE

performance merit (although some studies [15,17] try to minimize the total area cost, these approaches still suffer from limited area reduction since their area models do not consider resource sharing). Identified CIs can be pipelined to an optional I/O access serialization phase. In this phase, registers are inserted into the data-path to serialize the I/O access. As proposed in [23], this approach allows the CI identification process to identify CIs that require more I/O ports than the ones supported by the base processor and thus improves the overall performance. In the last step of the flow, resource sharing is performed to minimize the area cost.

Such a traditional CI generation flow helps minimize the total area cost of the design, but can still result in an overdesigned architecture taking too much area and possibly providing unnecessarily high performance for a given application. The main limitation of this flow is the absence of a feedback from the resource sharing phase to the CI identification phase. In particular, it ignores the fact that the total delay can be affected by the multiplexers and registers inserted into the data-path in the resource sharing process and so the performance merit of a CI can be affected. In such a case, the identified CIs from the first phase may not be the best possible solution. In addition, since the CI identification phase only selects CIs based on performance merit, it may miss the opportunity of selecting the CIs that consume less area after resource sharing. In the end, such a flow can hardly generate optimal CIs under a tight area constraint, especially when the CIs passed from the first phase are too large to be optimized to meet the constraint.

Based on our analysis, we claim that, to generate the optimal resource-shared CI, resource sharing problem should be considered during CI identification phase. In this paper, we propose a resource-shared CI generation approach to generate a CI under a performance constraint or an area constraint. The problem can be formally stated as follows:

Problem 1: *Performance Constrained Area Minimization*

Given a DFG G, find a subgraph C for CI such that
$Area\ (C)$ is minimized, and
$T_{SW}(C) - T_{HW}(C) \geq PERFORMANCE\_CONSTRAINT$

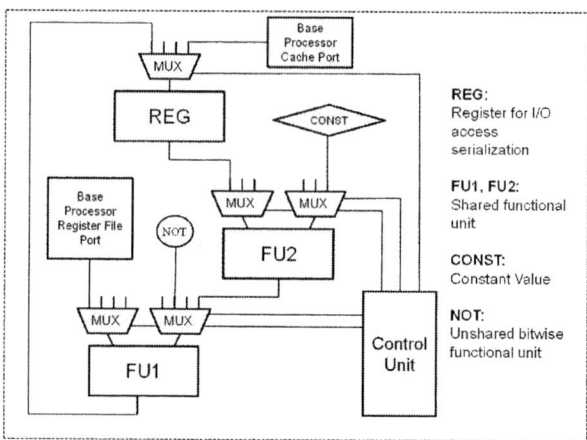

Figure 2. An example of resource-shared CI.

Problem 2: *Area Constrained Performance Maximization*

Given a DFG G, find a subgraph C for CI such that
$T_{SW}(C) - T_{HW}(C)$ is maximized, and
$Area(C) \geq AREA\_CONSTRAINT$

where $T_{SW}(C)$ is the number of cycles required to execute C on the base processor, and $T_{HW}(C)$ is the number of cycles required to execute C with the generated CI. Thus $T_{SW}(C) - T_{HW}(C)$ represents the *performance gain* obtained by the CI. $Area(C)$ is the area of C after resource sharing.

III. PROPOSED APPROACH

A. Area Model

For a CI without resource sharing, which is implemented with combinational logic, the area model can be as simple as the sum of the FU area (in the case of [23], area of registers for I/O access serialization should also be added). However, when considering a resource-shared CI, the area of multiplexers and the control unit must also be taken into consideration.

Figure 2 shows an implementation of a resource-shared CI. FU1 and FU2 indicate two different FUs, which are used at different control steps. REG is a register for I/O access serialization. Notice that this register can also be shared. Each input port of the FUs and REG has a multiplexer to select a proper input among different sources. The input of an FU can be a constant, an output of the base processor register file, or an output of another FU (in the case of FU chaining). The input of REG can be an output of an FU or the base processor cache port. The control unit sends signals for multiplexers' selection inputs and REG's write enable input.

In our approach, the area model is given by:

$$Area(C) = A_1*X_1 + A_2*X_2 + ... + A_n*X_n + A_{glue} + A_{control} \quad (1)$$
$$A_{glue} = A_{MUX}*X_{MUX} + A_{REG}*X_{REG} \quad (2)$$
$$A_{control} = A_{NAND}*\log_{BETA}((1/Kp)*X_{I/O}) + A_{1bit\_reg}*\log_2(t\text{-}cycle) \quad (3)$$

where A_i is the area of an FU of type i, X_i is the number of FU instances of type i, A_{glue} and $A_{control}$ are area of glue logic and control unit, respectively. A_{MUX} and A_{REG} indicate the area of a 32-bit 2:1 multiplexer and a 32-bit register for I/O access serialization, respectively, and X_{MUX} and X_{REG} are the number

---

Figure 1. Traditional flow for resource-shared CI generation.

978-1-4673-2895-1/12 $31.00 © 2012 IEEE

of multiplexers and registers, respectively. $A_{control}$ is broken down into combinational logic part and register part. Combinational logic can be further modeled with Rent's Rule. In formula I, *BETA* and *Kp* are constants for the Rent's rule, $X_{I/O}$ is the number of I/O ports of the control unit, and A_{NAND} is the area of a 2-input NAND gate. Finally, A_{1bit_reg} is the area of a 1-bit register and *t-cycle* is the total number of cycles needed to implement *C* as a CI.

B. Overall Flow

Figure 3 shows overall flow of our proposed approach. As well as the traditional flow, we take the target source code and an architecture description as the input. In the candidate enumeration phase, we adopt the branch-and-bound approach proposed in [18] to enumerate all candidate CIs from the DFG representation of a kernel of the input source code. All candidate CIs from this phase should satisfy a relaxed I/O constraint and a convexity constraint [23]. Notice that the relaxed I/O constraint can be larger than the actual number of I/O ports provided by the base processor, because our scheduling algorithm can also serialize the I/O accesses as the approach in [23] does.

We perform scheduling on each of the candidate CIs before sending them into the candidate CI library. In this phase, we provide unlimited FUs for the scheduling, but limit the number of I/O ports. Thus we perform list scheduling for I/O operations, but for other operations, we basically perform as-soon-as-possible (ASAP) scheduling. In this way, we can obtain an upper bound of the performance gain (*UB_P*) of each candidate CI for the given architectural constraints. Then we record the number of instances of each FU type actually used in the scheduling. It is the maximum number of instances that the candidate CI can utilize and can be viewed as the upper bound of the number of FU instances (*UB_F*). Both upper bounds will be used for pruning the *resource allocation space* during the exploration phase to speed up the process.

Among the candidate CIs collected in the library, we need to select the best CI, i.e., the CI that gives highest performance while satisfying the area constraint or the one that takes smallest area while satisfying the performance constraint. However, to select the best CI, we need to implement the CIs one by one to obtain the area and performance numbers. The

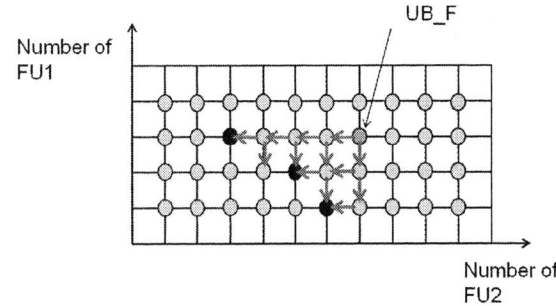

Figure 4. Resource allocation space exploration.

problem is that a candidate CI can be implemented in many different ways of allocating resources. For this, we perform *resource allocation space exploration* for each of the candidate CIs in descending order of their *UB_P*. For each point in the resource allocation space, we apply a complete HLS process, including FU scheduling, FU binding, and register binding. To improve the runtime, we use *UB_F* to prune the resource allocation space, and use *UB_P* to prune the candidate CIs. At the end of the process, we take the best solution obtained so far as the final solution.

C. Resource Allocation Space Exploration

Resource allocation space exploration is a process of determining the number of instances of each FU type to be used in the design. This is a key process of determining the performance and area of a CI. An example of resource allocation space with two types of FUs (FU1 and FU2) is shown in Figure 4. Each dot in the space represents a feasible resource allocation (FRA). For further discussion, let us define several terms.

Definition 1 (dominance): For two distinct FRAs, FRA1 and FRA2, FRA1 is said to dominate FRA2 if and only if every coordinate of FRA2 is greater than or equal to the corresponding coordinate of FRA1. For instance, (1, 1) is said to dominate (4, 2).

Definition 2 (direct dominance): For two distinct FRAs, FRA1 and FRA2, FRA1 is said to directly dominate FRA2 if and only if FRA1 dominate FRA2 and the sum of all coordinates' differences between FRA1 and FRA2 is one. For instance, (4, 1) and (3, 2) are said to directly dominate (4, 2).

Definition 3 (child): For two distinct FRAs, FRA1 and FRA2, FRA1 is said to be a child of FRA2 if and only if FRA1 directly dominates FRA2. For instance, (4, 1) and (3, 2) are children of (4, 2).

Depending on the number of desired types of FUs for resource sharing, the resource allocation space can be multi-dimensional. We first prune the design space with the *UB_F* obtained from the ASAP scheduling phase. Because this upper bound is obtained from an unlimited resource scheduling, there is no meaning of adding more FUs beyond this upper bound. Notice that not every CI contains all types of FUs for resource sharing. So this pruning approach can also reduce the dimension of the design space.

In addition, rather than exploring the entire remaining resource allocation space, we explore the space following the

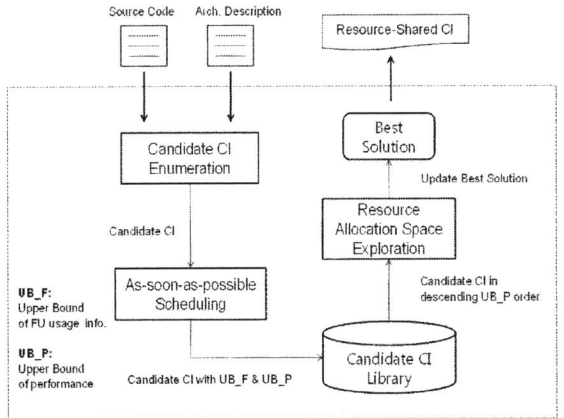

Figure 3. Overall flow of the proposed approach.

978-1-4673-2895-1/12 $31.00 © 2012 IEEE

dominance relationship to further prune the design space. Let us consider performance constrained problem first. In this case, the objective is to generate a minimum area CI that satisfies the performance constraint. So the objective of the resource allocation space exploration process is to obtain a minimum area cost implementation of the given CI while satisfying the performance constraint. The resource allocation space exploration starts from the UB_F point in the space as shown in Figure 5. In the beginning, it adds UB_F into the worklist. Then, every iteration of the while loop, it extracts one FRA from the head of the worklist and performs minimum latency list scheduling [27] with the FRA. The number of cycles obtained from the scheduling is taken as the $T_{HW}(C)$. Based on the scheduling, it binds the operations and their output data in the candidate CI to the FUs and registers, respectively. With the scheduling and binding information, it calculates the total area cost. If the performance constraint is satisfied by the current FRA, it adds the children of the current FRA at the end of the worklist as long as they are not dominated by any FRAs existing in the list or already examined FRAs. It also removes all the FRAs dominated by the newly added FRAs from the list. The whole process terminates when the worklist is empty.

For the case of area constrained problem, the objective is to generate a maximum performance CI that satisfies the area

```
 1: procedure RASE_PC(candidateCI, UB_F) // Resource
    allocation space exploration for performance
    constrained problem
 2:   bestSolution = NULL
 3:   worklist.insert(UB_F)
 4:   while worklist is not empty do
 5:     curFRA = worklist.extract_front()
 6:     curSolution = HLS(candidateCI, curFRA) // HLS includ
          es FU scheduling, FU binding and register binding
 7:     if curSolution.perf < CONSTRAINT then
 8:       continue
 9:     end if
10:     if curSolution.area < bestSolution.area then
11:       bestSolution = curSolution;
12:     end if
13:     for descendant in descendents(curFRA) do
14:       if descendent is not examined or does not exist in
            worklist then
15:         worklist.insert(descendent)
16:       end if
17:     end for
18:   end while
19:   return bestSolution
20: end procedure


21: procedure Generate_CI_PC(source code, architecture
    description)  // performance constrained CI generation
22:   canidateCiLibrary = Enumerate_Candidate_CI(source
        code, architecture description)
23:   bestCI = NULL
24:   for CI in candidateCiLibrary do
25:     CI.{UB_F, UB_P} = ASAP(CI) // apply as-soon-as-
          possible scheduling
26:   end for
27:   sort candidateCiLibrary with UB_F
28:   for CI in candidateCiLibrary do
29:     if CI.UB_P < CONSTRAINT then
30:       break
31:     end if
32:     curCI = RASE_PC(CI, CI.UB_F)
33:     if curCI.area < bestCI.area then
34:       bestCI = curCI
35:     end if
36:   end for
37:   return bestCI
38: end procedure
```

Figure 5. Pseudocode for performance constrained CI generation.

constraint. The only difference from the performance constrained problem is that children of the current FRA are added into the worklist only when the area constraint is violated (if the area constraint is no longer violated, then we do not need to further explore FRAs having smaller area than that).

D. Candidate CI Pruning

Although all of the candidate CIs are stored in the candidate CI library, the resource allocation space exploration process needs not be applied to all of them. In our proposed approach, the upper bound performance UB_P is used to prune some of the candidate CIs. In the performance constrained problem, for example, any candidate CI whose UB_P is smaller than the given constraint is pruned.

In the area constrained problem, the candidate CI pruning is slightly different. Figure 6 illustrates the basic idea of the pruning. The dashed line represents the UB_P of the candidate CIs sorted by the UB_P. For each CI, starting from the first CI in the sorted list, we perform resource allocation space exploration to obtain the performance merit under the area constraint. As we proceed with the CIs in the list, the UB_P decreases and eventually becomes less than the performance merit of the best solution obtained so far. Because the performance merit of a CI cannot exceed its UB_P, the remaining candidate CIs can be safely pruned. Then we take the best solution obtained so far as the final solution.

IV. EXPERIMENTS

We selected several basic blocks from Mibench [28] and DSPstone [29] as examples to demonstrate the effectiveness of our proposed approach. The base processor used in these experiments had an ISA that is compatible with ARM7TDMI-S. The register file in the base processor supported two read ports and one write port, which was an architectural constraint for the generation of a CI. The relaxed I/O constraint for candidate CI enumeration was set to 4-inputs/2-outputs in all of the experiments to take advantage of the I/O access serialization technique. Memory operations were also supported.

We synthesized FUs with their fastest implementations with Synopsys Design Compiler in TSMC 130nm process to obtain area and delay. We allowed sharing adders, multipliers, and shifters, but did not allow sharing other FUs such as bitwise AND, OR, XOR, and NOT due to high sharing cost

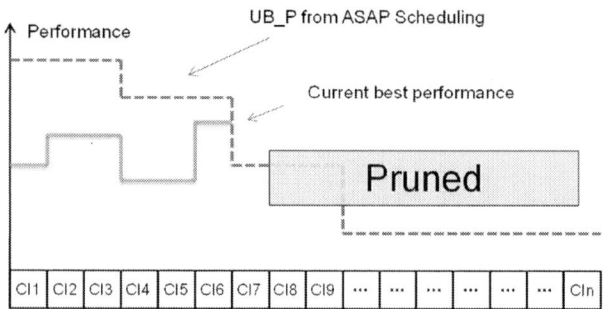

Figure 6. Illustration of candidate CI pruning for area constrained problem.

compared to their area.

A. Comparision with Traditional Approach

First, we took the area constrained problem to demonstrate the merit of our proposed approach compared to the traditional approach. For both approaches, we applied area constraints and compared the performance merit of the generated CIs. As the traditional approach, we took the one in [4], which considers I/O access serialization. For both approaches, we applied the same relaxed I/O constraint and the same HLS process mentioned in this paper. Recall that our approach considers resource sharing problem during the CI identification, whereas the traditional approach considers it after CI identification. The results are shown in Figure 7. Horizontal axis represents the given area constraint, and vertical axis represents the speedup of the benchmark example achieved by adding the generated CI. Notice that speedup of 1 indicates failure of generating a proper CI under the area constraint.

When the area constraint is tight, traditional approach fails to generate a proper CI in all examples. As analyzed in Section III, the CI identification phase of the traditional approach always returns the CI with largest performance merit. Such a CI requires too many resources, which cannot satisfy the given area constraint. However, our approach can successfully generate proper CIs under tight area constraints.

As the area constraint is loosened, both approaches can generate a proper CI that satisfies the constraint. However, compared with the traditional approach, proposed approach is able to return a CI with higher speedup. When the constraint is further loosened, sufficient numbers of resources can be used for implementing CIs with large performance merits. So the two approaches return the same result. In the end, the two approaches saturate at the upper bound speedup.

B. CI Generation under Performance/Area Constraint

We tested the proposed approach for both performance constrained area minimization problem and area constrained performance maximization problem. Experimental results are shown in Figure 8, where the two results are put together for comparison. For performance constrained problem, horizontal axis represents performance constraint, and vertical axis represents area cost. In contrast, for area constrained problem, vertical axis represents area constraint, while horizontal axis represents speedup. Notice that performance constrained CI generation always finds CIs with no larger area cost (red dots are not above any blue dots) for the same performance. Similarly, area constrained CI generation always finds CIs with no lower performance (blue dots are not on the left side of any red dots) for the same area. The experiment results demonstrate that the proposed approaches provide quite stable solutions over a wide range of area and performance. And regarding the runtime of our proposed methods, they take around 1 minute to find a solution for two largest examples having 62 and 64 nodes, respectively.

V. CONCLUSION

In this paper, we propose a new approach for resource-shared CI generation. Our approach considers resource sharing

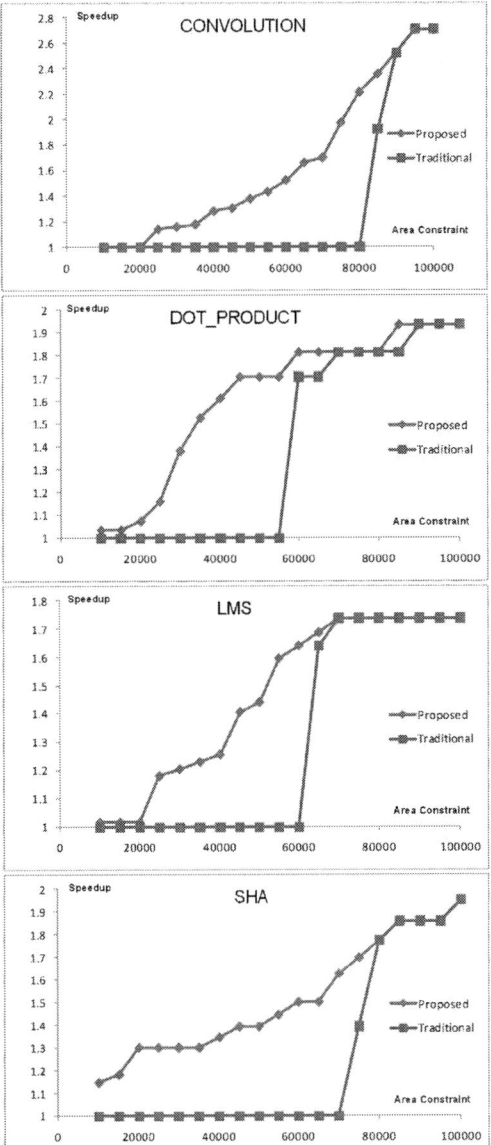

Figure 7. Comparison with the traditional approach.

problem during the CI identification process unlike traditional approaches, which consider the problem after CI identification. Experimental results show that our approach can efficiently generate proper CIs under a design constraint such as performance constraint or area constraint.

REFERENCES

[1] R.E. Gonzalez, "Xtensa: a configurable and extensible processor," *IEEE Micro*, vol. 20, no. 2, pp. 60–70, 2000.

[2] Altera Nios II. Available: http://www.altera.com/devices/processor/nios2/ni2-index.html

[3] Xilinx MicroBlaze. Available: http://www.xilinx.com/tools/microblaze.htm

[4] Synopsys ARC 600/700 Family. Available: http://www.synopsys.com/IP/ConfigurableCores/ARCProcessors/Pages/default.aspx

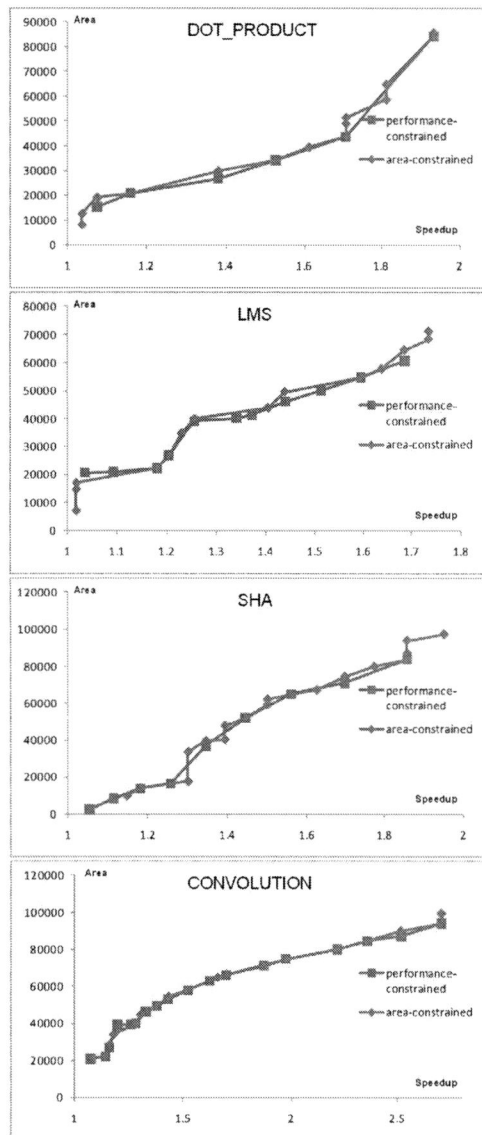

Figure 8. Resource-shared CI generation under performance/area constraints.

[5] MIPS Pro Series. Available: http://www.mips.com/products/cores/32-64-bit-cores/pro-series-family/

[6] "Xtensa, a new ISA and Approach" Available: http://bwrc.eecs.berkeley.edu/classes/cs252/Notes/xtensa_022400.pdf

[7] N. Cheung, S. Parameswaran, and J. Henkel, "A quantitative study and estimation models for extensible instructions in embedded processors," in *Proc. International Conference on Computer-Aided Design*, 2004.

[8] N. Pothineni, P. Brisk, P. Ienne, A. Kumar, and K. Paul, "A high-level synthesis flow for custom instruction set extensions for application-specific processors," in *Proc. Asia and South Pacific Design Automation Conference*, 2010.

[9] M. Zuluaga and N. Topham, "Resource sharing in custom instruction set extensions," in *Proc. Symposium on Application Specific Processors* 2008.

[10] P. Brisk, A. Kaplan, and M. Sarrafzadeh, "Area-efficient instruction set synthesis for reconfigurable system-on-chip designs," in *Proc. Design Automation Conference*, 2004.

[11] R. Kastner, A. Kaplan, S. Ogrenci Memik, and E. Bozorgzadeh, "Instruction generation for hybrid reconfigurable systems," *ACM Transactions on Design Automation of Electronic Systems*, vol. 7, pp. 605–627, Oct. 2002.

[12] P. Brisk, A. Kaplan, R. Kastner, and M. Sarrafzadeh, "Instruction generation and regularity extraction for reconfigurable processors," in *Proc. International Conference on Compilers, Architectures and Synthesis for Embedded Systems*, 2002.

[13] P. Bonzini and L. Pozzi, "A retargetable framework for automated discovery of custom instructions," in *Proc. International Conference on Application-specific Systems, Architectures and Processors*, 2007.

[14] P. Bonzini and L. Pozzi, "Recurrence-aware instruction set selection for extensible embedded processors," *IEEE Transactions on Very Large Scale Integration (VLSI) Systems*, vol. 16, pp. 1259–1267, Oct. 2008.

[15] F. Sun, S. Ravi, A. Raghunathan, and N. K. Jha, "Custom instruction synthesis for extensible-processor platforms," *IEEE Transactions on Computer-Aided Design of Integrated Circuits and System*, vol. 23, no. 2, pp. 216–228, Feb. 2004.

[16] J. Cong, Y. Fan, G. Han, and Z. Zhang, "Application-specific instruction generation for configurable processor architectures," in *Proc. International Symposium on Field-Programmable Gate Arrays*, 2004.

[17] K. Atasu, C. Özturan, G. Dündar, O. Mencer, and W. Luk, "CHIPS: custom hardware instruction processor synthesis," *IEEE Transactions on Computer-Aided Design of Integrated Circuits and Systems*, vol. 27, no. 3, pp. 528–541, March, 2008.

[18] K. Atasu, L. Pozzi, and P. Ienne, "Automatic application-specific instruction-set extensions under microarchitectural constraints," *International Journal of Parallel Programming*, vol. 31, no. 6, pp. 411–428, Dec. 2003.

[19] L. Pozzi, K. Atasu, and P. Ienne, "Exact and approximate algorithms for the extension of embedded processor instruction sets," *IEEE Transactions on Computer-Aided Design of Integrated Circuits and Systems*, vol. 25, no. 7, pp. 1209–1229, Jul. 2006.

[20] A. K. Verma, P. Brisk, and P. Ienne, "Fast, nearly optimal ISE identification with I/O serialization through maximal clique enumeration," *IEEE Transactions on Computer-Aided Design of Integrated Circuits and Systems*, vol. 29, pp. 341–354, Mar. 2010.

[21] K. Atasu, O. Mencer, W. Luk, C. Özturan, and G. Dündar, "Fast custom instruction identification by convex subgraph enumeration," in *Proc. International Conference on Application-specific Systems, Architectures and Processors*, 2008.

[22] J. Ahn and K. Choi, "An efficient algorithm for isomorphism-aware custom instruction identification for extensible processors," in *Proc. International Conference on Hardware/Software Codesign and System Synthesis*, Oct. 2011.

[23] L. Pozzi and P. Ienne, "Exploting pipelining to relax register-file port constraints of instruction-set extensions," in *Proc. International Conference on Compilers, Architectures and Synthesis for Embedded Systems*, 2005.

[24] I. Lee, D. Lee, and K. Choi, "Memory operation inclusive instruction-set extensions and data path generation," in *Proc. International Conference on Application-Specific Systems, Architectures, and Processors*, 2007.

[25] D. Wu, I. Lee, and K. Choi, "Fast custom instruction generation under area constraint," in *Proc. International SoC Design Conference*, Nov. 2010.

[26] D. Wu, I. Lee, J. Ahn, and K. Choi, "Fast generation of multiple custom instructions under area constraints," *Journal of Semiconductor Technology and Science*, vol. 11, no. 1, pp.51-58, Mar. 2011.

Comparative Analysis of Dynamic Task Mapping Heuristics in Heterogeneous NoC-based MPSoCs

Leandro Möller[1], Leandro Soares Indrusiak[2], Luciano Ost[3], Fernando Moraes[4], Manfred Glesner[1]

[1] Darmstadt University of Technology - Institute of Microelectronic Systems - Darmstadt, Germany
[2] Department of Computer Science - University of York - York, United Kingdom
[3] LIRMM – University of Montpellier II, France
[4] Faculty of Informatics - Catholic University of Rio Grande do Sul - Porto Alegre, Brazil

Abstract — **Dynamic mapping heuristics can cope with dynamic application scenarios by allocating tasks to cores of an MPSoC during runtime. In this paper, we compare eight heuristics in terms of the response time of application tasks - that is, the time between the issuing of a task and the time when it completes executing and communicating. By taking into account the task execution, communication and waiting times, we could better evaluate the quality of the different heuristics and show that there is room for improvement when it comes to heterogeneous platforms under high utilization.**

Keywords — *multiprocessor systems-on-chip, networks-on-chip, embedded systems, dynamic task mapping.*

I. INTRODUCTION

Applications running on Multiprocessor Systems-on-Chip (MPSoCs) may vary dynamically at execution time, according to user (e.g. load of new applications) and/or performance (e.g. change the frequency operation for optimizing battery lifetime) requirements, which leads in both time-changing processor workload and communication patterns [1][3][4]. Thus, offline-mapping techniques can be sub-optimal or inadequate in many scenarios. In this context, dynamic task mapping techniques have been used to achieve the required runtime adaptability demanded by such multiprocessing systems [6][7]. Such dynamic task mapping techniques are evaluated in both homogeneous and heterogeneous platform architectures.

More than avoiding congestion and placing communicating tasks near to each other, heterogeneous MPSoCs need to care about the affinity of tasks with the IP cores available on the platform. This is only true when the same task is developed for different IP cores and trading efficiency against utilization of these cores is left for the system to balance. The result is then a computing system that can analyze its own resources and allow the use of them in a more optimized manner. Therefore, smart implementations of dynamic mapping algorithms are vital for the MPSoC to execute applications with good performance figures and using as few resources as possible.

Our contribution is to evaluate quantitatively and comparatively dynamic task mapping using the affinity of tasks to the IP cores available on the heterogeneous MPSoC. Some of the presented algorithms are multi-objective, considering not only the affinity of the task and the congestion of the network, but also the utilization of the IP cores, the position on the network and the amount of communication among tasks.

This work is divided as follows. Section II presents the state of the art on dynamic task mapping for MPSoCs. The joint validation model composed by platform, application and task mapper used on this work is presented on section III. Section IV presents the dynamic mapping algorithms compared on this work. Section V presents the case studies and obtained results. Section VI concludes this work.

II. STATE OF THE ART

Examples of dynamic task mapping techniques explored in homogeneous architectures are [4][6][9]. In turn, dynamic task mapping on heterogeneous MPSoC platforms are investigated in [1][2][3][5][8][10][11]. Due to the distinct nature of processing elements (PEs) that can be integrated in such platforms, the mapping process is more complex when compared to the homogeneous case because additional constraints (e.g. the affinity of the task to a PE) must be considered at run-time. In this context, Carvalho et al. [1] proposed and evaluated the performance of six mono-task mapping heuristics considering different application workloads. Some of these heuristics were extended to consider multi-tasks mapping onto the same PE, while minimizing the commutation overhead in the same NoC-based platform [2]. Singh et al. [2] also proposed new heuristics that consider the power consumption as the product of number of bits to be transferred and distance between source-destination pair.

Faruque et al. [3] present a distributed agent-based mapping scheme. The proposed scheme divides the system into virtual clusters. A cluster agent (CA) is responsible for all mapping operations within a cluster. Global agents (GAs) store information about all the clusters of the NoC and use a negotiating policy with CAs in order to define to which cluster an application will be mapped. Another distributed approach is proposed in [4], which explores different implementations of a decentralized self-embedding algorithm, aiming to minimize network contention and latency while providing fault-tolerance support for NoC-based systems.

III. JOINT VALIDATION MODEL

Määttä et al. present in [12] the joint validation of an application mapped onto different platform models based on NoCs. This validation model enforces the use of a well-defined API among the main layers of the MPSoC: application, mapper and platform. An **application** is modeled by any number of concurrent tasks that communicate by explicitly exchanging

978-1-4673-2895-1/12 $31.00 © 2012 IEEE

messages. The communication dependencies between tasks are modeled using directed graphs with tasks represented by the nodes and messages by the edges. Each task is characterized by its computation time, and each inter-task communication message is characterized by its source and destination tasks, and its data volume. Many different models of computation can be used to further describe the concurrent behavior of tasks and messages. In this paper, we reuse the model described in [12], where messages are sent by a task only when they finish execution, and tasks can only be triggered by a timer or a predefined combination of messages (which may have to arrive in a predefined sequence). The **mapper** is responsible to map tasks to PEs on the platform. In turn, a **platform** is composed of PEs interconnected by a NoC. When tasks are triggered, the PEs onto which they are mapped are made busy for the task's execution time (or it enters a scheduling queue in case the PE was already busy). Once a task has finished its execution, the PE sends the respective messages to the NoC, which simulates its transmission towards the core where the destination task is mapped. The latency of the task execution on the PEs and the flit-by-flit message transmission over the NoC is then back-annotated to the application model, allowing for an accurate estimation of the response time of each task of the system, taking into account the contention for the PEs as well as NoC links and routers.

In this work, we extend all three layers described above: application, platform and mapper. One extension is that the PEs of the platform are now multi-tasking, using an Earliest Deadline First scheduling algorithm. Another extension that involved all three layers was to support the heterogeneity of the MPSoC. This involves setting the type of each PE (e.g. CPU1, DSP1, CPU2, ...) and setting for each task of the application which are the PEs that can execute it (i.e. the system contains the object code of the task for a certain PE). We also extended the characterization of application tasks by modeling the interplay between computation time and affinity. **Computation time** denotes how long does it take for a task to execute all its functionality. Sometimes the computation time may depend on the inputs of the task. In such cases, it is common in real time systems to define it as the worst case computation time to guarantee that the task will always be able to execute without missing the deadline. **Affinity** is measured in percentage and it is a multiplicative factor to increase or reduce the computation time depending on which PE the task is mapped. Every task must have its computation time defined in relation to the PE over which the task has greater affinity. So, the computation time of a task m mapped on a PE k (CT_{mk}) can be calculated by

$$CT_{mk} = \frac{CT_m}{Af_{mk}} \qquad (1)$$

where CT_m is the computation time of the task m when mapped on a PE which has 100% affinity and Af_{mk} is the affinity of the task m to the PE k.

IV. DYNAMIC MAPPING ALGORITHMS

Carvalho et al. [1] compare six dynamic mapping algorithms. The **First Free (FF)** simply selects the next compatible IP core to map a given task, thus walking sequentially through all IP cores before considering an IP core

again. **Nearest Neighbor (NN)** considers first the IP cores located near to the requesting task, and it maps the target task on the first compatible IP core found. **Minimum Maximum Channel load (MMC)** considers all possible mappings for a given task and chooses the one that increases the least the peak load of a channel of the NoC. **Minimum Average Channel load (MAC)** considers all possible mappings for a given task and chooses the one that increases the least the average load of the channels of the NoC. **Path Load (PL)** considers all possible mappings for a given task and chooses the one that increases the least the sum of the load of the channels between the requesting task and the target task. **Best Neighbor (BN)** considers first the IP cores located near to the requesting task, and if there is more than one candidate mapping at the same hop distance from the requesting task, the best alternative is selected according the PL algorithm.

Two dynamic mapping algorithms were developed in the frame of this work. **Minimum Data Exchange (MDE)** considers all possible mappings for a given task and computes for each of them the total amount of data that must be sent/received by the already mapped tasks. The PE with less communication load receives the target task. If more than one PE returns the same communication load (very likely to happen in the beginning of the execution of the system), the PE with minimum hops distance to the requesting task is selected. If again there is more than one candidate PE, the first candidate of an array of final candidates is selected. The **Cost Based (CB)** dynamic mapping algorithm considers all possible mappings for a given task and chooses the one with minimum cost according to the following equation

$$Cost = \frac{U_k \times H_{st} \times L_{st}}{Af_{tk}} \qquad (2)$$

where U_k is the current utilization of the PE under consideration for mapping k, H_{st} is the number of hops between the source task s and the target task t (considering t mapped on k), L_{st} is the load between s and t measured by the amount of bytes exchanged by them, and Af_{tk} is the affinity of the target task t to k. The utilization of a PE can be calculated by

$$U_k = \sum_{m=0}^{q} \frac{CT_{mk}}{P_m} \qquad (3)$$

where CT_{mk} is the computation time of task m when mapped to the PE under consideration k, P_m is the period of task m and q is the number of tasks mapped onto k.

Table 1 presents the metrics used by the cost functions of the dynamic mapping algorithms introduced on this work. FF is for sure the fastest algorithm, since it requires only to find the next compatible PE for a task. NN is also fast and tries to put the communicating tasks near to each other. BN comes next in terms of speed since it searches for possible PEs according to NN and only uses PL when more than one candidate PE is found. All the other algorithms consider all PEs for making a mapping decision, therefore, they become slower with the increase of the number of PEs. On the other hand, other algorithms can consider the channels of the NoC and the communication load of the tasks for preventing congestions. The computation load of the tasks mapped on a PE is also an

978-1-4673-2895-1/12 $31.00 © 2012 IEEE

important metric for avoiding the overload of the PE, and is considered by the CB algorithm.

All dynamic mapping algorithms, except the FF, require a requesting task to perform their mapping decision accurately. As the tasks that start the application do not dispose of a requesting task and such a decision can affect all subsequent decisions of the dynamic mapper, two initial mapping algorithms were developed to deal with this situation. One is the FF that was already presented, and the other is the **Cluster (CL)** initial mapping algorithm. This algorithm divides the PEs in clusters and maps each initial task to a different cluster. The size of the cluster depends on the amount of initial tasks and PEs the system contains. The goal of this algorithm is to separate the initial tasks, allowing tasks mapped later to be near to their corresponding initial tasks.

Table 1. Metrics used by the cost functions of different dynamic mapping algorithms used on this work.

	Network position	Channel load	Comm. load	Task affinity	PE utilization
FF	✓				
NN	✓				
MMC	✓	✓			
MAC	✓	✓			
PL	✓	✓			
BN	✓	✓			
MDE	✓		✓		
CB	✓		✓	✓	✓

V.   CASE STUDY

In order to evaluate the different dynamic mapping algorithms, one synthetic application was developed and executed over a 4x4 and a 5x5 heterogeneous platforms based on the HERMES NoC [12]. These heterogeneous platforms are configured with 2 DSPs, one on the upper left corner and another on the lower right corner. All the other PEs are GPPs, and both platforms reserve the PE on the lower left corner for the mapper. The application is composed by 30 tasks, where 12 use the initial mapper and 18 are dynamically mapped. These 30 tasks are used and reused by a total of 15 task graphs which describe different functionalities of the application. The computation time of the tasks range from 1,000 to 70,000 simulation cycles and the period of the tasks range from 200,000 to 500,000 simulation cycles. Each simulation cycle corresponds to one clock cycle of the real platform. The message sizes communicating the tasks range from 1,000 to 50,000 bytes. Six tasks are compatible to the 2 DSPs available on the platform, where these tasks have an affinity of 100% to them, while they have an affinity of 20% to the GPPs. Each platform-mapper combination was simulated for 10,000,000 simulation cycles and every time a task is mapped to a certain PE it remains there until the end of the simulation.

Figure 1 presents boxplots with the response times of 5 task graphs of the application mapped to the 4x4 platform (graphs A-E) and the 5x5 platform (graphs F-J). The numbers 1 and 2 used on the label of the plots indicate respectively the use of the CL and FF initial mappers. The plots aligned on the same row indicate the task graph. Each plot presents on the X-axis the dynamic mapping algorithm used by the application. The Y-axis refers to the response time of a complete task graph (i.e. the time between triggering the execution of the first task and

the time when the last task finishes executing and communicating). Boxplots show minimum, lower quartile, median, upper quartile and maximum response times of all executions of each task graph, measured on simulation cycles.

On the first glance to any of the graphs, it is possible to see that the worse response time of a task graph can be more than the double of the best response time, indicating that the mapping algorithm really influences the timing of the application. One expectation was that the dynamic mapping algorithm CB would always present better timing results, since it considers both communication and computation of the application. However, CB provided the best timing results on only 40% of the cases presented on

Figure 1. While it is enough to know that the CB was the dynamic mapping algorithm that performed better on most of the times, it cannot be forgotten that the CB is also the one that costs more in terms of communication overhead (i.e. information about communication load, task affinity and PE utilization, which are information that need to be transmitted through the network from all PEs to the mapper). This costs in terms of time and network usage for transferring this mapping information is currently not considered on this work.

Another expectation was that the initial mapping algorithm CL would always present better timing results, since it creates clusters to keep the initial tasks of different task graphs apart from each other, thus giving space for the subsequent tasks mapped with the dynamic mapping algorithm to be grouped together with their corresponding initial task. While this is true and visible when compare graph G1 with G2 and graph H1 with H2, the CL was only better than FF on 42.5% of the cases on the 4x4 platform and 82.5% of the cases on the 5x5 platform.

VI.   CONCLUSIONS

Eight dynamic mapping algorithms for heterogeneous MPSoCs were compared on two platform configurations. These dynamic mapping algorithms consider different cost functions like the position of the task on the platform, the congestions of the network, the amount of data transmitted by the tasks, the affinity of the tasks to the different types of PEs available on the platform and the utilization of the PEs. Hence, the timing characteristics of computation and communication of the application and platform were taken into account on the presented case studies.

From the results obtained we could see that in most of the cases the variance w.r.t. the timing to execute the application task graphs is enormous. This is mainly due to the fact that the baseline HERMES NoC is employed, which does not contain any mechanism that can enable Quality-of-Service (QoS) guarantees. The results also showed that the initial mapper CL was better than the FF on 42.5% of the times over the 4x4 platform and 82.5% of the times over the 5x5 platform. The dynamic mapping algorithm CB provided the best timing results on 40% of the cases. The main future work is to consider NoC architectures that provide QoS and to consider in a lightweight manner the costs of the mapper in terms of computation time and communication through the network to deliver the information required for the mapper. Another future work is to allow the definition of deadlines for each message and keep control if the deadlines are missed during simulation. Further comparisons regarding the dynamic mapping algorithms will then be performed to extract a more solid evaluation.

REFERENCES

[1] E. Carvalho, N. Calazans, and F. Moraes, "Dynamic Task Mapping for MPSoCs," *IEEE Design & Test of Computers*, vol. 27(5), 2010.

[2] Singh, A. K., at al. Communication-aware heuristics for run-time task mapping on NoC-based MPSoC platforms. Journal of Systems Architecture, 56(7), 2010.

[3] Faruque, M.A.; et al. ADAM: Run-time Agent-based Distributed Application Mapping for on-chip Communication. In: DAC'08, 2008.

[4] Weichslgartner, A., et al. "Dynamic Decentralized Mapping of Tree-Structured Applications on NoC Architectures". In: NoCS'11, 2011.

[5] Hölzenspies, P.K.F.; et al. Run-time Spatial Mapping of Streaming Applications to a Heterogeneous Multi-Processor System-on-Chip (MPSoC). In: DATE'08, 2008.

[6] Wildermann, S.; et al. Run time Mapping of Adaptive Applications onto Homogeneous NoC-based Reconfigurable Architectures. In: FPT'09, 2009.

[7] Molnos, A.; et al. Composable, energy-managed, real-time MPSOC platform. In: OPTIM'10, 2010.

[8] Smit, L.T.; et al. Run-time mapping of applications to a heterogeneous SoC. In: SoC'05, 2005.

[9] Chou, C-L. and Marculescu, R. "Run-time task allocation considering user behavior in embedded multiprocessor networks-on-chip". IEEE Transactions on Computer-Aided Design of Integrated Circuits and Systems, vol. 29(1), 2010.

[10] Ferrandi, F.; et al. "Ant colony heuristic for mapping and scheduling tasks and communications on heterogeneous embedded systems". IEEE Transactions on Computer-Aided Design of Integrated Circuits and Systems, vol. 29(6), 2010.

[11] Huang, L., et al. "Customer-Aware Task Allocation and Scheduling for Multi-Mode MPSoCs". In: Design Automation Conference, 2011.

[12] Määttä, S.; et al. "Joint Validation of Application Models and Multi-Abstraction Network-on-Chip Platforms". Int. J. of Embedded and Real-Time Communication Systems (IJERTCS), vol. 1(1).

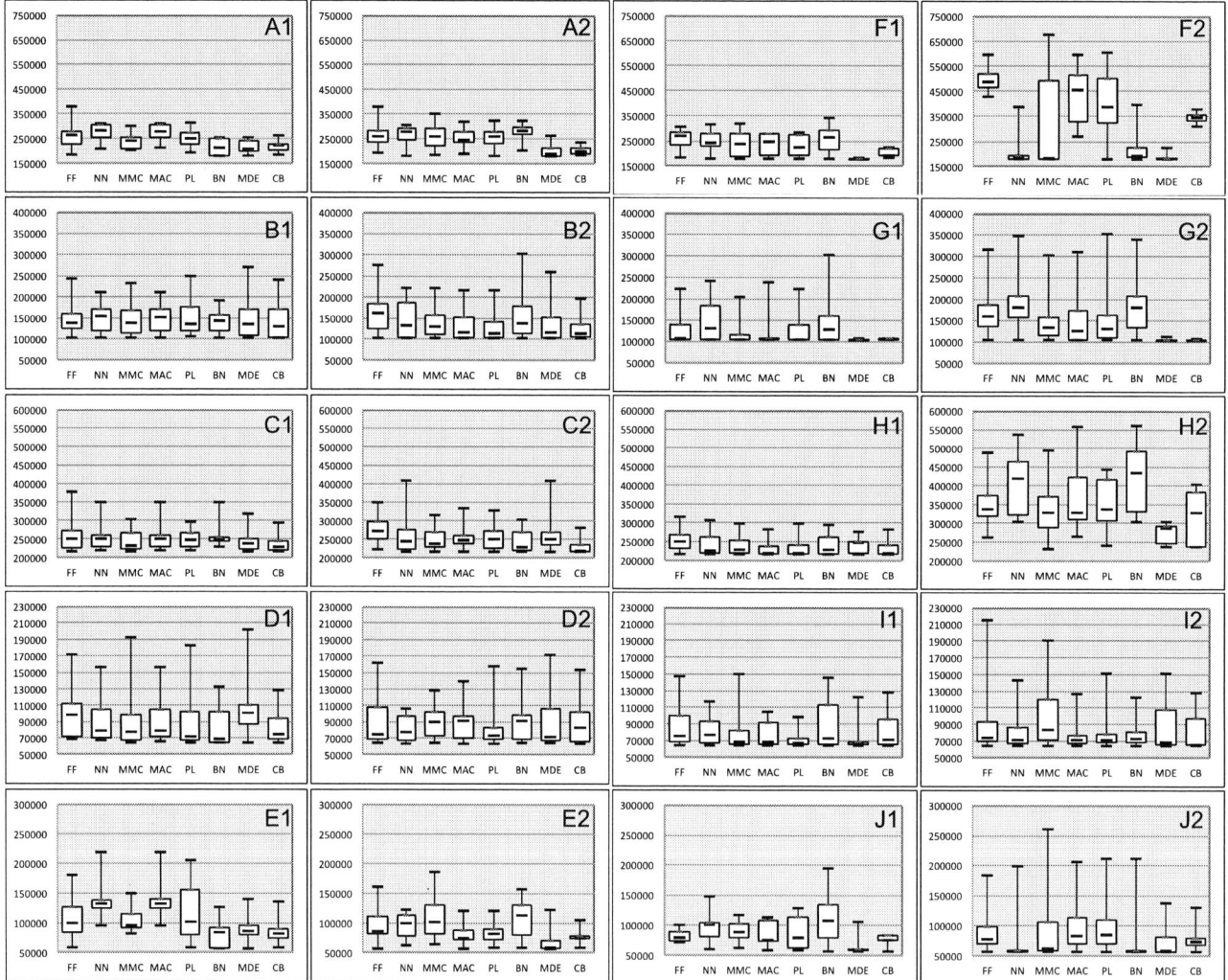

Figure 1. Response time of 5 application task graphs for the 4x4 platform (A-E) and the 5x5 platform (F-J). Graphs labeled with 1 use the CL initial mapper and graphs labeled with 2 use the FF initial mapper. The X-axis presents the dynamic mapping algorithm and the Y-axis presents the response times of all executed instances of each task graph in cycles (minimum, lower quartile, median, upper quartile and maximum response times).

A Hybrid Chip Interconnection Architecture with a Global Wireless Network Overlaid on Top of a Wired Network-on-Chip

Ling Wang and Zhen Wang
Dept. of Computer Science and Technology
Harbin Institute of Technology
Harbin, China

Yingtao Jiang
Dept. of Electrical and Computer Engineering
University of Nevada, Las Vegas
Las Vegas, USA

Abstract—**Multi-core platforms are emerging trends in the design of Systems-on-Chip(SoCs). Among a number of possible alternatives, the wireless network-on-chip (WNoC), enabled by the availability of miniaturized on-chip antennas, is envisioned as a revolutionary approach which may bring in significant performance gain. In this paper, we present a new WNoC architecture with a layered topology. In essence, the whole on-chip mesh-based network is divided into several subnetworks where each of these subnetworks has a wireless node sitting at its center. This architecture reduces the latency and the dynamic power consumption between two long distant nodes communication in the network and can improve the communications throughput in high traffic applications. Experiment results shows that the performance improvements are 8% and 13% respectively, compared with two other WNoC architectures.**

Keywords- Network-on-Chip; wireless; subnet; 2-Level Hybrid Mesh topology

I. INTRODUCTION

Network-on-Chip (NoC) has emerged as a communication backbone to enable a high degree of integration in multi-core System-on-Chips (SoCs) [1]. Despite their advantages, an important performance limitation in conventional NoCs arises from planar metal interconnect-based multi-hop communications, wherein the flits between two long distant nodes experience overly long latency and cause high power consumption. To alleviate this problem, long-range links using conventional metal wires are inserted in a standard mesh network [2-6]. Another effort to improve the performance of multi-hop NoC was undertaken by introducing ultra-low-latency and low power express channels between communicating nodes [7, 8]. These communication channels are yet made of metal wires as well. In the near future, improvements in metalic wires will no longer be able to keep up with the performance requirements and new interconnect paradigms are needed. As a result, different NoC approaches have been explored including on chip wireless communications using ultra-wideband (UWB) technology [9, 10] that is higher rate, lower delay and shorter communication distance.

In addition to the UWB technology, CMOS radio frequency (RF) [11] and carbon nanotube (CNT) antenna [12] are two

possible on chip wireless communication technologies, but their maximum communication distance is much longer than UWB. Yet the exponentially increase of wireless launch energy consumption, limited numbers of wireless channels and the serious congestion on wireless router becomes a greater limitation.

Meanwhile, the latest research [13, 14] mainly focuses on the so called mixed WNoC which deploys wired link between adjacent nodes while nodes that are far apart can talk to each other through a single hop wireless link. In the field of mixed WNoC research, at present, mainly put forward the concept of subnet, nodes in the local subnet is wired linked and each subnet has a wireless router to communicate with the other subnets with single-hop wireless way. However, all these wireless Network-on-Chip (WNoC) structures do not scale well with the network size, because of the likeliness of wireless router congestion and the exponential increase of wireless launch energy consumption determined by the distance between two wireless nodes. The mentioned two kinds of structure shared a serious problem that a large volume of data communications may go through the wireless nodes which will easily cause the wireless communication congestion by such single-hop wireless communication way between the subnets.

Here we propose a novel mixed wireless-multi-hop WNoC architecture which is referred as 2-Level Hybrid Mesh. It addresses simultaneously the latency, power consumption and interconnect routing problems. We integrate both the advantage of the existing pure wire WNoC and the mixed WnoC and proposed a new mixed WNoC structure based on multi-hop wireless communication and the structure we proposed can significantly dissipate less energy and to achieve notable improvements in throughput and latency compared to traditional wired NoCs and the existing WNoC structures.

II. PROPOSED WNOC STRUCTURE

A. Proposed topology

Modern complex network theory provides us with a powerful method to analyze network topologies and their properties. Networks with 2-Level Hybrid Mesh structure have a very short average path length, which is commonly measured as number of hops between any pair of nodes. Figure 1 shows the proposed network structure, which is divided into several subnets. Within each subnet, there is one wireless node which is connected with the subnet's wired nodes via wired links. This wireless node connects with other subnets' wireless nodes through wireless links. All the wireless nodes are equipped with wireless base stations (WBs) that can transmit and receive data packets over wireless channels. In Figure 1, red nodes represent wireless routers, blue nodes for wired routers; The red real line represents the wireless link, and the blue line represents the wired link. All the wired links constitute a wired mesh network and all of the wireless links constitute an overlay top-level wireless mesh network.

When a data packet needs to be sent to a core located in a different subnet, starting from the source node, the data packet travels through intermediate wireless and wire routers before reaching the destination. As wormhole routing is adopted, data packets are broken down into smaller entities called flow control units or flits. The header flit holds the routing and control information, and it establishes a path that all subsequent payload or body flits follow. The routing algorithm is described in Section B.

We consider an N × N mesh topology. Without loss of generality, we assume N is a multiple of 3 for convenience. First, the original N × N mesh is divided into 3×3 submeshes. Then, the central node of each 3 × 3 submesh is connected to each other in East, South, West, North four directions. The low-level mesh consists of nodes with wire connected. The central nodes of each submesh are used to construct an up-level mesh. In the up-level mesh, packets are transmitted in multi-hop wireless way. Central node is chosen so that the wireless node as the average distance from other eight nodes is minimal. As the central node is shared by both low and up level meshes, four wireless ports are added to the original for connecting up-level meshes.

Compared with other structures, this structure has the following advantages:

First, in this structure, wireless nodes in the network are uniformly distributed and communication way is multi-hop, it will greatly reduce the occur probability of wireless link congestion.

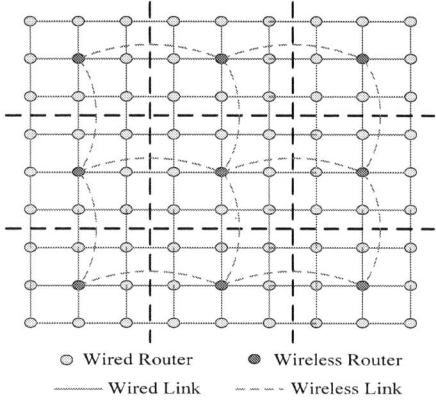

○ Wired Router   ● Wireless Router
—— Wired Link   - - - - Wireless Link

Figure 1. Proposed WNoC Topology

Second, in this structure, each subnet has a fixed size and moderate scale, which will both have a better expansibility and avoid the wireless router working pressure with the expanding network scale.

In further performance analysis, we will see a fact that our 2-Level Hybrid Mesh topology have a better performance than Hybrid Mesh topology while a little worse than Small-World topology in terms of maximum path length and average path length. However, in the way of wireless nodes and links, our 2-Level Hybrid Mesh topology is much more than those two topological structures, hence, it incorporated the path length and network congestion, that will greatly reduce the probability of data collision on wireless up-level subnet. In our following section of performance analysis, we will give a comparison of those three mixed WNoC topology on four indexes: maximum path length, average path length, numbers of wireless nodes and wireless links.

B. Routing algorithms

As we previously mentioned, the routing algorithm in the lower wired mesh can be extremely simple, while in the upper wireless mesh, the routing algorithm needs to handle massive data volume passing through the wireless nodes.

For the proposed wireless NoC architecture, we design a congestion-control routing algorithm (WFXY) combined with deterministic and adaptive characteristics:

First, packets routed in the wired mesh follow the deterministic XY routing algorithm, which has a low algorithm complexity and guarantees the shortest wired path;

Second, in the top wireless mesh, the partially adaptive routing algorithm West-First is used to route packets, so the congestion level of the wireless mesh can degrade, and the wireless path is the shortest.

Further, we define a threshold T for the routing distance. If a packet whose Manhattan distance between the source and the destination is greater than T, it will be classified as a long distance packets, otherwise it is a short distance packet. The long distance packets are routed through wireless mesh, while the short distance packets can only be transferred through wired mesh. In out experiment, the threshold T is set as 10.

(1) WFXY Routing Algorithm

WFXY routing algorithm is a combination of West-First routing algorithm and XY routing. As a distributed routing algorithm, WFXY is implemented at every router, and the routing decision is made collectively at every router on the path from the source to the designation. Whenever a packet arrives at a node, WFXY algorithm will determine one direction to switch the packet. This dedcision is based on the current node C, the target node D, the packet type and available buffer sizes of its neighbour nodes.

WFXY routing algorithm in wired nodes

Input: The coordinates of current node C and target node D, the packet type;

Output: Packet routing;

1) If the packet is a short distance packet, it will be routed towards D by wired links using XY strategy;

978-1-4673-2895-1/12 $31.00 © 2012 IEEE

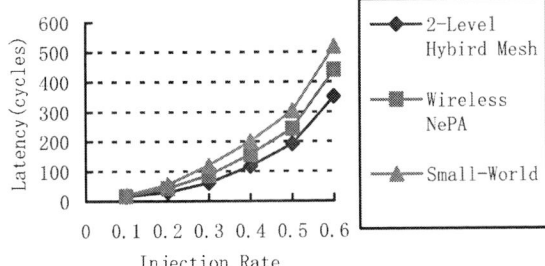

Figure 2. Latency under Uniform Random Model

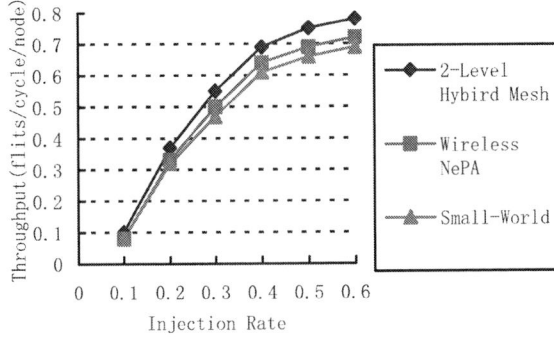

Figure 3. Throughput under Uniform Random Model

2) Else if the packet is a long distance packet, C and D are in different subnets, then the packet will be transferred towards the central node by wired links using XY strategy;

3) Else if the packet is a long distance packet, C and D are in the same subnet, then the packet will be routed towards D by wired links using XY strategy;

WFXY routing algorithm in wireless nodes

Input: The coordinates of current node C and target node D, the packet type, the buffer levels of neighbour wireless nodes;

Output: Packet routing;

1) If the packet is a short distance packet, it will be routed towards D by wired links using XY strategy;

2) Else if the packet is a long distance packet, C and D are in the same subnet, then the packet will be routed towards D by wired links using XY strategy;

3) Else if the packet is a long distance packet, and the subnet of D is at the west side of C, then the packet will be transferred by the X- wireless link;

4) Else if the packet is a long distance packet, and the subnet of D is not at the west side of C, then the packet will be transferred adaptively by X- or Y+ or Y- wireless link according to the buffer levels of the neighbor wireless nodes;

Considering an N×N network, time complexity of WFXY algorithm is O(N), and space complexity is $O(N^2)$.

When the long distance and short distance packets exist in the network at the same time, it is very likely to cause a deadlock with formed cyclic routing paths involving both wired and wireless links. To resolve this potential deadlock

problem, the routers employ two virtual channels in each port: VC0 for the long distance packets and VC1 for the short distance packets. Because the two types of packets are routed through different virtual channels, and no VC can dictate the switching fabric indefinitely, the possibility of having a deadlock can be eliminated.

III. PERFORMANCE ANALYSIS

To evaluate the performance of our proposed WNoC, a cycle-accurate WNoC simulator based on SystemC is used. In the experiment, we compare the performance (latency and throughput) of the proposed 2-Level Hybrid Mesh structure with that of two other WNoC architectures, Small-World and Wireless NePA. We assume that all three architectures are used to connect a system with 144 cores. Two traffic distribution models are adopted in the experiment: (1)Uniform Random model, where every source node has equal probability to communicate with all other nodes; and (2)Hotspot model, where 8 hotspot nodes are introduced and they receive 15% of the total network traffic. At 0.18 um technology, the area overhead of the virtual-channel router is 0.21 mm².

The average latency and throughput under Uniform Random pattern are shown in Figure 2 and 3, respectively. They are measured with respect to the traffic injection rate. At low traffic load, all of the three architectures perform well. When the injection rate rises, the 2-Level Hybrid Mesh structure has the lowest latency and the highest throughput (8% higher than Wireless NePA and 13% higher than Small-World). So it is evident that our proposed architecture outperforms the other two counterparts.

Under Hotspot pattern, a lot of packets are transmitted to the 8 hotspot nodes, so the network is more likely to become congested. Figure 4 shows the average latency under Hotspot model, and it is obvious that the 2-Level Hybrid Mesh architecture has the lowest latency. In the 2-Level Hybrid Mesh WNoC, as multi-hop wireless links are used to transmit packets, the data can be scattered in different wireless paths. Moreover, the adaptive routing algorithm "West-First" is introduced in our design, which can control the congestion. So the 2-Level Hybrid Mesh WNoC has the lowest latency under heavy traffic load.

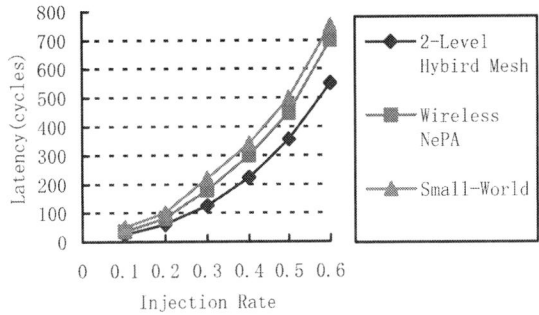

Figure 4. Latency under Hotspot Model

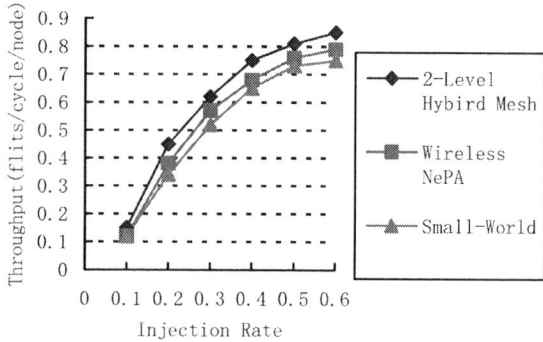

Figure 5. Throughput under Hotspot Model

In Figure 5, the throughput under Hotspot pattern is shown. Like the result under Uniform Random model, the 2-Level Hybrid Mesh architecture still has the highest throughput (7% higher than Wireless NePA and 13% higher than Small-World), proving it performs better than the other two architecture once again.

IV. CONCLUSION

In this paper, we have proposed a new Wireless NoC structure, and respective routing algorithm. The proposed architecture is an overlay of two networks. At the upper layer, nodes can communicate through a wireless mesh network. At the lower level, nodes can communicate by wired links. To avoid network congestion, packets are classified as long distance packet and short distance packet to reflect their lengths. These two kinds of packet will be routed at different virtual channels in the upper wireless network to avoid any possible deadlocks. Experiment results have shown that the proposed WNoC has lower latency than others, and the throughput improvements are 8% and 13% respectively.

ACKNOWLEDGMENT

This paper is supported by the Fundamental Research Funds for the Central Universities (Grant No.HIT.NSRIF.2012049).

REFERENCES

[1] A. Jantsch and H. Tenhunen (Eds.), Networks on Chip, Kluwer, 2003.

[2] P. P. Pande, et al., "Performance Evaluation and Design Trade-offs for Network-on-chip Interconnect Architectures," IEEE Transactions on Computers, vol. 54, No.8, August 2005, pp. 1025-1040.

[3] Q. Yang and Z. Wu, "An improved mesh topology and its routing algorithm for NoC," International Conference on computational intelligence and software engineering (CiSE), pp. 1-4, 2010.

[4] M. Saneei, A. Afzali-Kusha and Z. Navabi, "Low-latency multi-level mesh topology for NoCs," International Conference on Microelectronics (ICM '06), pp. 36-39, 2006.

[5] K. Chen, C. Peng and F. Lai, "Star-type architecture with low transmission latency for a 2D mesh NOC," IEEE Asia Pacific Conference on Circuits and Systems (APCCAS), pp.919-922, 2010.

[6] A. Tavakkol, R. Moraveji and H. Sarbazi-Azad, "Mesh connected crossbars: A novel NoC topology with scalable communication bandwidth," International Symposium Parallel and Distributed Processing with Applications (ISPA), pp. 319-326 , 2008.

[7] A. Kumar et al., "Toward Ideal On-Chip Communication Using Express Virtual Channels", IEEE Micro, vol. 28, issue 1, January-February 2008, pp. 80-90.

[8] T. Krishna et al., "NoC with Near-Ideal Express Virtual Channels Using Global-Line Communication", IEEE Symposium on High Performance Interconnects, HOTI, 26-28 August, 2008, pp. 11-20.

[9] D. Zhao and Y. Wang, "SD-MAC: design and synthesis of A hardware-efficient collision-free QoS-aware MAC protocol for wireless network-on-chip," IEEE Transactions on Computers, vol. 57, pp. 1230-1245, 2008.

[10] D. Zhao, Y. Wang, J. Li and T. Kikkawa, "Design of multi-channel wireless NoC to improve on-chip communication capacity," IEEE/ACM International Symposium on Networks on Chip (NoCS), pp. 177-184, 2011.

[11] M. Chang, J. Cong, A. Kaplan, M. Naik, G. Reinman, E.Socher and S. Tam, "CMP Network-on-Chip Overlaid With Multi-Band RF-Interconnect," IEEE 14th International Symposium High Performance Computer Architecture (HPCA), pp. 191-202, 2008.

[12] K. Kempa, et al., "Carbon Nanotubes as Optical Antennae," Advanced Materials, vol. 19, pp. 421-426, 2007.

[13] S. Deb, K. Chang, A. Ganguly and P. Pande, "Comparative performance evaluation of wireless and optical NoC architectures," IEEE International on SOC Conference (SOCC), pp. 487-492,2010.

[14] C. Wang, W. Hu and N. Bagherzadeh, "A wireless network-on-chip design for multicore platforms," 19th Euromicro International Conference on Parallel, Distributed and Network-Based Processing (PDP), pp. 409-416, 2011.

[15] W. Dajin and C. Jiannong, "On optimal hierarchical configuration of distributed systems on mesh and hypercube," Parallel and Distributed Processing Symposium, Proceedings. International, pp. 8, 2003.

978-1-4673-2895-1/12 $31.00 © 2012 IEEE

Statistical Timing Characterization

Z. WU, P. MAURINE, N. AZEMARD
LIRMM, University of Montpellier II
Montpellier, France
firstname.lastname@lirmm.fr

G. DUCHARME
Dept. Math University of Montpellier II
Montpellier, France
ducharme@math.univ-montp2.fr

Abstract—Monte Carlo (MC) method is the standard method to characterize statistical moments of cell delays and slopes. However, this method suffers from very high computational cost. In this paper, we propose a technique to quickly and accurately estimate Standard Deviation (SD) of standard cell delays and slopes. The proposed technique is based on the identification, performed with a reduced set of MC simulations, of delay and output slope SD functions that take input slope, output load and supply voltage as input arguments. These identified functions are then used to estimate SDs of delays and slopes at different operating conditions (input slope, output load, supply voltage). This proposed method provides at least 76% of CPU gains, with respect to MC, while keeping high accuracy.

Keywords-Monte Carlo method; Statistical Static Timing Analysis; Probability Density Functions

I. INTRODUCTION

Similar to traditional Static Timing Analyses (STA), Statistical Static Timing Analysis (SSTA) [1–4] aims at providing timing information allowing guaranteeing the functionality of a design. The main difference between STA and SSTA lies in the nature of the information reported. Indeed, SSTA propagates Probability Density Function (PDF) of timing performance rather than worst case timings. This constitutes a significant advantage allowing estimating the manufacturing yield. However, SSTA presents some drawbacks. First, the timing analysis itself is, in view of computational cost, more expensive than the traditional corner approach. Second, structural correlations introduced by the logic synthesis must be captured to obtain accurate PDF of the propagation delays. Third, statistical timing characterization of standard cell library, which is generally performed using a Monte Carlo (MC) method, remains extremely expensive.

To accelerate the procedure of characterization, we may reduce either the number of runs considered during MC simulation, or the number of operating conditions (input slope, output load, supply voltage) to be simulated, i.e. the size of timing Look-Up Tables (LUT). Concerning the first way, many variants of MC, such as importance sampling [5], Latin hypercube sampling [6], have been proposed to reduce the number of runs without losing too much accuracy. However, applying these techniques on dozens of random process parameters remains complicated. The second way is usually implemented by characterizing the timing moments for a reduced subset of operating conditions and deducing the others with some techniques. The contribution in this paper follows this latter idea.

Indeed, we propose a technique, called Semi-Monte-Carlo (SMC), to quickly and accurately estimate for each standard cell, the functions linking its operating conditions (input slope, output load, supply voltage) to the standard deviations of its timings. These functions are fitted on a reduced set of MC simulations performed for few operating conditions, and then are used to estimate the statistical timing values at all the other operating conditions considered the timing library file.

The rest of this paper is organized as follows. Section II presents the SMC method. Section III validates the proposed SMC method and shows the CPU time gains that can be obtained time with respect to MC. Finally, Section IV concludes the paper.

II. SEMI-MONTE-CARLO METHOD

To characterize the SD of a CMOS cell timing metric, a large number of electrical simulations are usually launched under specific environmental conditions (Supply Voltage V_{dd} and temperature T). But this requires very high computational cost if these MC simulations have to be repeated for various operating conditions (output load and input slope). In this section, we present a method allowing reducing significantly the number of MC simulations to be launched to fully fill such timing LUT.

A. SMC method for input slope τ_{in}

Given cell type, input pin and output edge, cell delay d and output slope τ_{out} can be respectively described as:

$$\begin{cases} d = f(P, \tau_{in}, C_{out}, T, V_{dd}) & (1) \\ \tau_{out} = g(P, \tau_{in}, C_{out}, T, V_{dd}) & (2) \end{cases}$$

where:

- the functions f, g are not explicitly known, but can be modeled by circuit simulations;
- $P = (p_1, p_2, \dots, p_l)$ is a vector of process parameters and $p_i \sim F_i(\mu_i, \sigma_i)$ with F_i, μ_i, σ_i known;
- τ_{in}, C_{out} represent the two operating conditions input slope and output load;
- T, V_{dd} are environmental parameters: temperature and supply voltage.

Suppose $d \sim F_d(\mu_d, \sigma_d)$, $\tau_{out} \sim F_{\tau out}(\mu_{\tau out}, \sigma_{\tau out})$, and τ_{in}, C_{out}, T, V_{dd} are bounded, i.e. $\tau_{in} \in (0, \tau_{max})$, $C_{out} \in (0, C_{max})$, $T \in [T_{min}, T_{max}]$ and $V_{dd} \in [V_{min}, V_{max}]$. Let us define the problem: given $\tau_{in} = \tau_0$, $C_{out} = C_0$, $T = T_0$, $V_{dd} = V_0$, estimate the SDs $\sigma_d, \sigma_{\tau out}$.

978-1-4673-2895-1/12 $31.00 © 2012 IEEE

According to (1) and (2), we know that the random variations of d and τ_{out} arise from P. The moments μ_d, $\mu_{\tau out}$, σ_d, $\sigma_{\tau out}$ are affected by the variational factors τ_{in}, C_{out}, T, V_{dd}. Here, we first consider the case where C_{out}, T, V_{dd} are constants and study the relationships between both σ_d, $\sigma_{\tau out}$ and τ_{in}, denoted respectively as $\sigma_d(\tau_{in})$ and $\sigma_{\tau out}(\tau_{in})$. Fig. 1 shows the curves of $\mu_d(\tau_{in})$ and $\mu_d(\tau_{in}) + \sigma_d(\tau_{in})$. These moments are estimated with data obtained by drawing a random sample of size $N = 10^4$ for the process vector P and running circuit simulation for each P_n, $n = 1, 2, \ldots, N$. The dotted curves are the delays simulated with the first five experiments of P, i.e. $d = f(P, \tau_{in})$, where $P = P_1, P_2, \ldots, P_5$. Note that $f(P, \tau_{in})$ is the same function as in (1); the omission of the terms C_{out}, T, V_{dd} indicates they are kept constants. From this figure, we see that the curves $\mu_d(\tau_{in})$ and $\mu_d(\tau_{in}) + \sigma_d(\tau_{in})$ are similar in form to the curves $f(P, \tau_{in})$.

In order to simplify the description of our method, from here on, we focus on estimating σ_d; the approach is the same for $\sigma_{\tau out}$.

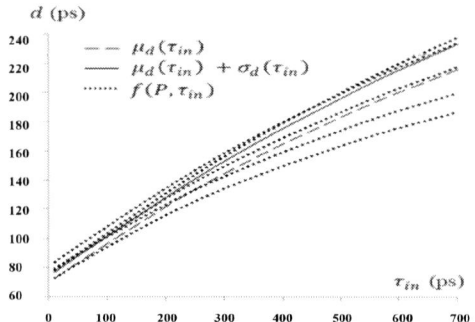

Figure 1. Comparison of curves $\mu_d(\tau_{in})$, $\mu_d(\tau_{in}) + \sigma_d(\tau_{in})$ and $f(P, \tau_{in})$

According to the feature deduced from Fig. 1, we propose the following empirical property: there exist two experiments P_K, P_L of the random vector P, so that

$$\begin{cases} \mu_d(\tau_{in}) = f(P_K, \tau_{in}) & (3) \\ \mu_d(\tau_{in}) + \sigma_d(\tau_{in}) = f(P_L, \tau_{in}) & (4) \end{cases}$$

If P_K, P_L are known, then for any value τ_0 of τ_{in}, $f(P_K, \tau_0)$ and $f(P_L, \tau_0)$ can be obtained by running one circuit simulation. So, we have:

$$\begin{cases} \mu_d(\tau_0) = f(P_K, \tau_0) & (5) \\ \sigma_d(\tau_0) = f(P_L, \tau_0) - f(P_K, \tau_0) & (6) \end{cases}$$

According to the way statistical model cards are constructed, the expectation, $\mathbb{E}\big(f(P, \tau_{in})\big)$, of $f(P, \tau_{in})$ can be obtained by passing the expectation of P, $\mathbb{E}(P) = (\mu_1, \mu_2, \ldots, \mu_I)$, inside $f(P, \tau_{in})$. This entails that P_K can be set to $(\mu_1, \mu_2, \ldots, \mu_I)$. Thus, only P_L needs to be identified. To this end, we used the following procedure:

1) For the same sample $P_1, P_2, \ldots, P_N$ of P, estimate $\sigma_d(\tau_1)$, $\sigma_d(\tau_2)$, $\sigma_d(\tau_3)$ with data from MC circuit simulations, where τ_1, τ_2, τ_3 are values of τ_{in} and chosen according to the sensitivity of μ_d, σ_d on τ_{in}.
2) Compute $q_j = \mu_d(\tau_j) + \sigma_d(\tau_j)$ with $j = 1, 2, 3$.
3) For each P_n, $(n = 1, 2, \ldots, N)$, compute:

$$e_n = \sum_{j=1}^{3} \left| f(P_n, \tau_j) - q_j \right| \qquad (7)$$

4) $P_L = P_S$ where $S = argmin_n(e_n)$.

With the above procedure and Eq. (5) and (6), we can obtain any $\sigma_d(\tau_{in})$ with $J = 3$ characterized points $\sigma_d(\tau_1)$, $\sigma_d(\tau_2)$ and $\sigma_d(\tau_3)$. That is to say, no matter how many $\sigma_d(\tau_{in})$ need to be characterized, they can always be deduced from $\sigma_d(\tau_1)$, $\sigma_d(\tau_2)$ and $\sigma_d(\tau_3)$.

The above criteria to find P_K, P_L and the empirical property in (3) - (4) are based on an empirical analysis. They will be proved by validating SMC in latter section. Note that we can have better accuracy if more than three values of τ_{in} are used to find out P_L, but this requires more CPU runtime to get the data. This conclusion will also be shown later.

B. Extensions for C_{out} and V_{dd}

The method presented in subsection II.*A* can be respectively extended to C_{out} and V_{dd}.

Consider C_{out} as an example. Suppose that τ_{in}, T, V_{dd} are constants. We propose a property similar to (3) - (4) as: there exist two experiments P_K^*, P_L^* of the random vector P, so that

$$\begin{cases} \mu_d(C_{out}) = f(P_K^*, C_{out}) & (8) \\ \mu_d(C_{out}) + \sigma_d(C_{out}) = f(P_L^*, C_{out}) & (9) \end{cases}$$

Similarly, we set $P_K^* = (\mu_1, \mu_2, \ldots, \mu_I)$ and follow the same four steps in subsection II.*A* to identify P_L^*.

But such an extension cannot be applied to T. As shown in Fig. 2, the forms of the curves $\mu_d(T)$, $\mu_d(T) + \sigma_d(T)$ and $f(P, T)$ are different. An empirical property as in (3) and (4) cannot be proposed. This can be explained by Fig. 3. Indeed, the function f in (1) is made up of two parts: f_1, f_2. Random process parameters $P = (p_1, p_2, \ldots, p_I)$ are first transformed to a group of electric parameters $E = (e_1, e_2, \ldots, e_M)$ by function f_1 as:

$$E = f_1(P, T) \qquad (10)$$

Then, cell delay d is a function of E described as below:

$$d = f_2(E, \tau_{in}, C_{out}, V_{dd}) \qquad (11)$$

Passing the expectation inside f_2 is still possible as before, however we cannot push further due to the exponential behavior of function f_1 [7].

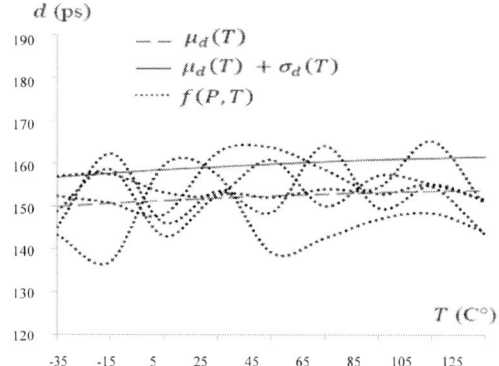

Figure 2. Comparison of curves $\mu_d(T)$, $\mu_d(T) + \sigma_d(T)$ and $f(P, T)$

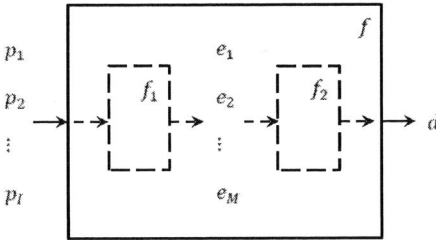

Figure 3. Decomposition of the function f.

C. Improved SMC method (ISMC method)

In fact, Eq. (4) can be generalized as:

$$\mu_d(\tau_{in}) + a \cdot \sigma_d(\tau_{in}) = f(P_L^{**}, \tau_{in}) \qquad (12)$$

where a is a real number. Then, Eq. (4) is a special case setting $a = 1$. Combining (3) with (12), we have:

$$\sigma_d(\tau_{in}) = \frac{1}{a} \cdot [f(P_L^{**}, \tau_{in}) - f(P_K, \tau_{in})] \qquad (13)$$

We define a finite sequence $a = A_1, A_2, \ldots, A_H$. Then, given τ_0, for each value A_h of a, compute $\hat{\sigma}_d(\tau_0, A_h)$ using the procedure in section II.A. Note that $\hat{\sigma}_d$ indicates the estimate of the real value σ_d. After that, the estimate $\hat{\sigma}_d(\tau_0)$ of $\sigma_d(\tau_0)$ is obtained by:

$$\hat{\sigma}_d(\tau_0) = \frac{1}{H} \cdot \sum_{h=1}^{H} \hat{\sigma}_d(\tau_0, A_h) \qquad (14)$$

It is clear that $\hat{\sigma}_d(\tau_0)$ is the average of all $\hat{\sigma}_d(\tau_0, A_h)$. This treatment can reduce the noise brought by a single value of a, therefore it offers a better accuracy.

III. VALIDATIONS AND DISCUSSIONS

We applied ISMC to characterized 9 types of cells with statistical model cards in the 65nm process provided by industry. These cards use the BSIM4 [8] model and define 51 random process parameters.

First, we ran 10^4 times MC under all possible combination of the following conditions:

$$\begin{cases} \tau_{in} = 10, 30, 60, 100, 150, 210, 280, 360, 450, 550 \text{ ps} \\ C_{out} = 5, 7, 10, 13, 16, 20, 25, 30, 35, 40 \text{ fF} \\ V_{dd} = 1, 1.05, \ldots, 1.35, 1.4 \text{ V} \\ T = -35, 25, 125 \text{ °C} \end{cases}$$

Then, $\sigma_{\tau out}$ and σ_d were estimated and used as references, or more precisely considered as real values. Next, for each value of T, we applied ISMC presented in section II.C to characterize $\sigma_{\tau out}$ and σ_d by considering one of τ_{in}, C_{out} and V_{dd} as a variable. As an example, the procedure to characterize σ_d for τ_{in} is as follows:

1) Run 1000 times MC for $\tau_{in} = 10, 210, 550$ ps and estimate $\hat{\sigma}_d(10), \hat{\sigma}_d(220), \hat{\sigma}_d(550)$.

2) Set $a = 0.8$ and apply the procedure in section II.A to find P_L^{**}.

3) Run circuit simulation with P_L^{**} and get $\hat{\sigma}_d(30, 0.8), \hat{\sigma}_d(60, 0.8), \ldots, \hat{\sigma}_d(450, 0.8)$ using (13).

4) Repeat the first three steps for $a = 1, 1.2, -0.8, -1, -1.2$ and compute their averages according to (14).

Finally, we computed the absolute error e_{abs} and relative error e_{rel} of $\hat{\sigma}_d, \hat{\sigma}_{\tau out}$ with respect to referenced values σ_d, $\sigma_{\tau out}$. Note that $\hat{\sigma}_d$ in the first step, which is the estimate of σ_d, are obtained by running 1000 instead of 10^4 times MC. Fig. 4(a) and 4(b) show the results. To describe the quality of estimates, we define:

$$\begin{cases} \text{zone 1: } e_{abs} < 0.5\text{ps and } e_{abs} < 10\% \\ \text{zone 2: } e_{abs} < 0.5\text{ps and } e_{abs} > 10\% \\ \text{zone 3: } e_{abs} > 0.5\text{ps and } e_{abs} < 10\% \\ \text{zone 4: } e_{abs} > 0.5\text{ps and } e_{abs} > 10\% \end{cases}$$

In Fig. 4(a), all the points are in zone 1, which demonstrates that ISMC gives very good accuracy on $\hat{\sigma}_d$. In Fig. 4 (b), most of the points are in zone 1. Those in zone 2 have relatively large relative errors, but their absolute errors are less than 0.5ps, which could be neglected in timing analysis. Points in zone 3 have absolute errors between 0.5ps and 1ps, while their relative errors are small, less than 3%. Finally, only four points in zone 4, but their absolute errors less than 1ps. In summary, Fig. 4(a) and 4(b) validate ISMC. Considering the good accuracy of $\hat{\sigma}_d$, from here on, we only compare the quality of $\hat{\sigma}_{\tau out}$ which is more dispersed.

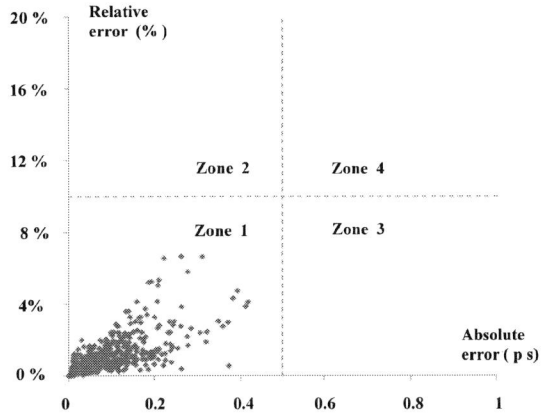

Figure 4(a). Absolute and relative errors of $\hat{\sigma}_d$.

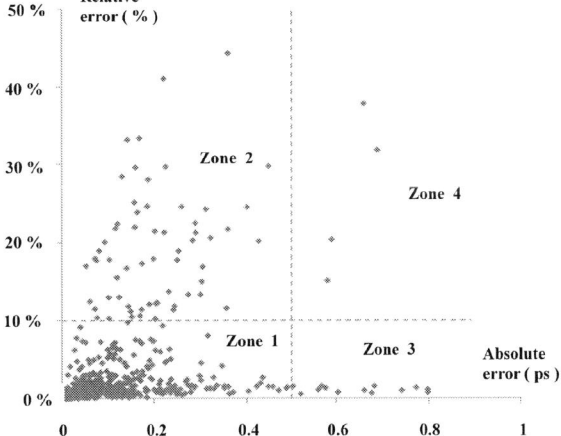

Figure 4(b). Absolute and relative errors of $\hat{\sigma}_{\tau out}$.

978-1-4673-2895-1/12 $31.00 © 2012 IEEE

To get more information from the results, we classified the points in Fig. 4 into three groups: values estimated by varying respectively τ_{in}, C_{out} and V_{dd}. Table I shows that ISMC gives better accuracy for C_{out} in zone 1 and for V_{dd} in zone 1, 2 and 3. The validation of ISMC is reconfirmed by the fact that 99.5% of all the points are in zone 1, 2 and 3.

TABLE I. QUALITY ON $\hat{\sigma}_{\tau out}$ FOR GROUPS CLASSIFIED BY VARYING RESPECTIVELY τ_{in}, C_{out} V_{dd} AND ALL

	% points in zone 1	% points in zone 1,2 and 3
τ_{in}	87.8%	99.6%
C_{out}	92.6%	98.9%
V_{dd}	89.3%	100%
All	90.0%	99.5%

Table II compares the accuracy of MC (1000 runs), SMC and ISMC. As shown, ISMC is slightly better than MC (90.0% vs. 86.2% in zone 1) and about 10% of gain with respect to SMC. We conclude from this table that ISMC can replace MC to do characterizations in view of accuracy.

TABLE II. COMPARING QUALITY ON $\hat{\sigma}_{\tau out}$ OF DIFFERENT METHODS.

		% points in zone 1	% points in zone 1,2 and 3
MC (1000 runs)		86.2%	99.9%
SMC	$a = 0.6$	73.3%	95.9%
	$a = 1$	80.8%	98.9%
	$a = 1.6$	79.3%	97.3%
ISMC		90.0%	90.0%

The results in Fig. 4 were obtained by ISMC setting $a = 0.8, 1, 1.2, -0.8, -1, -1.2$. These values were chosen according to an analysis shown in Fig. 5. In this figure, the quality of SMC is not good for small values of a. This is because a lot of experiments of the sample locate in the interval $(\mu - 0.8 \cdot \sigma, \mu + 0.8 \cdot \sigma)$, which leads to much difficulty to find an acceptable P_L with the procedure in section II.A. In contrast, when $a < -2$ and $a > 2$, there are not enough candidate experiments to choose, which introduces too many noises. In consequence, considering the flexibility, we have chosen the above six values of a. This flexibility means that these values of a will not introduce more noises even though the runs of MC are reduced, e.g. from 1000 to 500.

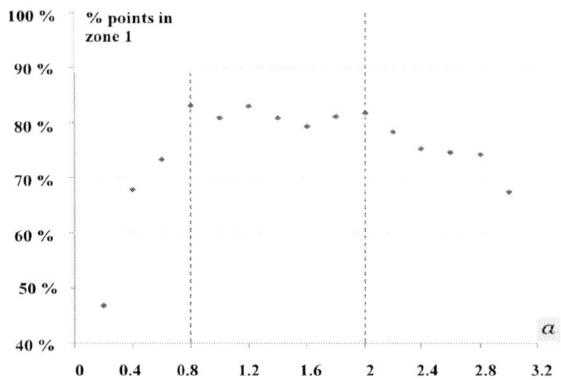

Figure 5. Percentage of points $\hat{\sigma}_{\tau out}$ in zone 1 for different values of a.

In the validation and Eq. (7), three values of each variable τ_{in}, C_{out} and V_{dd}, i.e. $J = 3$, are used to fit the functions and to find P_L. Table III compares the quality for different values of J. If we consider the quality of MC in Table II (86.2% of points in zone 1) as reference, so we should set $J_{\tau_{in}} = 2$, $J_{C_{out}} = 2$ and $J_{V_{dd}} = 3$. However, considering the fact that $\sigma_{\tau out}$ is sensitive to τ_{in}, therefore $J_{\tau_{in}} = 3$ is preferable.

According to the comparison in Table III, we only need to characterize SD for 6 couples of values of (τ_{in}, C_{out}), i.e. 3×2, to construct a LUT given a value of T and V_{dd}.

Table IV shows the significant gain of CPU time using ISMC with respect to MC. As an example, we have a gain of 76% when constructing a LUT of size 5×5, which requires doing characterization for 6 instead of 25 points.

TABLE III. QUALITY ON $\hat{\sigma}_{\tau out}$ OF DIFFERENT VALUES OF J.

		% points in zone 1	% points in zone 1,2 and 3
τ_{in}	$J = 2$	86.1%	99.6%
	$J = 3$	87.8%	99.6%
	$J = 4$	87.4%	99.6%
C_{out}	$J = 2$	87.0%	96.2%
	$J = 3$	92.6%	98.9%
	$J = 4$	93.7%	99.6%
V_{dd}	$J = 2$	82.7%	100%
	$J = 3$	89.3%	100%
	$J = 4$	90.5%	100%

TABLE IV. GAIN OF CPU TIME USING ISMC WITH RESPECT TO MC

size	5×5	7×7	9×9
gain	76%	87.8%	92.6%

IV. CONCLUSIONS

In this paper, we present a method to characterize standard deviations of cell delay and output slope. The main contribution is to propose a way to describe the functions of standard deviations depending respectively on input slope, output load and supply voltage. This proposed method, comparing with MC (1000 runs), has been proved to provide significant gains of CPU time, at least 76%, while keeping the same accuracy.

REFERENCES

[1] H. Chang, S. Sapatnekar, "Statistical Timing Analysis Considering Spatial Correlations Using a Single PERT-like Traversal", ICCAD, 2003, pages 621 – 625.

[2] C. Visweswariah, K. Ravindran, K. Kalafala, S. Waler, S. Narayan, "First-order Incremental Block-based Statistical Timing Analysis", DAC, 2004, pages 331 – 336.

[3] L. Cheng, J. Xiong, L. He, "Non-Linear Statistical Static Timing Analysis for non-Gaussian Variation Sources", DAC, 2007, pages 250– 255.

[4] Z. Feng, P. Li, Y. Zhan, "Fast Second-order Statistical Static Timing Analysis Using Parameter Dimension Reduction", DAC, 2007, pages 244 – 249.

[5] R. Kanj, R. Joshi, S. Nassif, "Mixture Importance Sampling and its Application to the Analysis of SRAM Designs in the Presence of Rare Failure Events", DAC, 2006, pages 69 – 72.

[6] V. Veetl, D. Blaauw, D. Sylvester, "Criticality Aware Latin Hypercube Sampling for Efficient Statistical Timing Analysis", TAU, 2007.

[7] B. Lasbouygues, R. Wilson, N. Azemard, P. Maurine, "Temperature- and Voltage-Aware Timing Analysis", IEEE Trans. on CAD of Integrated Circuits and Systems, vol.26, no.4, pages 801 – 815, 2007.

[8] BSIM4.2.1 MOSFET Model – User's Manual, University of California, 2001.

Hierarchical Control Flow Matching for Source-level Simulation of Embedded Software

Kun Lu, Daniel Müller-Gritschneder, Ulf Schlichtmann

Institute for Electronic Design Automation, Technische Universität München

http://www.eda.ei.tum.de

Abstract—Source-level simulation (SLS) of embedded software annotates the source code based on the matching of the control flow graphs (CFG) between the source code and the cross-compiled binary code. However, existing SLS approaches still can not guarantee to find a matching for a CFG that is optimized by the compiler. Further, they rely on debug information, which may be unreliable. In this paper, the authors propose a hierarchical CFG matching approach to reduce the influence of compiler optimization and ambiguous debug information. This approach divides the CFGs of the source and binary code into nested regions. Then the matching of those two CFGs is performed for the regions in a top-down manner. In this way, heavy optimization or debug misinformation of certain basic blocks will not have global impact on the matching of other basic blocks. Moreover, optimized loops and branches are matched with respect to the optimization techniques used by the compiler.

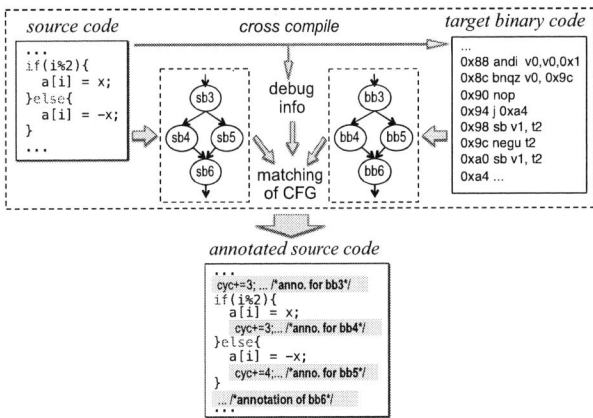

Fig. 1. Sketch of basic SLS annotation process

I. INTRODUCTION

For the development of system-on-chip (SoC), software development is causing more and more cost and effort. Virtual prototypes (VPs) have been widely used to enable early software performance estimation, debug and verification. There are two main computation models for VPs to simulate the SW programs. One is the instruction set simulator (ISS) that interprets the instructions of the *cross-compiled* program as the target machine would do. However, besides the high modeling effort, ISSs are very slow and thus not appropriate for simulating long software programs. The other is annotating the programs which are then directly compiled for the simulation host. Usually, the annotation information is obtained from the cross-compiled binary and models the program's property, such as timing. Because host-compiled simulation yields much faster simulation speed over ISS simulation, it has raised wide research interests in the last decade [1]–[12].

In general, it is more advantageous to the designer to annotate the original source code for source-level simulation (SLS) [5]–[7], [9]–[12]. However, annotating the source code becomes difficult in the presence of compiler optimization, which changes the control flows of the cross-compiled binary code. This requires to tackle the problem of resolving the position of annotation under compiler optimization. For illustration, we sketch the basic annotation process of SLS in Fig. 1. The source code is first cross-compiled. The control flow graph (CFG) of the binary code is then constructed. Then the basic blocks, i.e. nodes, in the CFGs of the source and binary code are mapped against each other. Information,

such as timing, is then extracted for each binary basic block. To handle timing for superscalers, timing in [9] is extracted for two consecutive basis blocks. Then the information is annotated based on the CFG mapping. Now, the annotated source code models the execution of the original program on the target machine. Finally, the annotated source code is directly compiled and executed on the simulation host. However, the core problem in this SLS annotation process is that the matching of the source and binary CFGs can not be easily found, because (1) the structure of CFG may be heavily changed due to compiler optimization and (2) the debug information required by the CFG matching may be erroneous. Recently, researchers start to tackle these problems by considering the domination concept [10], [12] which is used by the compiler in code optimization [13]. However, current SLS approaches still can not annotate the source code effectively and reliably. We exemplify these two problems in the following.

Problems of changed control flows: Consider an example in Fig. 2. In the source code, the branch condition at line 84 dominates the following code. In the binary code, due to branch elimination, the branch instruction compiled from line 84 is duplicated and pushed into its predecessors. This branch instruction does not dominate following binary basic block anymore. Thus, we will not find a matching for the following binary basic blocks based on dominator comparison. Many other optimization techniques can heavily change the control flows, such as loop transformations. Hence, it is insufficient

978-1-4673-2895-1/12 $31.00 © 2012 IEEE

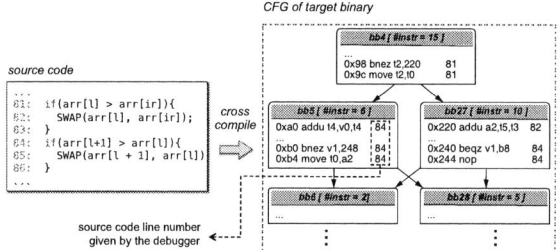

Fig. 2.   Example of changed domination and wrong debug information.

and unreliable to consider the whole CFG and apply the domination analysis for basic block matching.

Problems of erroneous debug information: CFG matching requires the debugger to provide information regarding the mapping from each target binary instruction to the source code line number. However, this information can be ambiguous. In Fig. 3, the binary instructions in the right part are compiled from different source code lines. However, all of them are reported in the debug information to be compiled from the source code line 133. Consequently, matching of these binary blocks can not be found. Besides, this misinformation will also change the domination relation for the following basic blocks.

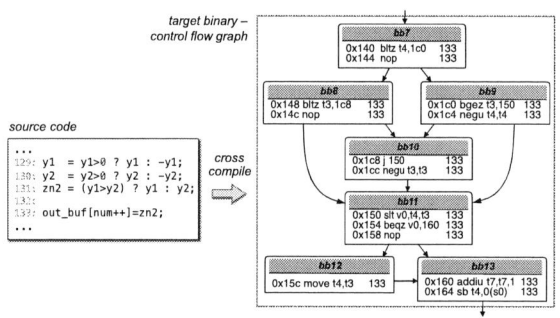

Fig. 3.   Example of erroneous line reference by the debugger.

Our contribution in this paper is a method to annotate the source code more reliably and accurately in the presence of compiler optimization and erroneous debug information. This method divides the CFGs of the source and binary code hierarchically into nested regions. The regions are then matched in a top-down manner. Thus the problem of matching the basic blocks of the original CFG is decomposed into matching the basic blocks of smaller sub-graphs. By doing this, the altered control flows and erroneous debug information in one region will not have global impact on matching the basic blocks in other regions. Thus, the influence of compiler optimization is reduced and the timing accuracy of the annotated source code is improved. Besides, we consider the optimization principles of loops and branches, which have been overlooked by previous SLS approaches. Hence optimized loops and branches can be annotated more effectively.

In the following, Sec. II surveys state of the art. Sec III presents our approach. Sec. IV gives experimental results

followed by the conclusion.

II. STATE OF THE ART

In the last few years, continuing research effort is given to source level simulation [5]–[7], [9]–[12]. At first, SLS approaches ignore compiler optimization [5]–[7], [9]. They extract information from the basic blocks in the control-flow of the cross-compiled binary code. Then they annotate the source code based on line references that are given by debuggers. For example, if the instructions of a basic block are compiled from certain lines in the source code, then the information of that basic block is annotated after the corresponding lines. These approaches can not resolve the annotation position if compiler optimization change the control flow of the original source code or the debug information is erroneous. Thus they may yield very inaccurate SLS results. Recently, approaches have been proposed to handle compiler optimization. Mueller et al. [10] consider the principles that are used by the compiler in optimization [10]. Based on those principles, they derive the *do*mination, post-domination and loop membership properties for the basic blocks in the control flow graphs (CFGs) of the source and cross-compiled binary code. They label each basic block with those properties and match the basic blocks by their properties. Although the use of dominance and loop membership can handle some optimizations such as like loop unswitching. However, some optimizations alter the control flows in a way that the domination relation is also changed. Besides, as the domination relation is obtained by analyzing the whole CFG, therefore debug misinformation of one basis block may have global impact and cause many other basic blocks to be unmatched. Stattlemann et al. [12] propose a similar approach as et al. [10], in that they also perform the matching based on domination relation. This approach shares the same limitation of [10]. The problem of debug misinformation is mentioned by Stattlemann et al. [14]. They re-map the wrongly mapped instructions to the source code lines. However, the correction of such debug misinformation may be even more complex than the matching of basic blocks. Because it requires both control flow and data flow analysis. Therefore, SLS approaches need to find a way of matching the basic blocks even in the presence of debug misinformation.

III. OUR SLS APPROACH

In the following, we first introduce some basic terms regarding the graph analysis. Then we present our approach to tackle the unsolved problems in the area of SLS. This approach hierarchically divides the CFG into sub-graphs and combines the domination in matching the sub-graphs and the basic blocks within them.

A. Preliminary

Compilers use certain principles to perform code optimization [13]. We explain several fundamental terms involved in the principles with a sample CFG in Fig. 4.

Dominator: Node bb_x dominates node bb_y if all paths from the *start* node to bb_y pass bb_x. In the figure, node $bb2$ dominates node $bb3, bb4$ and $bb5$. So $dom(bb2) = \{bb3, bb4, bb5\}$.

Post-dominator: Node bb_y post-dominates node bb_x if all paths from bb_x to the *end* node pass bb_y. In the figure, $pdom(bb2) = \{bb5, bb6\}$.

Immediate post-dominator: Node bb_y is the immediate post-dominator of node bb_x if bb_y is the first visited post-dominator in all paths from bb_x to the *end* node. In the figure, $ipdom(bb2) = bb5$.

Back edge is an edge going from a node to its dominator. Back edges are used to identify loops. In the figure, we can find the back edge $(bb5, bb2)$.

Loop header node is the tail node of a back edge. It is the entry node of a loop.

Loop latch node is the head node of a back edge.

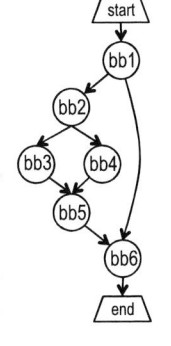

Fig. 4. Sample CFG.

B. Hierarchical CFG matching

We divide the CFG into sub-graphs successively in a top-down manner. This gives nested sub-graphs that represent the hierarchy of the CFG. For simplicity, we call a sub-graph as a code **region**. There are two types of divisions:

Division of branch region gives a sub-graph, in which all nodes are control-dependent on a branch node. Let node bb_x be a branch node and node bb_y its immediate post-dominator, then the branch region $RB(bb_x)$ of bb_x is computed as:

$$RB(bb_x) = dom(bb_x) - dom(bb_y). \qquad (1)$$

For example in Fig. 5, $RB(bb8) = \{bb8, bb9, bb19, bb20\}$.

Division of loop region gives a sub-graph, in which all nodes are contained in a loop. Let $RL(bb_x)$ denote the loop region of a loop header node bb_x. To find the nodes in $RL(bb_x)$, we check if a node is dominated by bb_x and reachable by going against the edges from the loop latch node. For example in Fig. 5, $RL(bb4) = \{bb4, bb5, bb17, bb6\}$.

Algorithm 1 hierarchicalCFGMatch(R_s, R_b). The CFG matching algorithm

---

Require: R_s/R_b := CFG of the source/binary code
1: matched = dominanceMatch(R_s, R_b)
2: **if** matched **then**
3:   annotate(R_s, R_b)   //matched found!
4: **else**
5:   $\mathbf{R}_s$ = getSubRegions(R_s)
6:   $\mathbf{R}_b$ = getSubRegions(R_b)
7:   $P = matchSubRegion(\mathbf{R}_s, \mathbf{R}_b)$
8:   **for** $(R'_s, R'_b) \in P$ **do**
9:     hierarchicalCFGMatch(R'_s, R'_b)
10:   **end for**
11: **end if**

Description of algorithm: The hierarchical CFG matching

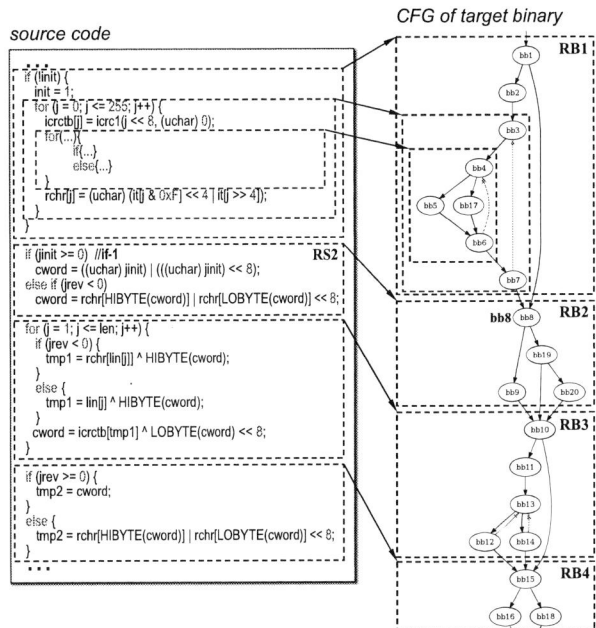

Fig. 5. Hierarchical division of CFGs.

algorithm is given in Alg. 1. The inputs are the source and binary CFGs to be matched. First, the basic blocks are matched strictly based on their dominators. If succeeded, the matching is found and annotation can be performed. However, with compiler optimization, this matching often fails on the overall CFG. If it fails, then the division rules are applied. For the example in Fig. 5, the resulted regions in the binary CFG after first division iteration are $RB1, RB2, RB3, RB4$. Each binary region is matched to the source region by matching its root node. For example, the branch instruction of the root node $bb8$ in $RB2$ is compiled from the statement $if-1$ in the source code, therefore $RB2$ is matched to $RS2$. After region matching, we have a set of pairs of sub-regions (denoted as P in the algorithm). Then each pair of sub-regions is matched in the same top-down manner until the basic blocks are reached. For unmatched region, the annotation solution is given in Sec. III-E.

Advantage: Hierarchical CFG matching decouples one region from another, thus altered control flow or erroneous debug information in one region will not have global effect.

C. Handling optimized loops

Compilers use certain rules to optimize loops [13]. We will conform to those rules in mapping the basic blocks of loops. These rules involve several terms:

Pre-test node is a branch node that controls whether the loop is entered. It is not necessarily the loop header node. A node is the pre-test node of a loop if its branch instruction is compiled from the loop condition and the loop header node is control-dependent on it.

Pre-header node is the only immediate predecessor of the loop header node. Compilers often create this node to hold the

978-1-4673-2895-1/12 $31.00 © 2012 IEEE

instructions that are moved outside form the loop body, or to be a common node for multiple branches that will otherwise branch to the loop header node.

For loop matching, the loop transformation patterns are first identified. Then the basic blocks of the loops are mapped accordingly. Consider the example in Fig. 6(a). *bb10* is the pre-test node of the loop. *bb11* is the pre-header node of the loop. The pre-test node is annotated before the loop (see comment 1 in figure). If the pre-header node is control-dependent on the pre-test node, then it means that the pre-header node is executed only in the first iteration. Instead of tracing the iteration count, we explicitly check the loop condition for the first iteration outside of the loop. This gives a place to annotate the pre-header node (comment 2).

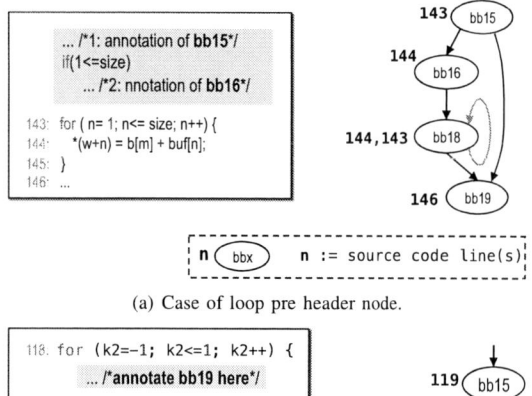

(a) Case of loop pre header node.

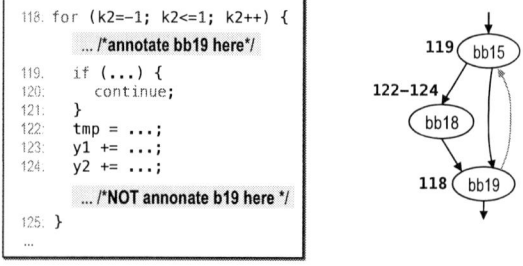

(b) Case of do-while transformation.

Fig. 6. Mapping basic block of loops.

In the presence of do-while transformation, the dominance relation is altered and can not be directly used for annotation. For example in Fig. 6(b), *bb19* is executed irrespective of the branch result of *bb15*, which is compiled from the `if` statement. However, if we annotate *bb19* after the `if` block in the source code, the annotated code will not be executed when the `if` branch is taken. Therefore, if the loop latch node post-dominates the loop header, then we annotate the binary block at the beginning of the `for` block. For loop splitting, the regions of the splitted binary loop bodies will be first mapped to their source code loop. Then mapping is performed for the basic blocks in each region respectively. For loop unswitching or overlapping, the use of the dominance and loop membership [10] can determine whether to map a binary basic block outside or inside a source code loop.

Fig. 7. Control node reconstruction for optimized branch.

D. Handling optimized branches

The original control node of a branch can be duplicated and eliminated by the compiler. In order to match the nodes that are dominated by this control node, we need to identify and reconstruct the control node. This is done by checking the following condition: If two branches in the binary are compiled from the same source code branch and they branch to the same target nodes, such as the branches of node $bb1, bb2$ in Fig. 7. then these two branches are duplication of the source code branch. For such branches, we add an intermediate node ($bb5$ in the figure) as a single control node for the branch target nodes. Now the resulted control flow will match that of the original source code. This method solves the problem discussed in Fig. 2.

E. Considering unmapped regions

If the basic blocks of a region can not be unambiguously mapped to the source code, then this region is considered as a single node. Its local worst case execution time is used for annotation. Consider the previously discussed case in Fig. 3, the binary CFG is divided into region $RB_{bb7} = \{bb7, bb8, bb9, bb10\}$ and $RB_{bb11} = \{bb11, bb12, bb13\}$. Because the debug information erroneously maps all instruction to line 133, no proper mapping is found by domination analysis. Thus, we take the longest path $bb7 \rightarrow bb9 \rightarrow bb10 \rightarrow bb11 \rightarrow bb12 \rightarrow bb13$ to calculate the worst-case execution cycle count. Suppose this count is 10, then we will annotate "`cycle+=10;`" after the source line 133. Although this method may cause over estimated timing, it is still a valid trade-off of accuracy for the automation of annotation.

IV. EXPERIMENTAL RESULTS

Several benchmarks (Tab. I) are simulated to validate and evaluate the proposed approach. These benchmarks are selected because their control flows are optimized so much that existing SLS approaches can not succeed in the CFG matching process. By applying the proposed approach, CFG matching is hierarchically performed and the source code can be annotated.

We use one case study to illustrate the necessity of using the proposed approach. Consider the example in Fig. 8. The source code in the left part is from the benchmark `select`. Because the variable `flag` is set to 0, the compiler finds that the `while` will always be entered and transforms it to a do-while loop. Besides, by performing data analysis on the index variable `flag`, the compiler finds that once the branch `if-1` is taken the loop will not be entered for the next iteration. In the compiled binary code, the binary instructions related to `if-2` block is pushed outside of the loop. Moreover, these instructions are placed after the loop body. Due to

978-1-4673-2895-1/12 $31.00 © 2012 IEEE

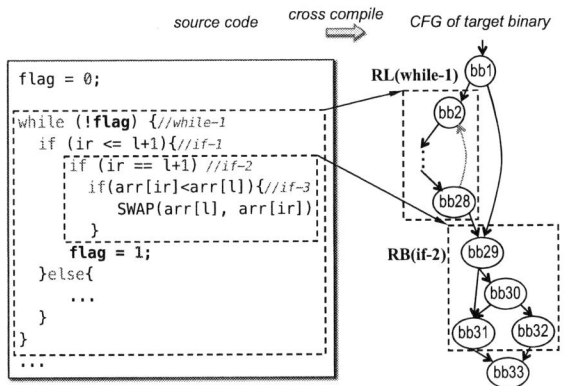

Fig. 8. Example of CFG matching.

TABLE I
COMPARISON OF SIMULATION ACCURACY AND SPEED

	crc	edgeDetect	prime	lzw	select
#. divsion	2	3	3	5	3
#. instr. (ISS)	21533	34552254	2409307	1585488	1992
#. instr. (SLS)	22305	34966371	2410509	1592424	1996
error(%)	3.6	1.2	<0.1	0.4	0.2
exe.time (ISS)	3.4ms	4.85s	314ms	265ms	343us
exe.time (SLS)	3.3us	5.6ms	270us	345us	0.5us
speedup	1020	870	1162	770	685

these optimization, the dominators for the basic blocks within and after this while loop are changed. As a result, existing SLS approach will not be able to find a matching for the basic blocks in the CFGs. With the proposed approach, region **RL(while-1)** and **RB(if-2)** in the binary are isolated with other regions. Matching of the basic blocks within these regions can be performed independently in a top down manner. Hence the change of control flow for this while loop does not affect the matching of the rest part of the CFG.

In the experiment, the benchmarks are cross-compiled with optimization level -*O2* and simulated by an interpretive ISS [15] as golden reference. The ISS is not a superscaler and has 5 pipeline stages and supports MIPS ISA. It uses a delay slot after each branch, hence no branch prediction is needed. The overall results are in Tab. I. The number of division means how many hierarchical division iterations are required to match the basic blocks within the divided regions. This number relatively indicates how intense the compiler optimization is. The overall estimation error is quite acceptably low for source-level simulation. In all programs, the compiler optimized the loops by adding pre-check nodes or using do-while transformation. To annotate these loops, the method in Sec. III-C is used. For the program `lzw` and `select`, the branch nodes are reconstructed prior to the CFG matching. For the program `crc` and `edge_detect`, the worst case timing is used for annotating certain unmatched regions. This causes an over estimation error, which is small enough to be accepted. As for the simulation speed, annotated source-level simulation provides a speed-up around three orders of magnitude.

Discussion: When function calls are used, the called functions also need to be annotated. When optimization level -O3 is used, certain small loops can be completely unrolled. The unrolled binary loop body can be divided by each iteration. The source code loop is annotated based on the iteration number. If this is not feasible due to heavy optimization within the unrolled loop body, then a worst case estimiation is used and annotated outside the source code loop. A large enough data cache in the experiment is used so that the error is mainly caused by CFG matching. In the future, timing due to cache conflicts will be examined.

V. CONCLUSIONS

The proposed hierarchical CFG matching decouples one region from another, thus minimizes the influence of altered control flows or erroneous debug information. Optimized loops are handled by considering the loop transformation patterns used by the compilers. Optimized branches are reconstructed in the binary code to retain the dominance relation. Several benchmarks are investigated , which can not be handled with existing SLS approaches. The annotated source level simulation shows adequate accuracy.

REFERENCES

[1] J. Y. Lee and I. C. Park, "Timed compiled-code simulation of embedded software for performance analysis of SOC design," in *ACM/IEEE Design Automation Conference (DAC)*, 2002.
[2] H. Posadas, F. Herrera, P. Sanchez, E. Villar, and F. Blasco, "System-level performance analysis in SystemC," in *Design, Automation and Test in Europe (DATE)*, 2004.
[3] K. Karuri, M.A.A.Faruque, S. Kraemer, R. Leupers, G. Ascheid, and H. Meyr, "Fine-grained Application Source Code Profiling for ASIP Design," in *ACM/IEEE Design Automation Conference (DAC)*, 2005.
[4] E. Cheung, H. Hsieh, and F. Balarin, "Framework for fast and accurate performance simulation of multiprocessor systems," in *IEEE International High Level Design Validation and Test Workshop*, 2007.
[5] J. Schnerr, O. Bringmann, A. Viehl, and W. Rosenstiel, "High-performance timing simulation of embedded software," in *ACM/IEEE Design Automation Conference (DAC)*, 2008.
[6] T. Meyerowitz, A. Sangiovanni-Vincentelli, M. Sauermann, and D. Langen, "Source-Level Timing Annotation and Simulation for a Heterogeneous Multiprocessor," in *Design, Automation and Test in Europe (DATE)*, 2008.
[7] P. Gerin, M. M. Hamayun, and F. Petrot, "Native MPSoC co-simulation environment for software performance estimation," in *International conference on Hardware/Software codesign and system synthesis*, 2009.
[8] L. Gao, J. Huang, J. Ceng, R. Leupers, G. Ascheid, and H. Meyr, "Totalprof: A Fast and Accurate Retargetable Source Code Profiler," in *CODES ISSS*, 2009.
[9] K.-L. Lin, C.-K. Lo, and R.-S. Tsay, "Source-Level Timing Annotation for Fast and Accurate TLM Computation Model Generation," in *IEEE/ACM Asia and South Pacific Design Automation Conference (ASP-DAC)*, 2010.
[10] D. Mueller-Gritschneder, K. Lu, and U. Schlichtmann, "Control-flow-driven Source Level Timing Annotation for Embedded Software Models on Transaction Level," in *EUROMICRO Conference on Digital System Design (DSD)*, Sep. 2011.
[11] S. Stattelmann, O. Bringmann, and W. Rosenstiel, "Fast and Accurate Source-Level Simulation of Software Timing Considering Complex Code Optimizations," *Design Automation Conference (DAC)*, 2011.
[12] S. Stattelmann, O. Bringmann, and W. Rosenstiel, "Dominator homo-morphism based code matching for source-level simulation of embedded software," in *International conference on Hardware/Software codesign and system synthesis (CODES+ISSS)*, 2011.
[13] A. V. Aho, R. Sethi, and J. D. Ullman, *Compilers Principles, Techniques and Tools*. Addison-Wesley Publishing Company, 1986.
[14] S. Stattelmann, A. Viehl, O. Bringmann, and W. Rosenstiel, "Reconstructing Line References from Optimized Binary Code for Source-Level Annotation," in *Forum on specification and Design Languages*, 2010.
[15] P. C. Model, http://www.opencores.org/projects/mips, 2011.

PowerMemo: A Power Profiling Tool for Mobile Devices in an Emulated Wireless Environment

Shiao-Li Tsao[+], Chih-Chen Kao, Ilter Suat, Yuchen Kuo, Yi-Hsin Chang, and Cheng-Kun Yu
Department of Computer Science, National Chiao Tung University
Hsinchu, Taiwan
sltsao@cs.nctu.edu.tw[+]

Abstract—In this paper, we present an architecture and implementation of a measurement-based energy profiling tool for mobile devices, called PowerMemo (<u>power</u> <u>me</u>ter for <u>mo</u>bile). The tool composes of a software event profiler and power measurement hardware to analyze process-level and function-level power consumption of mobile applications on Linux operating system and Dalvik virtual machine. Wireless signal attenuators and RF-shielded chambers are further integrated with the tool so that developers are able to emulate a real-life mobility scenario that a mobile device may encounter. The proposed tool overcomes the issue for profiling asynchronous I/Os and can correlate energy consumption of I/O events with software activities. This tool gives developers a broader view of software energy consumption in a mobile environment so that the developers can optimize the energy efficiency of their mobile applications.

Keywords- Embedded System, Mobile Device, Power Profiling, I/O Energy Consumption

I. INTRODUCTION

Embedded mobile devices such as mobile phone and tablets are getting more and more popular recently. However, limitation of standby and operation hours due to battery capacity has turn into a problem for mobile devices. This limitation even becomes worse after the coming of the powerful mobile platforms with multi-core CPUs and GPUs. Energy efficiency issues then have been increasingly important.

With the trend of cloud computing, complicated computation is expected to be accomplished by servers in a cloud center. Mobile applications, which require substantial wireless accesses and human-computer interactions, intend to be I/O-intensive rather than CPU-intensive. Energy consumption of wireless interfaces such as 3G, WLAN and few other I/O components have been proved to be much crucial for power consumption problem. To get detailed and useful information of energy consumption of I/O events, we, therefore, propose an energy profiling tool which focuses on I/Os for mobile devices.

Energy profiling tools can be categorized into two approaches: simulation-based and measurement-based tools. In this study, we consider a measurement-based energy profiling tool. The measurement-based energy profiling tools instrument the target hardware and collect power measurements along with the system event logs. Accuracy of this category of energy profiling tools is closely related to the time synchronization between a measurement device and a system profiler, sampling rate of a measurement device, and type of events collected

during profiling. One major problem that was criticized before is the need of an expensive and bulky measurement equipment. Nowadays, there have been relatively affordable, fast and portable digital signal and data acquisition cards with multiple measurement inputs, and therefore the measurement hardware platform can be resolved.

In this paper, a generic I/O energy profiling framework is proposed to associate the power consumption with processes and functions, and to map the power consumption of asynchronous I/O activities with software functions. The proposed tool is able to indicate the I/O energy consumption at process level and/or function level, and provides insights of energy consumption for different software designs. Moreover, we implemented a cross-mapping technique on Dalvik virtual machine in order to evaluate the power consumption of high level applications such as Java and Android programs. The proposed tool was implemented on Android/Linux operating system running on a TI OMAP3/Beagle board platform [9].

The power consumption behavior of wireless interfaces is quite different from other I/Os. The distance between wireless interfaces and base stations or access points, channel contentions, interferences, etc. significantly affect the transmission of packets and the power consumption. This exceptional I/O characteristic makes it quite a challenge to obtain a realistic energy consumption figure when an MS roams in a wireless network. This problem may especially restrict developers from optimizing and fine turning the energy efficiency of software in a mobile environment. To tackle the challenge, our tool profiles the energy consumption of applications while emulating the desired set of wireless channel behaviors in a controllable wireless environment.

The tool we described in this paper makes four main contributions. Firstly, we provide a measurement-based energy profiling tool for high-level application developers to examine the energy efficiency of applications and develop energy-efficient software. Secondly, the tool supports both static (compiling time, source codes needed), and dynamic (run-time, source codes not needed) energy profiling of software programs. Thirdly, we propose a technique to map the power consumption of I/O devices to the asynchronous I/O events in Linux kernel, processes running on Linux and also to applications such as Java and Android programs running on a virtual machine. Fourthly, the tool can emulate wireless environment so that developers can understand the power consumption behaviors of their software in real-life mobility scenarios.

978-1-4673-2895-1/12 $31.00 © 2012 IEEE

II. RELATED WORKS

PowerScope [1] is a well-known measurement-based tool and uses a programmable multi-meter to statistically measure the total energy drained from the external power source. The tool associates these measurements with active processes. The profiling results may become inaccurate since it measures only total energy and relies on an external trigger mechanism to synchronize the time between the host and target. Moreover, power consumption of I/O events are not supported by the tool.

Xian et al. further proposed an energy profiling tool that can perform measurements on every component such as processor, memory, and I/O devices [2]. Their tool is the first one that assigns measurements of I/O operations to the processes that generate the I/Os. This work also proposed an accurate synchronization method. However, the wireless network emulation and power consumption analysis in an emulated wireless environment were not yet been considered in their work.

More and more mobile devices embed powerful multi-core application processors and can run operating system and complicated mobile applications on Java virtual machine. The energy profiling of mobile software must consider not only OS kernel, user processes, but also virtual machines and high level mobile applications such as Android programs. In [3], the paper described the importance of energy efficiency of Android/Linux operating system. In [4], Andrew Rice and Simon Hay provided a measurement framework for Android-based mobile devices. The main difference between our energy profiling tool and their solution is that they only offer a total energy of the executing application, but we offer detailed energy consumption of I/O activities and the power consumption of software in process-level or function-level granularities.

An embedded system platform includes many I/O components, and a wireless interface is one of power-consuming and complex devices. Many factors impact the power consumption such as the packet size, the transmission rate, wireless access contentions, the RF power level, signal qualities, etc. Previous research considered various parameters to measure the wireless energy consumption [5]. However, very limited wireless environment emulation was used in the energy profiling. To our knowledge, noise from the environment and the distance between a mobile and base station both influence the results of the wireless energy consumption, but most of the former works ignored these effects [5]. Compared with laptops, hand held devices are often used in a dynamic mobile environment. The wireless environment emulation, therefore, is critical for the energy profiling.

Wireless environment simulation has been used for energy profiling on hand held devices [6]. Simulation, nevertheless, faces difficulties to interact with physical environment. Glenn Judd and Peter Steenkiste have presented a functional wireless emulator for few kinds of wireless behavior emulation [7]. We, then, first contribute to combine a complete emulation of heterogeneous wireless environment with the energy profiling of software.

III. DESIGN AND IMPLEMENTATION

Since the processes share hardware resources and operations of hardware resources consume energy. The main purpose of this tool is to map energy consumption of I/O devices to corresponding software functions.

A. System Architecture

As depicted in Figure 1, the tool consists of two different sets of hardware/software components, which could be categorized as target-side components and host-side components. The core of host side components is a graphical user interface (GUI) which acts as a control center in our tool. The GUI is mainly responsible for collecting power measurement results from a data acquisition (DAQ) card, controlling the signal attenuator to emulate the mobility that a user defined, and mapping the power measurements to calculate the total energy for each system activity. The main component at the target side is the kernel module that provides the instrumentation functions to collect Linux process and kernel level data logs. The kernel module utilizes the concept from Kernel Probes (Kprobes) [14] and User-space Probes (Uprobes) [15] so that it supports both static and dynamic energy profiling of software programs. The static energy profiling implies that a developer can insert our proposed profiling function calls to the processes, functions, and code blocks that they want to profile. The tool performs the profiling process and generates the report. In this case, source codes of mobile applications are required. On the other hand, dynamic profiling is for the cases that source codes of applications are not available. The tool can generate a list of processes and functions that the application uses from the symbol tables of the application. The developer can select the processes and functions that they would like to profile. Then, profiling codes are inserted before and after the profiling function so that the power consumption of the process and functions can be obtained. The kernel module stores the time of system activities and some other necessary parameters. A user-space daemon keeps transferring these profiling data from kernel-space to user-space as files, and periodically transfers these files to the host for post processing.

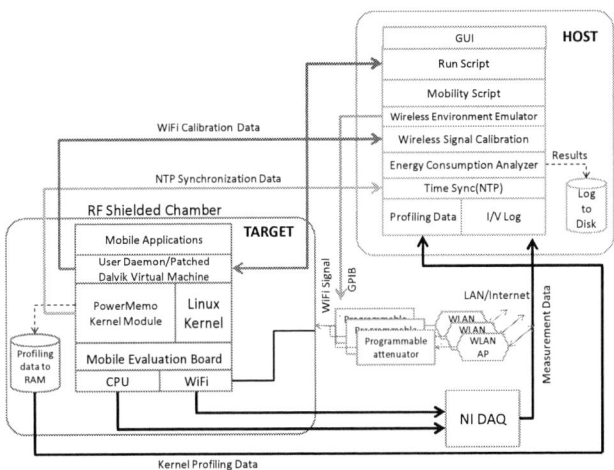

Figure 1. System architecture of the proposed tool.

B. Recording System Activities

1) Recording Process Activities

The begin and end time of each process activity running on a processor can be obtained by instrumenting the scheduler of the underlying kernel. The scheduler gives each process a time quantum to use processor exclusively. A process may or may not use the whole time quantum, and the interval it uses is defined as processor time slice. A process activity on a processor starts when this time slice is used. Therefore we add the instrumentation functions at the point where the current process is scheduled out and a new one is scheduled in. Because this is the exact point where the previous time slice ends and new one begins, we can correctly record begin and end time of activities.

2) Recording Function-level and Block-level Activities

PowerMemo is also capable of providing both function-level and code-block-level energy profiling results. According to our research, there are two types of techniques for obtaining the function-level and block-level profiling results. The first one is to record PC (Program Counter) in addition to PID (process identifier) when event occurs. These PCs can be used to compare memory addresses of functions with the binary image files of applications in order to correlate energy measurements to functions. The first approach is that the developers have to select the functions and processes they want to observe from our tool which can gather the function lists from the symbol tables of the program. These addresses of the functions are further processed and searched in the memory. The breakpoints are inserted to the locations of functions in the memory. When the processor runs the program and it is tripped at the locations of these functions, a profiling handler logs the time and related parameters which are processed later by PowerMemo. The source code of the program is not necessary for the energy profiling. The second technique takes advantages of special function calls and treats them as markers. A marker could be implemented as new system calls or I/O control, i.e. `ioctl()`. The developers have to insert these special function calls to the locations that they want to observe the power consumption. Then, the programs have to be recompiled and run. In this case, the source codes of the programs should be available. When the markers are executed, they trigger the kernel module to mark current system time. Later, the energy analyzer in PowerMemo Control Center uses these values to calculate energy for each function or code block. We have implemented the above two techniques in our tool.

3) Recording Java Program Activities

It is possible to record Java thread activities, such as loaded class information and method calls, within Dalvik virtual machine. To achieve this, first we have to generate time stamps for Java process events. Initially, we modified the Dalvik virtual machine and added several callback probes to trace the events with time information. In the new Android release, the Dalvik virtual machine has been implemented to support time logging. It is then possible to use Debug Java class to record event information and JDWP (Java Debug Wire Protocol) to transfer the event logs from Dalvik VM to a Linux process.

However, having traced all the Java process events with timestamps is not sufficient, since it only gives the start and the end time of a Java program. The Dalvik VM could be context-switched out and replaced by another process, which results in an inaccurate profiling report. To solve this problem, we adopt a methodology called "cross-mapping" which maps and compares the system logs generated by the kernel and program activity loggers generated by Dalvik VM. Figure 2 shows the concept of cross-mapping technique. Kernel and Dalvik VM activity loggers record the time that a specific Linux process, function calls or code blocks in the process, and the specific Java program or Java function calls are executed, respectively. Then, PowerMemo conducts a post process to associate the power measurements with the activities logs.

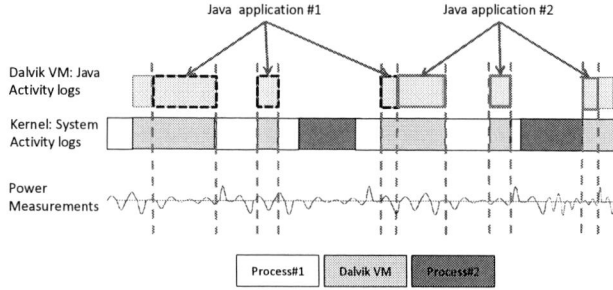

Figure 2. Cross-mapping technique for mapping Dalvik activities to power measurements.

C. I/O Energy Profiling

1) Java program mapping

Java programs use a mechanism called Java Native Interface (JNI) to transfer I/O requests to low-level system calls provided by operating system. To record the I/O requests, the tool modified the Dalvik virtual machine [10] to trace all JNI operations triggered by the Java classes. When a JNI is called, the tool records the caller information together with the current time stamp. After that, it combines the result to the system logs to obtain the power consumption of Java level I/Os.

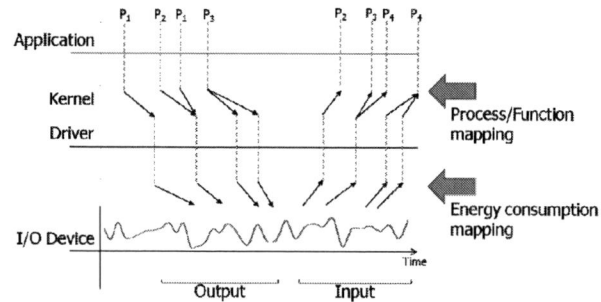

Figure 3. Asynchronous I/O issue.

2) I/O behavior mapping

Figure 3 shows the I/O behavior between application levels and actual I/O events. The association of I/O events and applications could be one-to-one, many-to-one, and one-to-many mappings. Operating system may perform scatter and gather in order to optimize the I/O performance. For example, a service data unit generated by an application I/O function may be split into a number of packet data units sent to I/O devices. On the other hand, a number of packet data units may be received and merged into a service data unit to an application

I/O function. According to I/O behaviors, our tool inserts probes in kernel and drivers to link the I/O events to applications; it also inserts probe points inside the JNI entry point in order to track the I/O activities performed by Java program in Dalvik virtual machine [10]. After analyzing the collected profiling information, the related I/O events can associate with processes, functions and Java classes.

3) Asynchronous I/O issue

For power profiling of a wireless interface, one essential information is the actual duration of receiving or transmitting a packet. This can be calculated by using bit rate, packet size and initial time of the receive/transmit event. We get the begin time of a transmit event before the underlying device driver begins to send out the first bit of the packet to the air, and in offline mapping stage we use bit rate information of this transmit event to calculate the end time. That is, the time when the last bit of the packet leaves the device is calculated by equation (1). Similarly, for a receiving event, we get the end time when the last bit of the packet reaches to the device driver, and then we use the bit rate information of this newly arrived packet to calculate the begin time. That is, the time when the first bit of the packet enters the device is obtained by equation (2).

$$t_{tx-end} = t_{tx-begin} + \frac{Packet\ size\ (in\ bits)}{TX\ bit\ rate}...(1)$$

$$t_{rx-begin} = t_{rx-end} - \frac{Packet\ size\ (in\ bits)}{RX\ bit\ rate}...(2)$$

With begin and end time of a packet, a duration value can be derived to calculate the power consumption of a system activity. However, it is still insufficient to associate the power consumption to a specific process. We have to identify the processes that are responsible for the energy consumption. Achieving this for CPU is as simple as reading the task data structure in the kernel, but for I/O components, it is more complicated due to the asynchronous nature of I/O operations. As illustrated in Figure 4, when a process writes data to a socket, this data travels down through the layers of network stack in kernel and finally reaches the device driver as a packet data unit. At some time during this flow, when we sniff the packet when it is formed, the current active process may have already been changed by kernel scheduler. If we simply record the current PID at this time, we may incorrectly identify the initiator of this system activity. To solve this problem, we add one more item, PID, to the socket data structure to identify the process that generates or receives the I/O events. When we record the time values for a packet, we simply access this item through the socket structure of packets.

D. Correlating Power Measurements with System Activities

One essential task on the host-side control center is to map, or correlate power measurements with system activities. After the profiling process is finished, the user-daemon program transfers profiling results to host for post processing. When the transfer ends, the energy consumption analyzer module in the control center analyzes the results file of system activities to check the process identify (PID) values that were collected during the profiling. Later on, the power consumption of CPU and I/O peripheral activities are associated with processes based on their PIDs.

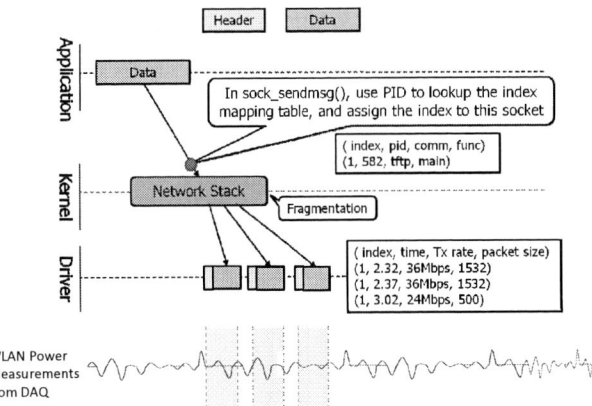

Figure 4. Correlating energy consumption with system activities (e.g. mapping power consumption of a TX packet to an applications).

E. Wireless Environment Emulator

Emulating wireless environment is achieved by using few wireless signal attenuators, RF-shielded chambers, and signal mixers. Developers can easily configure the location of each base stations or access points and the draw a mobility path of the target mobile device using our tool. Our current system only supports multiple WLAN APs shown but the architecture can be further extended to support different wireless access technologies such as 3G, etc.

A Wireless LAN channel model is employed to calculate the signal attenuation caused by the movement of a mobile device under profiling. Depending on the distance between the AP and mobile device, attenuator is set to a particular attenuation value. The mobile device gradually senses the attenuation as if it is moving away/close from/to wireless AP. We borrow the channel model presented in [8].

IV. EXPERIMENTS AND RESULTS

A. Experiment Setup

We use Beagleboard [9] as our embedded platform, and measure both voltage and current of each I/O component by using a National Instruments PCI-6115 data acquisition board with voltage probes and current clamps [11]. In order to get measurement point from WLAN interface, we choose a D-Link DWL-G122 USB Wireless NIC, rather than the WiFi module on Beagleboard. The wireless environment emulator is constructed with E-Instrument EPA-400BMG programmable attenuators [13], Angleton RF shielded boxes and D-Link DIR-600 WiFi APs and connect to the host with an NI PCI-GPIB Interface card. Figure 5 demonstrates the whole profiling system.

B. Experiments Result

We tested our energy profiling tool by measuring the energy of a simple email client application. In this test case, it connects to an email server on Internet and receives one email with a 10MB attachment. We found that the email client spends a significant amount of energy in receiving packets from WLAN. Also, we noticed that the energy consumption of the application increases considerable when the WLAN signal quality is poor. Therefore, we add a simple power management

978-1-4673-2895-1/12 $31.00 © 2012 IEEE   71

strategy in the email client that receives an email when the signal quality is higher than a pre-defined threshold. By using the proposed tool, experiment results can be easily collected, analyzed and reproduced. Results are shown in Figure 6.

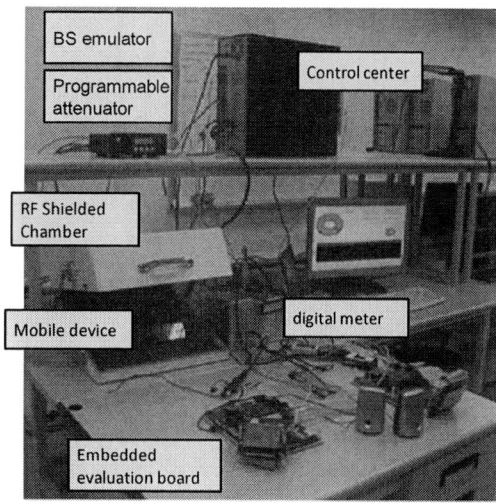

Figure 5.   Demonstration of the proposed power profiling system.

	Unmodified email client				Modified email client		
Experiment No	CPU Energy (Joule)	WNIC Energy (Joule)	Total Energy (Joule)	Experiment No	CPU Energy (Joule)	WNIC Energy (Joule)	Total Energy (Joule)
1	11.82	12.27	24.09	1	7.9	8.98	16.88
2	10.53	12.1	22.63	2	8.17	8.66	16.83
3	9.97	12.87	22.84	3	7.64	8.34	15.98
4	11.08	13.15	24.23	4	8.68	8.27	16.95
5	11.3	12.76	24.06	5	8.02	9.03	17.05
6	11.04	13.4	24.44	6	8.57	8.6	17.17
7	10.99	13.89	24.88	7	8.3	8.19	16.49
8	10.15	13.62	23.77	8	8.85	8.33	17.18
9	9.93	13.35	23.28	9	8.24	9.17	17.41
10	10.19	13.11	23.3	10	8.54	8.58	17.12
Average			23.752	Average			16.906

Figure 6.   Comparison between the orginal and modified email client.

As can be seen, unmodified version of the email client does not observe the wireless signal level. It immediately begins to download the email. On the other hand, our modified version of the email client periodically observes wireless signal levels to receive the email at good channel quality. If the WLAN signal level is below a pre-defined threshold (-82dBm), it sleeps for a while. When it detects the level is higher than the threshold, it begins to download the email. In total, we ran 10 sets of experiments and compared the unmodified and modified versions of the email client. Our experimental results demonstrate that about 30% energy consumption can be reduced but additional 380 seconds delay for email downloads is introduced. The case study shows that the proposed tool is useful and practical for developers to tradeoff the performance and the energy efficiency of mobile applications.

V.   CONCLUSIONS

In this paper, we introduced a measurement-based energy profiling tool combining with an emulated wireless environment. This tool is capable to profile the power consumption of application programs and processes at process-

level and function-level granularities, and to correlate energy consumption of I/O events to software activities. The tool was verified and implemented on a Beagleboard running Android/Linux system. The proposed tool facilitates developers to optimize energy efficiency of their mobile applications.

ACKNOWLEDGMENTS

The authors would like to thank MediaTek Inc. and National Science Council of the Republic of China for financially supporting this research under Contract No. 101-2219-E-009-010-, 101-2220-E-009-036-, 101-2918-I-009-004-, 101-2915-I-009-022, 101-3113-P-006-020-, 101-2219-E-009-001-, and Institute for Information Industry under the "Advanced Sensing Platform and Green Energy Application Technology Project" which is subsidized by the Ministry of Economy Affairs of the Republic of China.

REFERENCES

[1] J. Flinn, and M. Satyanarayanan, "PowerScope: A tool for profiling the energy usage of mobile applications," in Proceedings of the 2nd IEEE Workshop on Mobile Computing Systems and Applications, 1999.

[2] Changjiu Xian, Le Cai, and Yung-Hsiang Lu, "Power Measurement of Software Programs on Computers With Multiple I/O Components," IEEE Transactions on Instrumentation and Measurement, Vol. 56, pp. 2079-2086, 2007.

[3] K. Paul and T.K. Kundu, "Android on Mobile Devices: An Energy Perspective," IEEE 10th International Conference on Computer and Information Technology , 2010.

[4] A. Rice, and S. Hay, "Decomposing power measurements for mobile devices," IEEE International Conference on Pervasive Computing and Communications, 2010.

[5] L.M. Feeney and M. Nilsson, "Investigating the energy consumption of a wireless network interface in an ad hoc networking environment," in Proceedings of the Twentieth Annual Joint Conference of the IEEE Computer and Communications Societies, 2001.

[6] Mark STEMM and Randy H. KATZ, "Measuring and Reducing Energy Consumption of Network Interfaces in Hand-Held Devices," IEICE Transactions on Communications, Vol. E80-B, No.8, pp.1125-1131, Aug.1997.

[7] Glenn Judd and Peter Steenkiste, "Using emulation to understand and improve wireless networks and applications," in Proceedings of the 2nd conference on Symposium on Networked Systems Design & Implementation, Vol. 2, 2005.

[8] T. S. Rappaport, Wireless communications principles and practices, Prentice-Hall, 2002.

[9] "beagle board-xM platform," http://beagleboard.org/

[10] http://developer.android.com/index.html

[11] "NI   PCI-6115   DAQ   Card", http://sine.ni.com/nips/cds/view/p/lang/en/nid/11886

[12] "D-Link DWL-G122 USB Wireless NIC - FW Version 3.0", http://www.dlink.com/products/?pid=334

[13] "EPA-400   Programmable   Attenuator",   http://e-channel.com.tw/zencart/index.php?main_page=product_info&products_id=1175

[14] A. Mavinakayanahalli, P. Panchamukhi, J. Keniston, A. Keshavamurthy, and M. Hiramatsu, "probing the guts of kprobes", in Ottawa Linux Symposium, pp. 101–115, 2006.

[15] K. Jim, M. Ananth, P. Prasanna, and P. Vara, "Ptrace, Utrace, Uprobes: Lightweight, Dynamic Tracing of User Apps", in Proceedings of the Linux Symposium, Volume One, June 27th–30th, 2007.

Architecture Efficiency of Application-Specific Processors: a 170Mbit/s 0.644mm^2 Multi-standard Turbo Decoder

Rachid Al-Khayat, Amer Baghdadi, Michel Jézéquel

Institut Mines-Telecom; Telecom Bretagne; Lab-STICC CNRS UMR 6285
Electronics Department, Telecom Bretagne, Technopôle Brest Iroise CS 83818, 29238 Brest
Université Européenne de Bretagne, France
E-mail: {firstname.surname}@telecom-bretagne.eu

Abstract—Architecture efficiency, in terms of performance/area, of application-specific processors is directly related to the devised instruction set and pipeline stages usage. Most of recently proposed works on application-specific instruction-set processors (ASIP) do not consider or present this key point explicitly.

In this paper, we consider the challenging turbo decoding application where many recent implementations have been proposed to accommodate the related large flexibility and high throughput requirements. The paper demonstrates how the architecture efficiency of instruction-set based processors can be considerably improved by minimizing the pipeline idle time. A complete ASIP-based turbo decoder is proposed with further contributions on interleaving generators, extrinsic information exchange, and rapid reconfiguration. While supporting 3GPP LTE, WiMAX and DVB-RCS turbo codes, the proposed implementation achieves a throughput of 170Mbps with 0.644mm^2 @65nm CMOS technology. The proposed ASIP-based turbo decoder exhibits a high architecture efficiency of 3.12 bit/cycle/iteration/mm^2.

Index Terms—SoC design, Architecture efficiency, ASIP, Pipeline, Turbo codes, WiMAX, LTE, DVB-RCS, QPP interleaver, ARP interleaver.

I. INTRODUCTION

Application-specific processors are being widely investigated these last years in System-on-Chip design. The main reason behind this emerging trend is the increasing requirements of flexibility and high performance in many application domains. Digital communication domain is very representative of this trend where many flexible designs have been recently proposed for the challenging turbo decoding application. For this application, there is a large variety of coding options specified in existing and future digital communication standards, besides the increasing throughput requirement (Table I).

Standard	Codes	Rates	States	Block size	Channel Throughput
IEEE802.16 (WiMax)	DBTC	1/2 - 3/4	8	.. 4800	.. 75 Mbps
DVB-RCS	DBTC	1/3 - 6/7	8	.. 1728	.. 2 Mbps
3GPP-LTE	SBTC	1/3	8	.. 6144	.. 150 Mbps

TABLE I: Selection of standards supporting turbo codes [1]. DBTC: Double Binary Turbo Code, SBTC: Single Binary Turbo Code

In this context, many recent works have been proposed targeting flexible, yet high throughput, implementations of turbo decoders [2][1][3][4][5]. The flexibility varies from supporting

different modes of a single communication standard to the support of multi-standards multi-modes applications. Other implementations have even increased the target flexibility to the support of different channel coding techniques. Regarding the architecture model, besides the conventional parametrized hardware models, recent efforts have targeted the use of application-specific instruction-set processor models (ASIP). Such an architecture model enables the designer to freely tune the flexibility/performance trade-off as required by the considered application requirements. Related contributions are emerging rapidly seeking to improve the resulted architecture efficiency in terms of performance/area and in addition to increase the flexibility support. However, the architecture efficiency of application-specific processors is directly related to the devised instruction set and pipeline stages usage. Most of recently proposed works do not consider or present this key point explicitly.

In this paper, and considering the challenging turbo decoding application, we demonstrate how the architecture efficiency of instruction-set based processors can be considerably improved by minimizing the pipeline idle time. A complete ASIP-based turbo decoder is proposed with novel contributions on interleaving generators, extrinsic information exchange, and rapid reconfiguration.

The rest of the paper is organized as follows. Section II gives a brief introduction on the considered application of turbo decoding for both DBTC and SBTC modes. Section III presents the proposed ASIP component decoder architecture with a maximized architecture efficiency. Sections IV and V present the other contributions regarding interleavers design, extrinsic exchange management, and rapid reconfiguration. The synthesis results and comparisons w.r.t. state of the art implementations are given in Section VI and finally the paper concludes with Section VII.

II. DECODING ALGORITHMS

This section gives a brief introduction on the considered application of turbo decoding [1]. The typical turbo decoding system consists of two component decoders exchanging extrinsic information via an interleave (Π) and deinterleave (Π^{-1}) processes. One component decoder receives Log-likelihood ratio Λ^k (1) for each bit k of a frame of length N in the natural order while the other component decoder is initialized in interleaved order.

$$\Lambda^k = log\frac{Pr\{d^{k=0}|y^{0..N-1}\}}{Pr\{d^{k=1}|y^{0..N-1}\}} \quad (1)$$

For efficient hardware implementation Max-Log MAP algorithm is used, as described in [6]. For DBTC, the three normalized extrinsic information are defined by (2) where $i \in (01, 10, 11)$ of the kth symbol while s' and s are the previous and current corresponding trellis state respectively.

$$Z_k^{n.ext}(d(s',s) = i) = Z_k^{ext}(d(s',s) = i) - Z_k^{ext}(d(s',s) = 00) \quad (2)$$

The extrinsic information defined by (3) is calculated from the aposteriori probability given by (4), wherein $\alpha_k(s)$ and $\beta_k(s)$ are the state metrics in forward (5) and backward recursion (6) respectively and $\gamma_k(s',s)$ are the branch metrics (7). The $\gamma_k^{sys}(s',s)$ and $\gamma_k^{par}(s',s)$ are the systematic and parity symbol LLRs. Finally, when the required number of iterations N_{iter} are completed the hard decision is calculated as given by (9).

$$Z_k^{ext}(d(s',s) = i) = \Gamma \times (Z_k^{apos}(d(s',s) = i) - \gamma_k^{int}(s',s)) \quad (3)$$

$$Z_k^{apos}(d(s',s) = i) = \max_{(s',s)/d(s',s)=i}(\alpha_{k-1}(s)+$$
$$\gamma_k^{n.ext}(s',s) + \beta_k(s)), i \in \{00, 01, 10, 11\} \quad (4)$$

$$\alpha_k(s) = max_{s',s}(\alpha_{k-1}(s) + \gamma_k(s',s)) \quad (5)$$

$$\beta_k(s) = max_{s',s}(\beta_{k+1}(s) + \gamma_{k+1}(s',s)) \quad (6)$$

$$\gamma_k(s',s) = \gamma_k^{int}(s',s) + \gamma_k^{n.ext}(s',s) \quad (7)$$

$$\gamma_k^{int}(s',s) = \gamma_k^{sys}(s',s) + \gamma_k^{par}(s',s) \quad (8)$$

$$Z_k^{Hard.dec} = sign(Z_k^{apos}) \quad (9)$$

For SBTC, the trellis length is reduced by half through applying the one-level look-ahead recursion [7]. The modified α and β state metrics for this Radix4 optimization are given by (10) and (11) where $\gamma_k(s'',s)$ is the new branch metric for the combined two-bit symbol (u_{k-1}, u_k) connecting state s'' and s.

$$\alpha_k(s) = max_{s'',s}\{\alpha_{k-2}(s'') + \gamma_k(s'',s)\} \quad (10)$$

$$\beta_k(s) = max_{s'',s}\{\beta_{k+2}(s'') + \gamma_k(s'',s)\} \quad (11)$$

$$\gamma_k(s'',s) = \gamma_{k-1}(s'',s') + \gamma_k(s',s) \quad (12)$$

The extrinsic information for u_{k-1} and u_k are computed as:

$$Z_{k-1}^{n.ext} = \Gamma \times (max(Z_{10}^{ext}, Z_{11}^{ext}) - max(Z_{00}^{ext}, Z_{01}^{ext})) \quad (13)$$

$$Z_k^{n.ext} = \Gamma \times (max(Z_{01}^{ext}, Z_{11}^{ext}) - max(Z_{00}^{ext}, Z_{10}^{ext})) \quad (14)$$

III. ASIP ARCHITECTURE FOR TURBO DECODING

The proposed ASIP architecture considers the base design presented in [1]. In this design, the turbo decoder system architecture consists of 2 ASIPs interconnected as shown in Fig. 1. It exploits the various parallelism levels available in turbo decoding including shuffled decoding [8] where the two ASIPs operate in a 1×1 mode. In this mode, one ASIP ($ASIP1$) processes the data in natural order while the other one ($ASIP2$) processes it in interleaved order. The generated extrinsic information are exchanged between the two ASIP decoder components via an *Extrinsic Exchange Module*. Furthermore, the system control and configuration are managed by a (*Global Controller & Input Interface Module* and a *Configuration Module*).

The rest of this section will present a new contribution which maximizes the architecture efficiency the ASIP component decoder. Subsequent Sections IV and V will present the other contributions regarding interleaving, extrinsic exchange, and rapid reconfiguration.

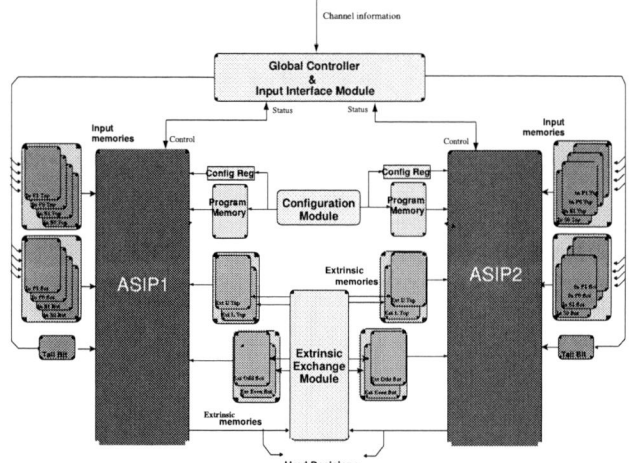

Fig. 1: Turbo decoder system architecture

A. Considered ASIP base design example

In order to illustrate our approach we considered the ASIP base design for turbo decoding proposed in [1]. This architecture exploits the BCJR computation parallelism by using two recursion units to implement the butterfly scheme [6] (Fig. 6). It specifies 9 pipelines stages and three computational instructions (DATA LEFT.., DATA RIGHT.., EXTCALC..) which are executed per symbol for each turbo decoding iteration.

Using the butterfly scheme, and for each turbo decoding iteration, processing a sub-frame (window) is done in two phases. In the first phase, only one instruction ("DATA LEFT") is repeated over the sub-frame to execute the left-butterfly recursion calculating the α/β metrics (10,11) and buffering them in a stack memory called the cross-metric memory. The second phase includes two instructions ("DATA RIGHT" and "EXTCALC") which are repeated to execute the right-butterfly recursion. "DATA RIGHT" continues the computation of α/β

978-1-4673-2895-1/12 $31.00 © 2012 IEEE

metrics while "EXTCALC" them along with the corresponding buffered ones from cross-metric memory to calculates and outputs the extrinsic information (13).

Analyzing the pipeline stages usage for these three instructions gives the results illustrated in Table II. The table indicates the average Pipeline Usage Percentage: $PUP = \frac{\frac{7}{9}+\frac{7}{9}+\frac{6}{9}}{3} \cong$ 74%. This sub-optimal usage is caused by the idle state of instructions "DATA LEFT" and "DATA RIGHT" in pipeline stages (MAX, ST) and instruction "EXTCALC" in (BM1, BM2, EX).

INSTRUCTION	Pipeline Stages									Usage
	PFE	FE	DEC	OPF	BM1	BM2	EX	MAX	ST	
DATA LEFT	X	X	X	X	X	X	X	-	-	7/9
DATA RIGHT	X	X	X·	X	X	X	X	-	-	7/9
EXTCALC	X	X	X	-	-	-	X	X	X	6/9

TABLE II: Pipeline usage percentage for [1]

INSTRUCTION	Pipeline Stages								Usage
	PFE	FE	DEC	OPF	BM1	BM2	EX	ST	
MetExtCALC LEFT	X	X	X	X	X	X	X	-	7/8
MetExtCALC RIGHT	X	X	X	X	X	X	X	X	8/8

TABLE III: Pipeline usage percentage for the proposed work

B. Proposed ASIP Architecture

Fig. 2 illustrates the overall architecture of the proposed ASIP component decoder with the memory structure and pipeline organization in 8 stages. The numbers in brackets indicate the mapping of decoding equations (referred in Section II) on the corresponding pipeline stage. The extrinsic information format, at the output of the ASIP, is also depicted in the same figure for the two modes SBTC and DBTC.

This proposed architecture maximizes the pipeline usage (PUP) by merging the two instructions DATA and EXTCALC, thus merging pipeline stages EX and MAX. So in right-butterfly both state metric and extrinsic information calculations are processed in the same pipeline stage (EX). Table III presents the usage of the new instructions (MetExtCALC LEFT & RIGHT) through the pipeline of the proposed work. We find that $PUP = \frac{\frac{7}{8}+\frac{8}{8}}{2} \cong 94\%$ which decreases the pipeline idle time of about 20%.

The challenge in this proposal is to keep the same balance of the functional units (Adders, Registers, Multiplexers) to avoid increasing the complexity and to maintain the same critical path. In fact, the proposed merge causes to duplicate part of the functional units. On the other part, it allows to remove several registers and multiplexers which were used for buffering and functional units sharing. Fig. 4 illustrates the architecture of the proposed recursion unit. In this architecture, each one of the 32 Adder Node Functional units (ANF) integrates 2 adders (instead of one) to compute the two additions ($\alpha + \gamma$ or $\beta + \gamma$) then ($\alpha + \beta + \gamma$) in the same clock cycle. On the other hand, per ANF, the following logic has been removed: 2 10-bit registers used to buffer $\alpha + \gamma$ or $\beta + \gamma$ (RADD Reg. and RC Reg.) and 2 multiplexers used to share the single adder. Furthermore, additional 8 comparators (each of 4 inputs) have been added instead of sharing the existing ones. This allowed removing

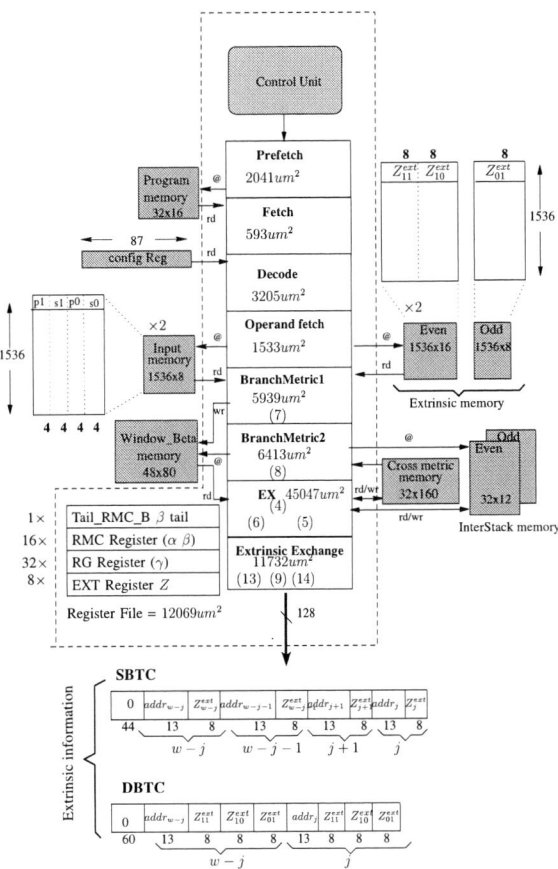

Fig. 2: ASIP pipeline architecture

the corresponding multiplexers besides 4 10-bit registers (EXT Reg.). The summary of the removed and added logic for the two recursion units of the ASIP is presented in Table IV which shows that the level of complexity is almost remained the same. Synthesis results of the overall ASIP logic present a slight increase of 0.004mm² in CMOS 65nm technology.

Logic removed			Logic added		
Logic	Data width (bits)	#	Logic	Data width (bits)	#
RADD Reg	10	64	2-input Adder	10	64
RC Reg	10	16	2-input Adder	10	48
EXT Reg	10	8	2-input MUX	10	48
2-input MUX	10	152	-	-	-

TABLE IV: Comparison table for added and removed logic for the proposed ASIP architecture

Fig. 3 compares the critical path of the proposed architecture with the initial base one [1] which is in the EX pipeline stage. The compare units are equivalent to the combination of 1 adder and 1 multiplexer. Fig. 3 shows that the critical path for work [1] integrates 6 multiplexers + 3 adders while that of the proposed work integrates 4 multiplexers + 5 adders. This illustrates how the proposed optimization does not impact the ASIP maximum clock frequency.

Regarding the turbo decoder throughput, it can be computed

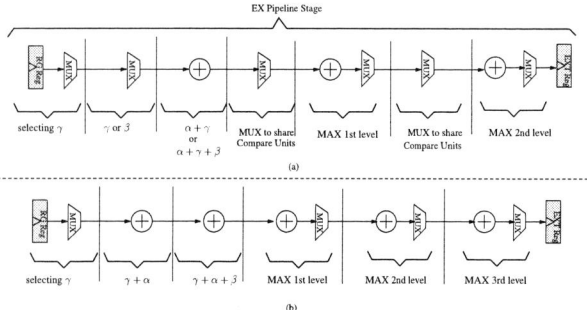

Fig. 3: Critical path comparison: (a) Critical path of the initial base architecture [1], (b) Critical path of the proposed architecture

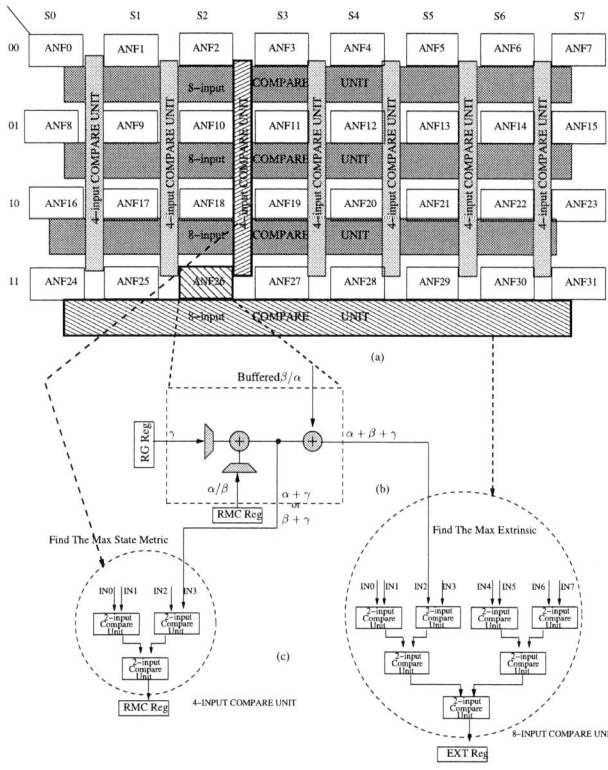

Fig. 4: (a) Forward recursion unit composed of 32 ANF (b) 8 4-input Compare Units used for state metric computation and 4 8-input Compare Units used for extrinsic computation(c) Adder Node Forward

through (15). With the proposed architecture, $N_{instr} = 2$ instructions per iteration are needed to generate the extrinsic information for $N_{sym} = 2$ symbols, where a symbol is composed of $Bits_{sym} = 2$ bits. This is true in DBTC and SBTC modes as RADIX4 is adapted in SBTC. Considering $N_{iter} = 6$ iterations, the maximum throughput achieved is $170 Mbps$ in both modes.

$$Throughput = \frac{N_{sym} * Bits_{sym} * F_{clk}}{N_{instr} * N_{iter}} \quad (15)$$

	PUP	logic area mm^2 @65nm	Decoding speed Bits/clk/iter	Throughput Mbps
Work [1]	74 %	0.1	1.33	115.5 @6iter
This work	94 %	0.104	2	170 @6iter

TABLE V: Results comparison in terms of PUP, logic area, decoding speed and throughput

Table V summaries and compares the results in terms of pipeline usage PUP, logic area, throughput, and decoding speed. It is worth noting that the memory area remains identical with the proposed optimization, and thus not considered in this table.

IV. INTERLEAVING AND EXTRINSIC INFORMATION EXCHANGE

This section presents the related contributions to interleavers design and parallel memory access management in extrinsic information exchange.

A. Interleaver Generator

Conventional designs utilize interleaver memory table in order to fetch interleaving address for each extrinsic information value. The draw-back for this method is the high area occupation due to the need of storing long address tables (max frame size in LTE is 6144 bits). However, advanced wireless communication standards have taken care of this issue and have specified interleaving rules that enable to generate addresses recursively. In this case, optimized interleaver address generators can be designed instead of using interleaving memories. In this context, the DBTC proposed in WiMAX and DVB-RCS uses Almost Regular Permutation (ARP) interleaving rule while the SBTC proposed in LTE uses Quadratic Permutation Polynomial (QPP) interleaver.

1) QPP Interleaver Generator: For the design of QPP interleaver generator, we adapted the efficient method proposed in [9]. Since our proposed work adapts RADIX4 technique and two extrinsic information are generated and should be addressed simultaneously, two interleaving addresses (for two consecutive bits) should be generated. To achieve this task, two QPP generators are designed, one for odd interleaving addresses and the other for even ones. In this case, the step size of every generator is set to 2 (rather that 1 in conventional generators).

2) ARP Interleaver Generator: The ARP interleaving addresses can also be computed recursively. In this regard, we propose a different formulation of the mathematical expressions in order to optimize the underlined hardware architecture. The basic formula of ARP interleaving is given in (16) for index j.

$$\prod_1(j) = (P_0 \times j + P_j + 1)\%N \quad (16)$$

where

$$
\begin{array}{ll}
P = 0 & \text{if } j\%4 = 0 \\
P = \dfrac{N}{2} + P_1 & \text{if } j\%4 = 1 \\
P = P_2 & \text{if } j\%4 = 2 \\
P = \dfrac{N}{2} + P_3 & \text{if } j\%4 = 3
\end{array}
\tag{17}
$$

For index $j + 1$, the above equation can be written as follows:

$$
\begin{aligned}
\prod\nolimits_1 (j + 1) &= (P_0 \times (j + 1) + P_{j+1} + 1)\%N = \\
&= (P_0 \times (j + 1) + P_j - P_j + P_{j+1} + 1)\%N
\end{aligned}
\tag{18}
$$

This expression leads to the following recursive equation:

$$
\prod\nolimits_1 (j + 1) = \prod\nolimits_1 (j) + Seed
\tag{19}
$$

where $Seed = (P_{j+1} - P_j + P_0)\%N$. Using the formula in (17), there are four cases for $Init$ as follows:

$$
\begin{array}{ll}
Seed0 = (P_0 - N/2 - P_3)\%N & \text{if } (j+1)\%4 = 0 \\
Seed1 = (P_0 + N/2 + P_1)\%N & \text{if } (j+1)\%4 = 1 \\
Seed2 = (P_0 + P_2 - N/2 - P_1)\%N & \text{if } (j+1)\%4 = 2 \\
Seed3 = (P_0 + P_3 + N/2 - P_2)\%N & \text{if } (j+1)\%4 = 3
\end{array}
\tag{20}
$$

Using this formulation, the proposed ARP interleaving generator is illustrated in Fig. 5. This architecture presents lower complexity that the one proposed in [10] (use of two adders rather than four).

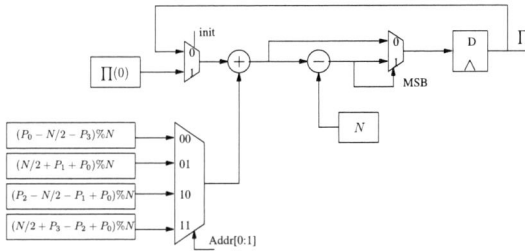

Fig. 5: ARP interleaver generator

3) Interleaving address generator with butterfly scheme:
Since the proposed turbo decoder uses butterfly scheme, the interleaving addresses are required in the right-butterfly when sending the extrinsic information in both backward and forward directions.

Having address generators in both directions will incurs significant area and control overheads due to the discontinuity of generating addresses (generating addresses only in right butterfly of every sub-block). This discontinuity multiplies the number of required initial values (Seed values). To avoid this issue, we propose to use only interleaving address generators in forward direction, so they generate address continuously in both left and right butterfly. In this case, the interleaving addresses generated in left butterfly can be used later for backward direction of the right butterfly. To that end, those addresses are buffered in small size stack memories (InterStack) of size 12×32. Fig. 6 illustrates the proposed interleaving address generation and stack memories in butterfly

scheme. It is worth to note that in LTE there is a need for two stack memories (InterStack1, InterStack2) for even and odd addresses respectively because of using RADIX4 which decode two bits simultaneously. As for WiMAX and DVB-RCS only InterStack1 is used.

As a conclusion a conventional interleaving memory of size $6144 \times 13 \cong 80$Kbits is replaced by $2 \times 12 \times 32 = 768$bits and additional very low complexity logic for address generators.

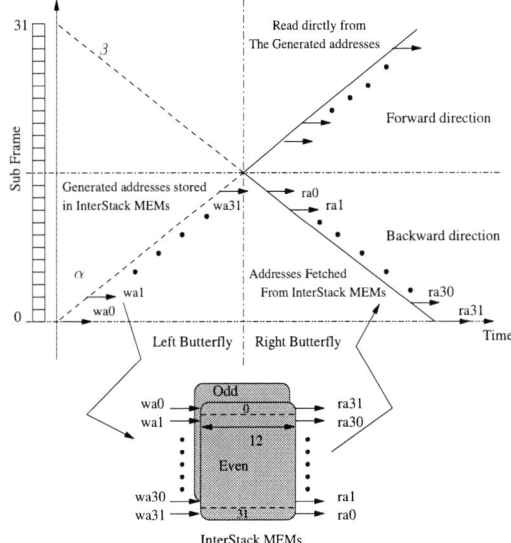

Fig. 6: Interleaving address generator with butterfly scheme

B. Extrinsic Exchange Module

In SBTC, Radix4 with butterfly scheme imply the generation of four extrinsic information every clock cycle in right butterfly. These four extrinsic information should update the extrinsic memories of the other component decoder. One of QPP features is that the generated interleaved address $\prod(j)$ used in LTE has the same even/odd parity as j. For that reason, extrinsic memories are split to four banks (TOP1, TOP2, BOT1, BOT2). Fig. 7 explains the functionality of the Extrinsic Exchange Module to manage parallel memory access conflicts. If LLRs of odd addresses 1 and 3 are not in conflict then they are sent to memories TOP1 and BOT1 accordingly. Otherwise, LLR with address 3 is stored in the FIFO2. Similarly, LLRs of even addresses are handled. Since no extrinsic information is generated during left butterfly of next processed window, the Extrinsic Exchange Module is free to update the remaining LLRs from the FIFOs.

Similar technique is used in DBTC turbo decoding, but only one FIFO is used since just two extrinsic information are generated every cycle in right butterfly.

V. ASIP DYNAMIC RECONFIGURATION

The proposed turbo decoder supports multi-standards (DVB-RCS, LTE, WiMAX) with wide range of block-sizes. In order to speed up the reconfigurability of the proposed

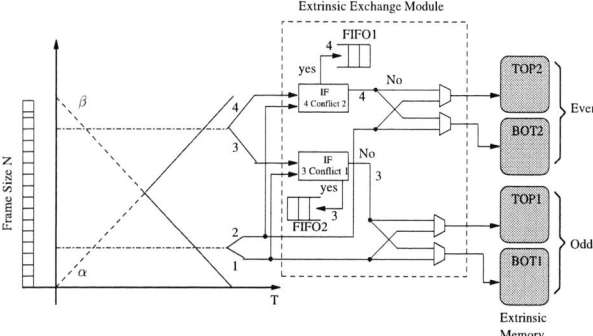

Fig. 7: Extrinsic exchange module

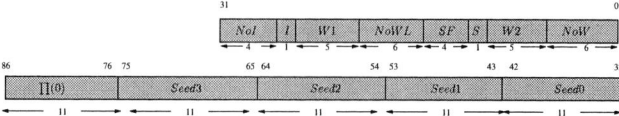

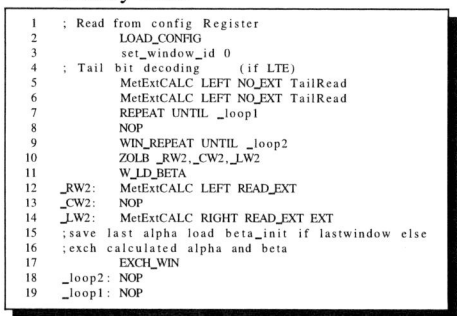

Fig. 8: Config_Reg reserved bits

architecture, we propose to unify SBTC and DBTC instruction program. To achieve that, all necessary parameters are placed in special register (Config_Reg) of 87 bits. Fig. 8 details the reserved bits for the needed parameters, and Table VI explains how each parameter can be calculated where FS stands for Frame size in symbols. The unified assembly code is shown in

Parameter	Description	Formula	Size in Bits
NoI	N# of iter.	Given by the user	4
S	Standard	$S = \begin{cases} 1 & \text{if (LTE)} \\ 0 & \text{otherwise} \end{cases}$	1
I	Intra-symbol permut.	$I = \begin{cases} 1 & \text{if (DVB)} \\ 0 & \text{otherwise} \end{cases}$	1
SF	Scaling factor	$SF = \begin{cases} 0.4375 & \text{if (LTE)} \\ 0.75 & \text{otherwise} \end{cases}$	4
$W1$	Window Size $-1*$	$W1 = \begin{cases} 31 & \text{if}(FS \geq 64) \\ (FS/2) - 1 & \text{otherwise} \end{cases}$	5
$W2$	Last Window Size $-1*$	$W2 = \begin{cases} 31 & \text{if}(FS\%64=0) \\ \frac{(FS - Floor(\frac{FS}{64}) \cdot 64)}{2} - 1 & \text{otherwise} \end{cases}$	5
NoW	N# of Windows	$Ceiling(\frac{FS}{64})$	6
$NoWL$	N# of Windows $-1*$	$NoW - 1$	6

* The "-1" is required because of the REPEAT & WIN_REPEAT instructions that execute for the "given value $+1$" times

TABLE VI: Parameters definition, method of calculation, and number of reserved bits

Listing 1. It is worth to note the reduced number of instructions and the short code size of the proposed ASIP (no high-level language programming support is needed). LOAD_CONFIG instruction reads from Config_Reg register to define the standard's mode, number of iterations, extrinsic scaling factor, window size, and last window size. The window maximum size supported is 128 bits, so any frame bigger than 128 bits is divided to windows of size 128 bits except for the last window which will be the remaining. ZOLB is the zero overhead loop instruction, instruction at @12 "MetExtCALC LEFT.." executes $W1$ times (half of window size) to compute the state metrics in left butterfly. The instruction at @14 "MetExtCALC RIGHT.." executes $W1$ times to compute the state metrics and extrinsic information in right butterfly. The only exception is when decoding the last window, ZOLB will iterate $W2$ times. Instructions at @5 and @6 "MetExtCALC...TailRead" read the tail bits in LTE mode while it reads zero in other modes.

Listing 1: STBC/DBTC Unified Assembly Code

```
1   ; Read from config Register
2           LOAD_CONFIG
3           set_window_id 0
4   ; Tail bit decoding      (if LTE)
5           MetExtCALC LEFT NO_EXT TailRead
6           MetExtCALC LEFT NO_EXT TailRead
7           REPEAT UNTIL _loop1
8           NOP
9           WIN_REPEAT UNTIL _loop2
10          ZOLB _RW2, _CW2, _LW2
11          W_LD_BETA
12  _RW2:   MetExtCALC LEFT READ_EXT
13  _CW2:   NOP
14  _LW2:   MetExtCALC RIGHT READ_EXT EXT
15  ;save last alpha load beta_init if lastwindow else
16  ;exch calculated alpha and beta
17          EXCH_WIN
18  _loop2: NOP
19  _loop1: NOP
```

In addition, the rest of the Config_Reg register is divided to 5 parts (11bits each) to store the initial seed values for interleaver generators depending on the decoding standard, besides the initial interleaving address ($\prod(0)$).

VI. SYNTHESIS RESULTS

The ASIP was modeled in LISA language using Synopsys (ex. CoWare) Processor Designer tool. Generated VHDL code was validated and synthesized using Synopsys tools and 65nm CMOS technology. Obtained results demonstrate a logic area of 0.104 mm^2 per ASIP with maximum clock frequency of $F_{clk} = 520MHz$. Table VII lists the required memories for each ASIP. Thus, the proposed turbo decoder architecture with 2 ASIPs occupies a logic area of 0.208 mm^2 with total memory area of 0.436 mm^2.

Memory name	#	depth	Width (bits)	Type
Ext Mem (odd)	2	1536	16	DP
Ext Mem (even)	2	1536	8	DP
Input Mem	2	1536	16	SP
Window Beta Mem	1	48	80	SP
Cross-metric Mem	2	32	80	SP
InterStack Mem	2	32	12	SP

TABLE VII: Memories configuration used for one ASIP decoder component

Table VIII compares the obtained results of the proposed architecture with other related works. For a fair comparison, we normalized the occupied areas to 65nm technology using the conversion formula (21):

$$NA = A \times \left(\frac{f_N}{f}\right)^2 \quad (21)$$

where,

f_N - Feature size of target technology for normalization

	This Work	[3]	[2]	[4]	[5]
Standard compliant	LTE, WiMAX, DVB-RCS	WiMAX, LTE	3GPP, LTE	LTE, WiMAX	LTE
LTE Mode supported	188	188	188	18	188
WiMAX Mode supported	17	17	-	17	0
Tech (nm)	65	130	65	90	65
Core area (mm^2)	0.644	10.7	0.5	3.38	2.1
Normalized Core area @65nm (mm^2)	0.644	2.67	0.5	1.76	2.1
Throughput (Mbps)	170 @6iter	187 @8iter	21 @6iter	186 @6iter	150 @6.5iter
Parallel MAPs	2	8	1	8	1
Decoding speed Bits/clk/iter	2	6	0.42	7.3	3.25
F_{clk} (MHz)	520	250	300	152	300
AE (bit/cycle/iter/mm^2)	3.12	2.24	0.84	4.2	1.56

TABLE VIII: Comparison with state of the art implementations

(65nm),

f - Feature size of used technology,

NA - Normalized Area,

A - Occupied Area.

The architecture efficiency (bit/cycle/iteration/mm^2) is defined by the expression (22):

$$AE = \frac{T \times I}{NA \times F} \qquad (22)$$

where,

AE - Architecture Efficiency,

F - Operational frequency,

NA - Normalized Area,

T - Throughput,

I - Number of turbo decoding iterations.

This table illustrates how the proposed implementation outperforms state of the art in terms of architecture efficiency with respect to flexibility. The presented ASIP in [2] supports wide range of 3gpp standards and convolutional code (CC). Although they reserved 6bits for channel input quantization, still the occupied area 0.5mm^2 is big when it is compared to the achieved throughput of 21Mbps. Thus, a low architecture efficiency of 0.84 is obtained. The work in [4] gives a good architecture efficiency result of 4.2. However, the design flexibility efficiency is low. They adapt vectorizable and contention-free parallel interleaver, so they support only 18 modes out of the 188 specified in LTE. Moreover the maximum throughput achieved is only for 9 modes otherwise for frame-sizes less than 360 bits has throughput less than 93Mbps.

The parameterized architecture of [3] supports both turbo modes (DBTC and SBTC) and achieves a high throughput of 187Mbps. However, the occupied area is more than 4 times compared to our implementation and it achieves an architecture efficiency of 2.24. The LTE-dedicated architecture proposed in [5] achieves high throughput of 150Mbps but at a high cost of almost 3 times the occupied area with architecture efficiency of 1.56.

VII. CONCLUSION

In this paper we have proposed a complete ASIP-based flexible turbo decoder supporting all communication modes of 3GPP LTE, WiMAX and DVB-RCS standards. We have illustrated how the architecture efficiency of instruction-set based processors can be considerably improved by minimizing the pipeline idle time. Furthermore, the paper has presented low complexity interleaver designs supporting QPP and ARP interleaving in butterfly scheme together with an efficient parallel memory access management. Results show that the ASIP pipeline usage percentage have been maximized to reach 94%, achieving a throughput of 170Mbps with 0.644mm^2 @65nm CMOS technology and an architecture efficiency of 3.12 bit/cycle/iteration/mm^2.

REFERENCES

[1] R. Al-Khayat, P. Murugappa, A. Baghdadi, and M. Jezequel, "Area and throughput optimized ASIP for multi-standard turbo decoding," in *Proc. of the IEEE International Symposium on Rapid System Prototyping (RSP)*, may 2011, pp. 79–84.

[2] C. Brehm, T. Ilnseher, and N. Wehn, "A scalable multi-ASIP architecture for standard compliant trellis decoding," in *Proc. of the International SoC Design Conference (ISOCC)*, nov. 2011, pp. 349–352.

[3] J.-H. Kim and I.-C. Park, "A Unified Parallel Radix-4 Turbo Decoder for Mobile WiMAX and 3GPP-LTE," in *Proc. of the IEEE Custom Integrated Circuits Conference (CICC)*, sept. 2009, pp. 487–490.

[4] C.-H. Lin, C.-Y. Chen, E.-J. Chang, and A.-Y. Wu, "A 0.16nJ/bit/iteration 3.38mm2 turbo decoder chip for WiMAX/LTE standards," in *Proc. of the International Symposium on Integrated Circuits (ISIC)*, dec. 2011, pp. 168–171.

[5] M. May, T. Ilnseher, N. Wehn, and W. Raab, "A 150 Mbit/s 3GPP LTE Turbo Code Decoder," in *Proc. of the Design, Automation and Test in Europe Conference & Exhibition (DATE)*, march 2010, pp. 1420–1425.

[6] O. Muller, A. Baghdadi, and M. Jezequel, "From Parallelism Levels to a Multi-ASIP Architecture for Turbo Decoding," *IEEE Transactions on Very Large Scale Integration (VLSI) Systems*, vol. 17, no. 1, pp. 92–102, 2009.

[7] Y. Zhang and K. Parhi, "High-Throughput Radix-4 logMAP Turbo Decoder Architecture," in *Proc. of the Asilomar Conf. Signals, Systems and Computers (ACSSC)*, nov. 2006, pp. 1711–1715.

[8] O. Muller, A. Baghdadi, and M. Jezequel, "Parallelism Efficiency in Convolutional Turbo Decoding," *EURASIP Journal on Advances in Signal Processing*, 2010.

[9] Y. Sun and J. R. Cavallaro, "Efficient hardware implementation of a highly-parallel 3GPP LTE/LTE-advance turbo decoder," *Integration, the VLSI journal*, vol. 44, no. 4, pp. 305–315, 2011.

[10] J.-H. Kim and I.-C. Park, "Double-Binary Circular Turbo Decoding Based on Border Metric Encoding," *IEEE Transactions on Circuits and Systems II, Express Briefs*, vol. 55, no. 1, pp. 79–83, jan. 2008.

Improving Logic-to-Memory Ratio in an Embedded Multi-Processor System via Code Compression

Roberto Airoldi, Piia Saastamoinen and Jari Nurmi
Tampere University of Technology
Department of Computer Systems
P.O. BOX 553, FIN-33101, Tampere, Finland
firstname.lastname@tut.fi

Abstract—This paper presents the evaluation of a homogeneous Multi-Processor (MP) architecture equipped with a code compression system. The MP architecture is composed by a 3x3 mesh of processing elements interconnected via a hierarchical Network-on-Chip. Each processing element hosts a RISC processor, data and instruction memories, a network interface and an independent code compression system. The so composed system was evaluated in performance, power consumption, logic utilization and obtained compression ratio, while executing significant kernels of wireless communication systems. Results show that the introduction of the code compression system improves the computational density of the architecture (e.g. $GOPS/mm^2$) due to a higher logic-to-memory ratio, while approximately retaining other performance figures in a worst-case analysis.

I. INTRODUCTION

In the past few years, multi-core and multi-processor (MP) architectures have been widely studied by both academia and industry as a feasible way to achieve high performance while retaining a high degree of flexibility, in order to support a wide set of applications [1]. In fact, these properties are becoming more and more important, allowing companies to develop a single Integrated Circuit (IC) as building block for many different families of products, redistributing the design costs over a wider set of products. Furthermore, the flexibility allows systems to follow the updates of protocols / algorithms without redesigning the hardware part of the system but only updating the software layer.

The flexibility is most often obtained by an extension of the software layer, from a simple control system to the actual implementation of the kernels in software. However, such extension does not come for free. Indeed, it introduces "new challenges": a more software centric design stresses more the memory system, which in an embedded application is characterized by limited amount of resources. In order to cope with these issues researchers have focused on many different aspects of the software layer in order to implement algorithms in a more efficient way and consuming the given resources in the most efficient way. Also on the hardware layer efforts have been made to provide efficient solutions for the utilization of the memory system. If these issues were important already in the context of single-core architectures, their importance has grown even more in the era of multi-processors.

Code compression techniques have been widely studied as a feasible way to reduce the memory footprint of applications. As a result, it is possible to either reduce the amount of on-chip program memory, or to fit more code to the original memory. In MP systems, the amount of memory usually scales up with the number of processing elements, becoming the dominant part of the chip. To implement computationally powerful and efficient systems it is important to increase the logic-to-memory ratio in order to raise, for example, the $GOPS/mm^2$. However, improving the logic-to-memory ratio of such systems is challenging. While increasing the size of the system is not a viable option, also decreasing for instance the program memory size is difficult, since the application size is not getting smaller, and also the time consumed in task distribution and context swapping should be minimized. Therefore, utilizing code compression techniques can equally multiply the benefits when moving from single-core to multi-core chips, but this issue has not been addressed by academia this far.

In this paper we analyze the performance of a multi-processor architecture equipped with a compression/decompression system. The paper is organized as follows: in Section II brief overviews of the studied multi-processor architecture, of the code compression scheme utilized, and of the used benchmarking applications are given; Section III evaluates the performance of the system when applying the code compression scheme, in particular compression ratio, computation time, power consumption and logic utilization are analyzed; finally Section IV discusses the achieved results and gives conclusions.

II. SYSTEM OVERVIEW

A. Ninesilica multi-processor architecture

Ninesilica is a homogeneous multi-processor architecture developed on the basis of the Silicon Cafè template [2]. Ninesilica is composed of nine computational nodes, arranged in a 3x3 mesh topology. Each node is composed of a processor core (COFFEE RISC core), data and instruction memories, and a network interface. The communication between nodes and within a single node is supported by a hierarchical Network-on-Chip [3]. The only node with access to the I/O is the node

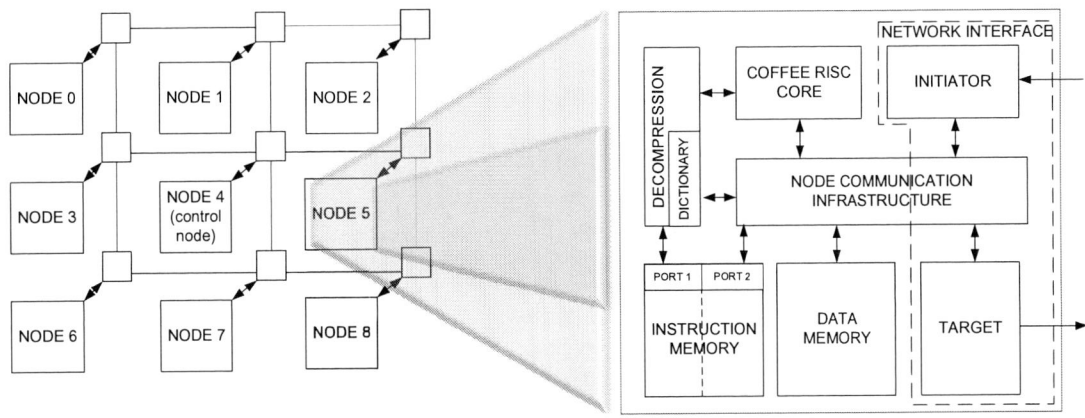

Fig. 1. Schematic view of CoNinesilica multi-processor architecture and its node structure, appended with the decompression hardware.

located in the central position of the mesh. This capability to access the I/O combined with its position in the architecture makes the central node a good candidate to be the controller node of the system, acting as data and task scheduler. The central location allows the control node also to have a uniform access time to all other nodes for updating/distributing data and tasks, keeping the workload on the NoC at the same time balanced. Furthermore, this enables effective utilization of the broadcasting features provided by the NoC, reducing the communication overhead by 30% [4].

B. Code compression/decompression system

Analyzing and compressing the application code is done off-chip after compilation, and the compressed code is then stored to an on-chip instruction memory (IMEM). All necessary modifications, for instance patching branch addresses to correspond to the compressed address space and aligning branch targets to memory word boundaries, as well as parameter explorations, such as optimizing the compression results and resulting decompression hardware properties, are done automatically by the compression tools. The resulting compressed code is processor specific, since it takes into account instruction set architecture (ISA) specific details. However, compressing the code again for a different processor type is merely a question of listing the ISA properties for the tools.

A stand-alone decompression engine, containing a small dictionary structure, is located between IMEM and processor core and handles restoration of the code on-the-fly. Therefore, appending the existing system with the compression/decompression system does not require modifying the processor core or the memory interface. In an MP system this feature enables for instance utilizing the code compression only in some of the nodes, and/or tailoring the system specifically towards certain application or task distribution plan.

The Ninesilica system appended with the decompression hardware is called CoNinesilica. For this analysis setup the decompression system has been integrated into each node independently. A detailed view of the architecture and its node structure is given in Figure 1. The dictionary structure in the decompression engine can be accessed through the network infrastructure. This feature enables the master node to update the content of the dictionaries as well as the local memories, if necessary. However, the decompression hardware is capable of decoding any code which has been compressed for the processor type in use, and therefore updating the dictionary is not necessary. In order to fully utilize the achievable compression ratio, the performance can still be optimized by updating the dictionary when changing application code.

Both the code compression method and the decompression hardware which are utilized here have been presented earlier for single-core system in [5], [6], [7], therefore a more detailed description of those is not included in this paper. The system used in this multi-processor study is fully compatible on algorithmic part with the one presented in the aforementioned publications, only the processor-specific parts have been fitted for COFFEE RISC core.

C. Benchmark applications

The used benchmark applications are two significant kernels of wireless communications. In particular, we have considered a radix-2 Fast Fourier Transform (FFT) [8], which is for example utilized in the demodulation step of OFDM systems as well as in many other signal processing domains. In addition to FFT, we utilized the cell search algorithm for W-CDMA systems [9]. The cell search algorithm can be further divided into 3 sequential tasks: slot synchronization, frame synchronization and scrambling code identification. However, each of these steps is based on the computation of correlations. More details about the implementation of the benchmark algorithms and their parallelization on the Ninesilica architecture can be found in [10], [11].

III. RESULTS

In order to evaluate the performance of CoNinesilica architecture over Ninesilica, different parameters/system properties were considered, in particular: achieved compression ratio, resource utilization, system performance and power consumption.

A. Compression ratio

The benchmark applications are partitioned for utilizing the parallel multi-core structure. In Ninesilica (and therefore in CoNinesilica as well), this is done by distributing the code between master and slave nodes. The compression ratios (CR), which are calculated as compressed code size divided by original code size, settle between 61.4 and 63.5%, as presented in Table I. The results already include all overheads of branch target aligning, address calculation, and dictionary structures. The only missing figure is the actual decompression engine area, which will be depicted in the following subsection with area results.

B. Resource Utilization

Table II reports the FPGA prototyping results of the original Ninesilica, the decompression hardware block, and the CoNinesilica system. The two architectures were synthesized targeting an Altera Stratix IV FPGA device. The utilization of the decompression hardware introduces about 33% of logic overhead, without taking into account the memory reduction.

Considering an ASIC implementation (standard-cell at 65nm technology) of the original Ninesilica architecture, the memories consume altogether 76% of the whole chip area [12], even when using moderate size memories of 16KB for both data and instructions. Portion of the instruction memories alone is 49%.

To compare the logic-to-memory figures of the original Ninesilica and compressed CoNinesilica, the systems were synthesized with standard cell technology. The data memory sizes in both systems were kept as 16KB. Instruction memories were fixed respectively at 64KB/node and 32KB/node for the Ninesilica and CoNinesilica systems. Considering Ninesilica, the ratio of the original useful logic (processing units and network structures) over total area of the system is 9,7%. In the CoNinesilica, the area of the decompression logic is counted on the 'negative' side together with the 32KB IMEM. However the final logic-to-memory ratio is 16%, which leads into an improvement of the computational density of 65%, noting that even though the memory sizes are different, they can host the same application codes.

TABLE II
FPGA SYNTHESIS RESULTS OF NINESILICA ARCHITECTURE, DECOMPRESSION HW AND CONINESILICA

Component	Adapt. LUT	Registers
COFFEE RISC	7862	4945
Local network node	346	232
Ninesilica Node	8237	5177
Network-on-Chip	2813	3548
Ninesilica	76780	50482
Decompression HW	2820	494
CoNinesilica node	9587	7563
CoNinesilica	90336	71078

C. Computation time

The benchmark applications were profiled on the Ninesilica and CoNinesilica architectures through RTL clock cycle accurate simulations. The computation times for each of the applications were normalized against the performance of the original Ninesilica. The system does not include instruction caches, which would complicate the performance analysis due to unknown amount of cache misses. Therefore, some performance loss will be introduced by the decompression hardware due to extra branch execution delay, which cannot be won back by decreased cache miss penalty. On this aspect, this analysis gives worst-case run-time results.

The first two columns of Table III collect the performance results for Ninesilica and CoNinesilica running at the same operating frequency. As it can be seen from the table, without any optimization the decompression HW degrades the execution time by 7–31%, depending on the application. To overcome such performance loss the decompression HW was then run at double clock frequency compared to the core, so that the branch calculation bottleneck could be removed or reduced. The last column of Table III reports the performance of CoNinesilica in this case. As shown, the performance loss due to the introduction of the decompression hardware has dropped down to a couple of percents, while for instance power consumption is not significantly compromised, as will be shown in the next section. However, this analysis does not take into account the task distribution time which is shorter due to a reduced amount of data to be transferred.

TABLE III
NINESILICA AND CONINESILICA NORMALIZED COMPUTATION TIME

algorithm	Ninesilica	CoNinesilica	CoNinesilica x2
2048-point FFT	1	1.07	1.02
Slot Synch.	1	1.31	1.05
Frame Synch.	1	1.18	1.06
Fsc. Identification	1	1.24	1.06

D. Power Consumption

Power consumption estimations of Ninesilica and CoNinesilica systems prototyped on the Altera Stratix IV FPGA and for the profiled benchmark applications were obtained utilizing the power analyzer tool of Altera Quartus II and the switching activity files generated during clock accurate simulations. For CoNinesilica architecture we evaluated the power consumption for both the version with uniform clock frequency and for the version in which the decompression hardware is clocked at double the frequency of the core. However, the difference in the power consumption results is not very significant, since increasing the decompression frequency improves branching characteristics, but most of the time the decompression hardware is idle. In order to follow the worst case analysis scenario, the double clocked system is the one offering higher performance as well as the one requiring higher power consumption, and therefore the most interesting case to be analyzed.

TABLE I
COMPRESSION RATIOS OF THE BENCHMARKED APPLICATIONS

Application	orig. number of IMEM words	compr. number of IMEM words	Compression Ratio [%]
2048-point FFT			
– Control node	6691	4222	63.5
– Computational node	6950	4392	63.2
W-CDMA			
– Slot Synch.			
—- Control node	3598	2249	62.5
—- Computational node	3050	1873	61.4
– Frame Synch.			
—- Control node	3214	1999	62.2
—- Computational node	2941	1809	61.5
– sc. Identification			
—- Control node	2961	1827	61.7
—- Computational node	2964	1820	61.4

Table IV presents the power consumption breakdown as comparison between the two architectures. As can be seen, the power consumption of the compressed system compared to the original one is improved with all applications codes, even in the case where the decompression hardware is clocked at double the core frequency.

TABLE IV
NORMALIZED POWER CONSUMPTION BREAKDOWN FOR NINESILICA AND CONINESILICA ARCHITECTURES

algorithm	Ninesilica	CoNinesilica x2
2048-point FFT	1	0.95
Slot Synch.	1	0.90
Frame Synch.	1	0.90
Fsc. Identification	1	0.84

IV. CONCLUSIONS

In this paper we presented the performance and evaluation of key parameters like compression ratio, resource utilization and power consumption of a multi-core architecture appended with a code compression system. The code compression system was independently designed and utilized as an add-on IP block for each node in the multi-processor architecture. The profiling results showed that the introduction of the code compression/decompression system improved the power consumption of the architecture by 5–16%. Without instruction caches, and therefore creating a worst-case performance analysis, the computation time suffered a couple of percents drawback, when excluding task distribution time. On the other hand, the CR settled around 62%, allowing to utilize smaller memory for the implementation of the benchmarked applications, leading into a higher logic-to-memory ratio and therefore a higher computational density of the system, which was improved by 65%. In addition, the CR could be used to trade the higher memory density to fit more applications on the same memory footprint leaving the logic-to-memory ratio fixed but requiring for example a reduced swapping time between applications and/or improving the overall system performance in scenarios where different applications are running on the multi-core system.

ACKNOWLEDGMENTS

The authors would like to thank GETA doctoral program and Nokia Foundation for the financial support. The research leading to these results has also been partially funded by SYSMODEL project (http: // www.sysmodel.eu).

REFERENCES

[1] W. Wolf, A. A. Jerraya, and G. Martin, "Multiprocessor System-on-Chip (MPSoC) Technology," *Computer-Aided Design of Integrated Circuits and Systems, IEEE Transactions on*, vol. 27, no. 10, pp. 1701–1713, Oct. 2008.

[2] J. Nurmi, "Silicon Caf: a Heterogeneous Multi-Processor Platform based on Coffee (RISC Core)," in *8th International Forum on Application-Specific Multi-Processor SoC*, 2008.

[3] T. Ahonen and J. Nurmi, "Hierarchically Heterogeneous Network-on-Chip," in *Proc. International Conference on "Computer as a Tool" EUROCON*, 9–12 Sept. 2007, pp. 2580–2586.

[4] R. Airoldi, F. Garzia, T. Ahonen, D. Milojevic, and J. Nurmi, "Implementation of W-CDMA Cell Search on a FPGA based Multi-Processor System-on-Chip with Power Management," in *Proceedings of the IX International Symposium on Systems, Architectures, MOdeling and Simulation (SAMOS IX). Springer Verlag (LNCS series)*, July 2009, pp. 88–97.

[5] P. Saastamoinen, I. Saastamoinen, and J. Nurmi, "Code compression in DSP processor systems," in *International Journal of Embedded Systems 2008 - Vol. 3, No.4 pp. 256 - 262*, 2008.

[6] P. Saastamoinen, I. Saastamoinen, M. Laiho, and J. Nurmi, "Minimizing Area Costs in GPS Applications on a Programmable DSP by Code Compression," in *Proc. Int. Symp. System-on-Chip SOC 2009*, 2009, pp. 91–94.

[7] P. Saastamoinen and J. Nurmi, "Parameterized Decompression HW for a Program Memory Compression System," in *Proc. Int. Symp. System-on-Chip SOC 2010*, 2010, pp. 63 – 67.

[8] J. W. Cooley and J. W. Tukey, "An Algorithm for the Machine Calculation of Complex Fourier Series," *Mathematics of Computation*, vol. 19, pp. 297–301, 1965.

[9] Y.-P. E. Wang and T. Ottosson, "Cell search in w-cdma," *IEEE Journal on Selected Areas in Communications*, vol. 18, no. 8, pp. 1470–1482, 2000.

[10] R. Airoldi, F. Garzia, and J. Nurmi, "FFT Algorithms Evaluation on a Homogeneous Multi-processor System-on-Chip," in *Proc. 39th Int Parallel Processing Workshops (ICPPW) Conf*, 2010, pp. 58–64.

[11] F. Garzia, R. Airoldi, T. Ahonen, J. Nurmi, and D. Milojevic, "Implementation of the w-cdma cell search on a mpsoc designed for software defined radios," in *Proc. IEEE Workshop Signal Processing Systems SiPS 2009*, 2009, pp. 030–035.

[12] R. Airoldi, F. Garzia, and J. Nurmi, "Implementation of a 64-point fft on a multi-processor system-on-chip," in *Proc. Ph.D. Research in Microelectronics and Electronics PRIME 2009*, 2009, pp. 20–23.

978-1-4673-2895-1/12 $31.00 © 2012 IEEE

Effects of Scaling a Coarse-Grain Reconfigurable Array on Power and Energy Consumption

Waqar Hussain, Tapani Ahonen, Jari Nurmi

Department of Computer Systems, Tampere University of Technology

P. O. Box 553, FIN-33101, Tampere, Finland

Email: firstname.lastname@tut.fi

Abstract—In recent past, we scaled a 4×8 processing element (PE) template-based Coarse-Grain Reconfigurable Array (CGRA) to a 4×4, 4×16 and 4×32 PE CGRA and generated matrix-vector multiplication (MVM) accelerators from each one of them. Furthermore, on each of the accelerators, MVM kernels of order $N = 4, 8, 16, 32$ were mapped. In this paper, we have estimated the power and energy consumption by generating the postfit gate-level netlist of each accelerator for a Field Programmable Gate Array as target platform. Based on our measurements, we have studied the effects of scalability of a CGRA on power and energy consumption.

I. INTRODUCTION

Embedded systems are required in almost every field of science and engineering. The requirements of the users may demand different types of embedded systems ranging from single processor system to multiprocessors. They also may vary from single accelerator system to multiple accelerator systems. One of the important system types is processor/ coprocessor model in which the general purpose processing is performed by the processor and the coprocessor accelerates the computationally intensive tasks. One of the important class of accelerators is a Coarse-Grain Reconfigurable Array (CGRA) which by its structure offers high hardware level parallelism and throughput. A number of CGRAs have been developed so far, for example ADRES [2], Morphosys [1], PACT-XPP [3] and BUTTER [4]. The general-purpose CGRA required an area of few million gates and their presence in the systems becomes expensive unless they are extensively used. To avoid this problem, we developed template-based CGRAs like CREMA [5] and [6]. Using template-based CGRAs, the user was able to generate special purpose accelerators on a specified mapping which reduces resource utilization. Many computationally intensive kernels were mapped on CGRAs in past, for example Wideband Code Division Multiple Access (WCDMA) cell search [7], image and video processing [4], [9] and Viterbi decoders [8]. Execution time constraints for processing Fast Fourier Transform (FFT) were also achieved for many wireless standards like IEEE-802.11a/g, 3GPP-LTE and IEEE-802.11n [7], [6].

Scaling the hardware is important from execution and resource utilization point-of-view. Especially in mobile devices, we may not need a high processing bandwidth at all times. For example, our mobile device may not need resources to carry out processing for a stream generated in multiple-input multiple-output Orthogonal Frequency-Division Multiplexing (OFDM) environment if the user ports it to a single-input single-output OFDM environment. Run-Time Partial Reconfigurable (RTPR) Field Programmable Gate Arrays (FPGA) allow reconfiguration of a fraction of the fabric at run-time. Using the RTPR FPGA, we can increase or decrease the resources on FPGA at run-time and therefore save the resources and power consumption. In this context, we need to know how scaling will effect the execution time, resource utilization, power consumption and also the application developement time [15].

In the next section, we will discuss the processing model based on a scalable CGRA template. In Section III, we will discuss the matrix-vector multiplication (MVM) accelerators generated from the scalable CGRA templates. In Section IV, we will focus on power and energy estimation of the generated accelerators. In the next section, we will discuss the scalability analysis based on the findings in the previous sections. Finally, we will present conclusions.

II. SCREMA BASED PROCESSING MODEL

Structure of SCREMA template is similar to CREMA template except that SCREMA can be scaled to different CGRA templates which can then generate accelerators of different sizes. SCREMA is written in VHDL and can be scaled to different sizes of templates by changing just a few parameters in the definition package VHDL file. After changing the definition package file, a parameter package VHDL file can be used which is generated on user specifications by a graphical tool. The parameter package file contains the information to craft the template to an accelerator on compilation.

The structural unit of SCREMA is a processing element (PE) like in CREMA. Each PE has two inputs (a, b) and two outputs (A, B). Output A carries the result based on the operand received by (a, b) and the output B receives the operand b so that it can transfered to other PE as required. Each PE can perform 32-bit integer and floating-point operations in IEEE-754 format. The internal structure of the PE is well defined in [5] and also the way it exchanges data with the neighbouring PEs in point-to-point fashion.

The generated accelerator is equiped with two local memories and the data to be processed is loaded in these local memories with the help of a Direct Memory Access (DMA) device. The DMA transfers the data from the main memory of the system to the local memories in a specific

978-1-4673-2895-1/12 $31.00 © 2012 IEEE

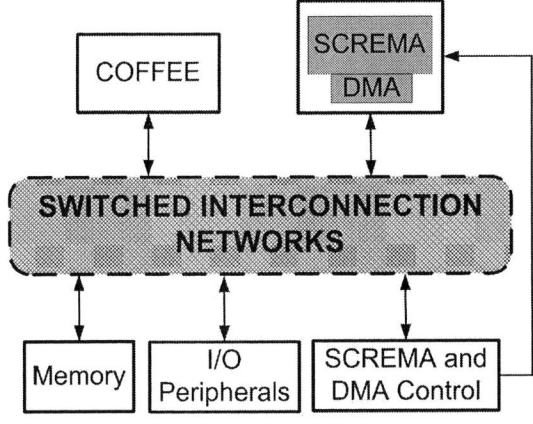

Fig. 1.   SCREMA based Embedded Processing Model

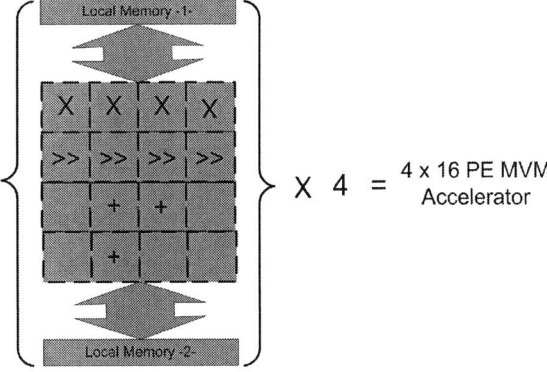

Fig. 2. 4×4 PE MVM accelerator generated by 4×4 PE SCREMA shown in braces. Four of such accelerators working in parallel will be equal to MVM accelerator generated by a 4×16 PE CGRA template.

pattern introduced by the user. Before the system start-up, the configuration data is loaded in the configuration memory of each PE. The configuration words are injected in the array using a pipelined infrastructure [14]. These words are used to select an operation to be performed by a PE and also the interconnections among PEs. The operations to be performed and the pattern of interconnection among all PEs is called a context. Different contexts can be designed by the user at compile-time and can be enabled at run-time based on the flow of an algorithm. The application mapping on SCREMA versions and the execution flow to be written in C is similar to the one explained in [13] for designing FFT accelerators.

SCREMA generated accelerators work as coprocessors of COFFEE RISC [10]. The program is written in C and compiled for COFFEE RISC which controls the processing of the accelerator in a polling mechanism. COFFEE writes the control words in the control registers of the generated accelerator which in return performs cycle-accurate processing. COFFEE and SCREMA generated accelerator interact with each other by a network of switched interconnections that provide dedicated connections for faster communication. Fig. 1 shows the overall system.

III. MVM ACCELERATORS

In this paper, we concentrate on Matrix-Vector Multiplication (MVM) applications as they are largely used in different science and engineering applications. We can define MVM process mathematically by considering a matrix $a = [a_{i,j}]$ and vector $\overrightarrow{b} = [b_i]$ of Nth-order and they are supposed to be multiplied to produce a product vector $\overrightarrow{p} = [p_i]$. Then the multiplication process can be defined as

$$[p_i] = \sum_{j=1}^{N} [a_{j,i}] \times [b_j] \tag{1}$$

where i, j = 1, 2, 3,...,N.

MVM accelerators were generated from CGRA-templates of sizes 4×4, 4×8, 4×16 and 4×32 PEs which were developed in recent past. The basic MVM accelerator was

generated from 4×4 PE CGRA-template and its structure is shown in Fig. 2. The first row consist of multipliers, the second performs the shift operation to avoid any possible overflows at the later stages. The third and fourth row perform the addition operations required for MVM process. The MVM accelerator generated by 4×8 PE CGRA template is similar to two 4×4 PE MVM accelerators working in parallel. Similarly 4×16 PE MVM accelerator contains four of such accelerators and 4×32 will have eight of those working in parallel. As each of the PEs has two inputs, the first input can receive a matrix element and the second input will receive the related vector element to be multiplied. The matrix and vector data to be multiplied is distributed over the MVM accelerator's local memory in a specific pattern. The data placement pattern in the local memories of the MVM accelerators is same for all the accelerators to carry out justified comparisons. The execution time shown in Table III does not include the time required by the DMA to load the data in the local memories. It is the time, when the data start processing over the array until it completes the processing and stores the data in the local memory as we are only interested in the performance measurement of the array.

The MVM accelerators were synthesized for Altera's Stratix-IV FPGA device (EP4SGX70HF35C2) and the resource utilization can be oberved from Table I for each of the accelerators. As the size of the MVM accelerator increases, the resource utilization increases almost proportionally. As all the accelerators have 32-bit processing so the multiplication of two 32-bit numbers will result in a 64-bit number. To fit a 64-bit number, four 18-bit DSP elements are required. As a 4×4 MVM accelerator shown in braces in Fig. 2 has four multipliers in the first row, it will need 16 of 18-bit DSP elements on the FPGA as shown in Table I.

IV. POWER AND ENERGY ESTIMATION

To estimate the power consumption, we generated the postfit gate-level netlist for each of the MVM accelerators and performed the timing simulation which is the most accurate method for estimating the power conumption. Table II shows

978-1-4673-2895-1/12 $31.00 © 2012 IEEE

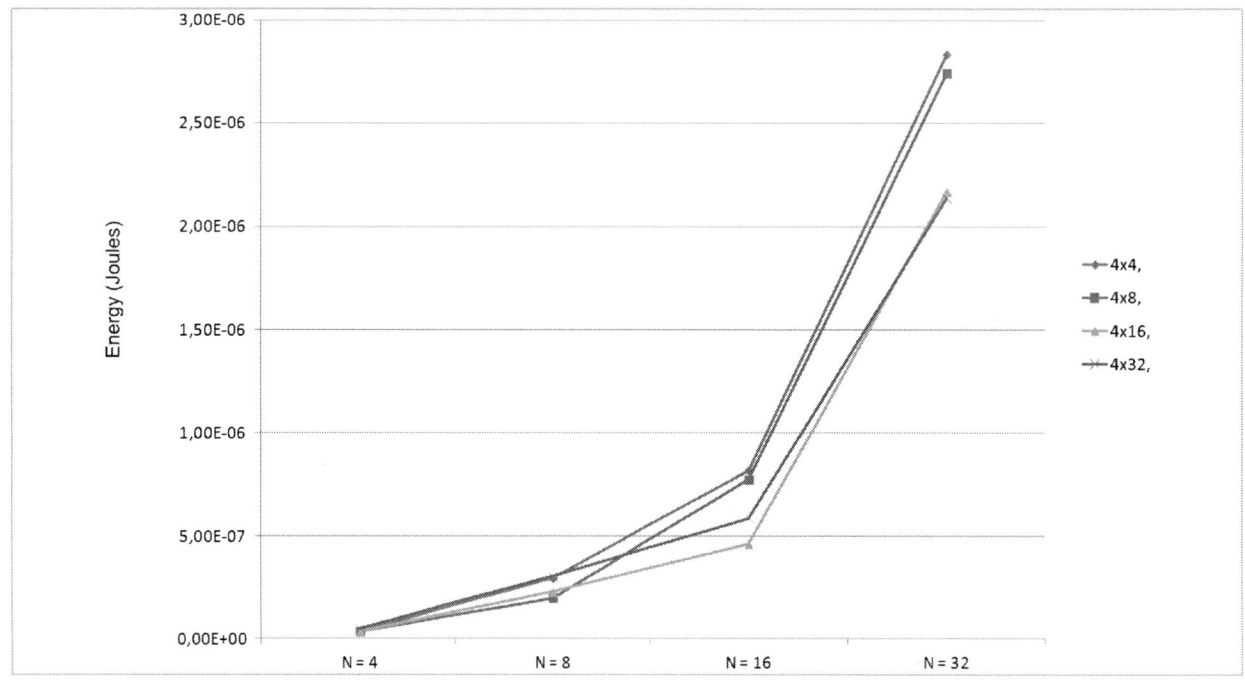

Fig. 3. Curves showing 4×16 PE MVM accelerator as the most energy efficient relatively. X-axis shows the order of MVM process and Y-axis shows the energy consumption in Joules.

MVM Accelerator Size	Comb ALUTs	Logic Registers	DSPs
4×4 PE	2,566	2,766	16
4×8 PE	4,805	3,820	32
4×16 PE	8,259	6,784	64
4×32 PE	15,522	12,057	128

TABLE I

RESOURCE UTILIZATION BY MVM ACCELERATORS ON STRATIX-IV

(EP4SGX70HF35C2) DEVICE

the static and dynamic power consumption of the four MVM accelerators at 85°C and 900mV. At these conditions, the four accelerators achieved the operating frequency between 169-172 MHz. From the table, it can be observed that there is no significant change in the static power consumption. A large offset approximately equal to 427 mW is visible due to the unused portion of the FPGA chip. If this offset is subtracted, we can observe that the static power consumption increases almost by a factor of two as the size of the accelerator doubles. Significant difference in dynamic power consumption can be observed as it depends on the number of signals having switching activity at a particular time instant. As the size of the MVM accelerator increases, the dynamic power consumption increases but the processing time for MVM decreases. Depending on the requirements for resource utilization, execution time and power consumption, it may be desireable to make a choice among these four MVM accelerators. An optimal choice can be based on energy consumption which is the product of total power consumption and the execution time

but in that case a comparison should also be made to the resouce utilization of the choosen MVM accelerator. The energy consumption is estimated only for the data processing by the accelerators and not for the process of transferring the data from the main memory of the system to the local memories of the accelerators.

V. SCALABILITY EFFECTS

The digital hardware due to its binary characteristics will tend to scale by a factor of 2^m where $m \in Z^+$. The application to be mapped should be built on the same scale to have best-fit compatibility. This is why we have chosen the CGRA sizes of 4×2^n and the matrix and vector of sizes $2^n \times 2^n$ and 2^n respectively, where $n = \{2, 3, 4, 5\}$.

From the curves shown in Fig. 3, we can easily observe that 4×8 and 4×16 PE MVM accelerators are relatively the most energy efficient for processing MVM of $N = 4, 8$ and $N = 16, 32$ relatively. The 4×4 PE MVM accelerator is the least energy efficient relatively otherwise while processing MVM of $N = 4$. Its relatively low dynamic power consumption was not enough to overcome the number of clock cycles required to process MVM kernels. The 4×32 PE MVM accelerator is not relatively as energy efficient as others than it has shown to be slightly more energy efficiency while processing MVM of order $N = 32$ then 4×16 PE MVM accelerator. From the curves in Fig. 3, it can be observed that an MVM accelerator is the most energy efficient when its order is matching with the order of the MVM process. As the experimental data set is kept fixed upto only MVM of order $N = 32$, we can speculate

MVM Accelerator	Static Power	Dynamic Power	I/O Power	Total Power
4×4 PE	428.48 mW	127.27 mW	58.07 mW	613.82 mW
4×8 PE	430.41 mW	208.63 mW	56.53 mW	695.57 mW
4×16 PE	435.11 mW	387.21 mW	56.95 mW	879.27 mW
4×32 PE	448.51 mW	728.25 mW	51.04 mW	1227.80 mW

TABLE II

POWER CONSUMPTION BY MVM ACCELERATORS OF DIFFERENT SIZES

MVM Accelerator	MVM Order	Execution Time	Total Power	Energy
4×4	4	57 ns	613.82mW	0.034 μJ
	8	479 ns		0.29 μJ
	16	1334 ns		0.818 μJ
	32	4617 ns		2.83 μJ
4×8	4	52 ns	695.57 mW	0.036 μJ
	8	285 ns		0.19 μJ
	16	1111 ns		0.773 μJ
	32	3940 ns		2.74 μJ
4×16	4	41 ns	879.27 mW	0.036 μJ
	8	259 ns		0.228 μJ
	16	525 ns		0.462 μJ
	32	2466 ns		2.17 μJ
4×32	4	41 ns	1227.80 mW	0.05 μJ
	8	248 ns		0.304 μJ
	16	478 ns		0.587 μJ
	32	1740 ns		2.14 μJ

TABLE III

ENERGY CONSUMPTION BY DIFFERENT SIZE OF ACCELERATORS WHILE EXECUTING DIFFERENT ORDERS OF MVM

that 4×32 PE MVM accelerator will be significantly more energy efficient than all others while processing $N \geq 32$.

An important dimension of comparison will be to evaluate energy consumption with the resource utilization of the MVM accelerators. A rough estimate about the resource utilization can be the average of the number of ALUTs and logic registers. We can ignore the DSP elements as their power consumption is almost negligible compared to the reconfigurable fabric of the FPGA. From the data in Table I, we calculated the averages and found that the cost in terms of resources is almost $1.7X$ which is required to scale-up an MVM accelerator to the next level. As the size of MVM accelerator is scaled-up, the dynamic power consumption increases by a factor of almost $1.8X$.

From the curves shown in Fig. 3, the investment of $1.7X$ resources was generally the most cost effective for scaling-up 4×8 PE MVM accelerator to 4×16 PE MVM accelerator while processing $N = 16, 32$ MVM and even for processing $N = 4, 8$, it is very close to 4×8 PE MVM accelerator curve.

VI. CONCLUSION

We estimated the power and energy consumption of 4×4, 4×8, 4×16 and 4×32 processing element (PE) matrix-vector multiplication (MVM) accelerators. We found that scaling-up MVM accelerators results in a constant increase of dynamic power consumption of almost $1.8X$ and requires around $1.7X$ of more resources when doubling the accelerator array width. The 4×8 and 4×16 PE MVM accelerators are relatively the most energy efficient while processing MVM of orders $N = 4, 8$ and $N = 16, 32$, respectively. The 4×32 PE MVM

accelerator is not as energy efficient as any of the other MVM accelerators.

REFERENCES

[1] H. Singh, M.-H. Lee, G. Lu, F. J. Kurdahi, N. Bagherzadeh, and E. M. C. Filho, "Morphosys: An integrated reconfigurable system for data-parallel and computation-intensive applications". IEEE Trans. Computers, vol. 49, no. 5, pp. 465-481, 2000.

[2] B. Mei, S. Vernalde, D. Verkest, H. D. Man, and R. Lauwereins, "ADRES: An architecture with tightly coupled VLIW processor and coarse-grained reconfigurable matrix", Field-Programmable Logic and Applications, vol. 2778, pp. 61-70, September 2003, ISBN 978-3-540-40822-2.

[3] V. Baumgarte, G. Ehlers, F. May, A. Nuckel, M. Vorbach, and M. Weinhardt, "PACT XPP-A Self-Reconfigurable Data Processing Architecture", The Journal of Supercomputing, vol. 26, no. 2, pp. 167-184, September 2003.

[4] C. Brunelli, F. Garzia, and J. Nurmi, "A Coarse-Grain Reconfigurable Architecture for Multimedia Applications Featuring Subword Computation Capabilities", in Journal of Real-Time Image Processing, Springer-Verlag, 2008, 3 (1-2): 21-32. doi:10.1007/s11554-008-0071-3.

[5] F. Garzia, W. Hussain and J. Nurmi, "CREMA, A Coarse-Grain Reconfigurable Array with Mapping Adaptiveness", in Proc. 19th International Conference on Field Programmable Logic and Applications (FPL 2009). Prague, Czech Republic: IEEE, September 2009.

[6] W. Hussain, F. Garzia, T. Ahonen and J. Nurmi, "Designing Fast Fourier Transform Accelerators for Orthogonal Frequency-Division Multiplexing Systems", in the Springer's Journal of Signal Processing Systems, December 2010.

[7] F. Garzia, W. Hussain, R. Airoldi, J. Nurmi, "A Reconfigurable SoC tailored to Software Defined Radio Applications", in Proc of 27th Norchip Conference, Trondheim (NO), 2009.

[8] Y. Kishimoto, S. Haruyama, H. Amano, "Design and Implementation of Adaptive Viterbi Decoder for Using A Dynamic Reconfigurable Processor" in Proc. Reconfigurable Computing and FPGAs, 2008. ReConFig '08, pp 247-252, doi=10.1109/ReConFig.2008.39, ISBN: 978-1-4244-3748-1.

[9] Chia-Cheng Lo, Shang-Ta Tsai, Ming-Der Shieh; , "A reconfigurable architecture for entropy decoding and IDCT in H.264", VLSI Design, Automation and Test, 2009. VLSI-DAT '09. International

Symposium on , vol., no., pp.279-282, 28-30 April 2009, doi: 10.1109/VDAT.2009.5158149, ISBN: 978-1-4244-2781-9.

[10] J. Kylliainen, T. Ahonen, and J. Nurmi, "General-purpose embedded processor cores - the COFFEE RISC example", In Processor Design: System-on-Chip Computing for ASICs and FPGAs, J. Nurmi, Ed. Springer Publishers, June 2007, ch. 5, pp. 83-100, ISBN-10: 1402055293, ISBN-13: 978-1-4020-5529-4.

[11] W. Hussain, F. Garzia and J. Nurmi, "Exploiting Control Management to Accelerate Radix-4 FFT on a Reconfigurable Platform", in Proc. International Symposium on System-on-Chip 2010. Tampere, Finland: IEEE, pp. 154-157, September 2010, ISBN: 978-1-4244-8276-4.

[12] C. Brunelli, F. Garzia, C. Giliberto and J. Nurmi, "A Dedicated DMA Logic Addressing a Time Multiplexed Memory to Reduce the Effects of the System Buss Bottleneck", in Proc. 18th International Conference on Field Programmable Logic and Applications, (FPL 2008), Heidelberg, Germany, 8-10 September 2008, pp. 487-490.

[13] W. Hussain, F. Garzia, and J. Nurmi, "Evaluation of Radix-2 and Radix-4 FFT Processing on a Reconfigurable Platform," in Proceedings of the 13th IEEE International Symposium on Design and Diagnostics of Electronic Circuits and Systems (DDECS'10). IEEE, pp. 249-254, April 2010, ISBN 978-1-4244-6610-8.

[14] F. Garzia, C. Brunelli and J. Nurmi, "A pipelined infrastructure for the distribution of the configuration bitstream in a coarse-grain reconfigurable array", in Proceedings of the 4th International Workshop on Reconfigurable Communication-centric System-on-Chip (ReCoSoC'08). Univ Montpellier II, July 2008, pp. 188-191, ISBN:978-84-691-3603-4.

[15] W. Hussain, T. Ahonen F. Garzia and J. Nurmi, "Application-Driven Dimensioning of a Coarse-Grain Reconfigurable Array", in Proc. NASA/ESA Conference on Adaptive Hardware and Systems (AHS-2011), San Diego, California, USA.

CRAVE: An Advanced Constrained RAndom Verification Environment for SystemC

Finn Haedicke[1]   Hoang M. Le[1]   Daniel Große[1]   Rolf Drechsler[1,2]

[1]Institute of Computer Science, University of Bremen, 28359 Bremen, Germany

[2]Cyber-Physical Systems, DFKI GmbH, 28359 Bremen, Germany

{finn, hle, grosse, drechsle}@informatik.uni-bremen.de

Abstract—A huge effort is necessary to design and verify complex systems like System-on-Chip. Abstraction-based methodologies have been developed resulting in Electronic System Level (ESL) design. A prominent language for ESL design is SystemC offering different levels of abstraction, interoperability and the creation of very fast models for early software development. For the verification of SystemC models, Constrained Random Verification (CRV) plays a major role. CRV allows to automatically generate simulation scenarios under the control of a set of constraints. Thereby, the generated stimuli are much more likely to hit corner cases. However, the existing SystemC Verification library (SCV), which provides CRV for SystemC models, has several deficiencies limiting the advantages of CRV. In this paper we present CRAVE, an advanced constrained random verification environment for SystemC. New dynamic features, enhanced usability and efficient constraint-solving reduce the user effort and thus improve the verification productivity.

I. INTRODUCTION

Creating a new *System-on-Chip* (SoC) involves many tasks. Once the specification is agreed on, the modeling phase starts to derive a potential solution. To manage the complexity of today's SoCs high-level languages are used for the first models. For this *Electronic System Level* (ESL) design phase [1] a widely accepted approach is SystemC [2], [3], [4], [5]. Integrated hardware and software models can be developed, exchanged and refined based on the SystemC IEEE standard [6]. In particular, *Transaction Level Modeling* [7], [8] allows to create high performance virtual platforms for early software development and architectural analysis. In addition, the high-level models serve as reference to verify the behavior of the more detailed descriptions built in the following stages.

In general, facing today's verification challenges a verification environment needs to be constructed. From a high-level perspective three major components are necessary: stimuli, assertions and coverage. Assertions are used to check the functional correctness and therefore monitor design variables [9]. The task of functional coverage is to measure which design functionality has been exercised during simulation [10]. Both are not in the focus of this work. Here, we target the problem of stimuli generation. But instead of deterministic values as defined in traditional directed testbenches we make use of *Constrained Random Verification* (CRV) [11], [12]. CRV applies input stimuli to the design that are solutions of constraints. These solutions are determined by a constraint-solver. CRV offers two key benefits: First, CRV enables to find

This work was supported in part by the German Federal Ministry of Education and Research (BMBF) within the project SANITAS under contract no. 01M3088 and by the German Research Foundation (DFG) within the Reinhart Koselleck project DR 287/23-1.

unexpected assertion violations since scenarios are simulated which the verification engineer might have not thought of. Second, the stimulus generation process is automated and hence a huge set of scenarios can be executed leading to higher coverage. Hence, for large and complex systems the confidence in the correct functionality significantly increases.

For SystemC CRV is available through the *SystemC Verification* (SCV) library [13], [14], [15]. However, the SCV library has several deficiencies:

1) Poor dynamic constraint support, hence no control of constraint effects at run-time
2) No constraint specification for dynamic data-structures
3) Low usability when specifying constraints for composed data structures
4) Poor information in case of over-constraining
5) Limits in complexity of constraints, since constraint-solving is based on *Binary Decision Diagrams* (BDDs) [16] only

In this paper we present CRAVE, an advanced *Constrained RAndom Verification Environment* for SystemC.[1] To overcome the limitations of the SCV library CRAVE provides the following features:

- New constraint specification API
  An intuitive and user-friendly *Application Programming Interface* (API) to specify random variables and random objects has been developed.
- Dynamic constraints and data structures
  Constraints can be controlled dynamically at run-time. Moreover, constraints for elements of dynamic data structures like e.g. STL vectors can be specified.
- Improved usability
  Inline constraints can be formulated and changed incrementally at run-time. Furthermore, automatic debugging of unsatisfiable constraints is supported.
- Parallel constraint-solving
  BDD-based and SAT/SMT-based techniques have been integrated for constraint-solving. A portfolio approach is used to enable very fast generation of constraint solutions.

Please note the usage of CRAVE is not limited to pure hardware designs. For example, the constraint solutions can also be used to describe software tests running on a SoC. In the experiments such an example is presented.

For an industrial application of CRAVE we refer to [17].

[1]CRAVE is freely available (w/ source code) under MIT license at www.systemc-verification.org.

The rest of this paper is structured as follows: Related work is discussed in Section II. Section III presents the API of CRAVE. Then, in Section IV the dynamic features of CRAVE are introduced. The usability aspects are described in Section V. Section VI presents the constraint-solving approach of CRAVE and Section VII compares CRAVE to the SCV library in an experimental evaluation. Finally, the paper is concluded in Section VIII.

II. RELATED WORK

As mentioned above, the CRV techniques of the SCV library have several weaknesses which limits their use in practice. Therefore, several improvements for the SCV library have been developed. In [18] bit-vector operators have been added and the uniform distribution among all constraint solutions is ensured in all cases. An approach to determine the exact reasons in case of over-constraining has been presented in [19]. In [20] the BDD-based constraint-solver is replaced by a method which uses a generalization of *Boolean Satisfiability* (SAT).

However, all these approaches compensate only some of the SCV weaknesses. In particular, no constraints on dynamic data structures can be specified, constraints cannot be controlled dynamically during run-time, references to the state of constraints are not available, and no inline constraints are possible restricting the usability. In addition, the integration of different constraint-solvers working in parallel is mandatory to reduce the time for stimuli generation to a minimum.

To standardize verification processes the so-called *Universal Verification Methodology* (UVM) has been developed [21]. Essentially, UVM is a methodology and a class library for building advanced and reusable verification components. The initial implementation has been done for SystemVerilog. Meanwhile UVM also provides a SystemC class implementation. However, it does not include CRV (which is the core verification technique in the UVM methodology).

III. CONSTRAINT SPECIFICATION

In this section we describe the basics of CRAVE. In particular this includes the APIs to create random variables and constrained random objects.

A. Random Variable

From the user's point of view, the most elementary entity is the template class *randv<T>*, which corresponds to a random variable of the C/C++ or SystemC built-in type *T*. All standard applicable operators (arithmetic, comparison, logical, etc.) are overloaded so that an instance *x* of *randv<T>* behaves as if it were a variable of type *T*. A call *x.next()* assigns a random value in the range of *T* to *x*. While the focus of CRAVE is on complex constraints involving many variables, it also supports simple constraints on a single variable. Two member functions *addRange* and *addWeightedRange* can be used to refine the distribution of *x.next()*. Furthermore, *x()* returns a symbolic link to the value of *x* to be used to specify constraints in conjunction with other instances of *randv<T>* as shown in the next section. Table I summarizes exemplarily the API of *randv<int>*.

```
1   struct packet : public rand_obj {
2     randv< unsigned int > src_addr;
3     randv< sc_uint<16> > dest_addr;
4
5     packet() : src_addr(this), dest_addr(this) {
6       constraint(src_addr() <= 0xFFFF);
7       constraint("diff", src_addr() != dest_addr());
8       soft_constraint(dest_addr() % 4 == 0);
9     }
10  };
```

Figure 1: A basic constrained random packet

```
1   struct packet1 : public packet {
2     randv< char > data;
3     packet1() : data(this) {
4       constraint('a' <= data() && data() <= 'z');
5       constraint(dest_addr() % 2 == 1);
6     }
7   };
```

Figure 2: An inherited packet

B. Random object

Complex constrained random objects can also be specified. They must inherit from the class *rand_obj* provided by CRAVE. Such an object can contain several instances of *randv<T>* and *rand_obj*. Constraints for each of these instances as well as constraints between them can be specified in a constructor of the object. For the instance *x* of *rand_obj*, *x.next()* randomizes all belonging instances of *randv<T>* and *rand_obj*, respecting the specified constraints.

We demonstrate the specification of constraints using an example. Figure 1 shows a constrained random packet consisting of two integers to be randomized: a source address as *randv<unsigned int>* and a destination address as *randv<sc_uint<16>>*. The member instances are forced to register themselves to the *rand_obj* as shown in line 5. The source address is constrained to be in the range [0x0, 0xFFFF] (line 6) and the source address and destination address must not be the same (line 7). Both constraints are so-called hard constraints, i.e. they must be satisfied otherwise *next()* should fail. The second constraint is also a named constraint, which enables dynamic management of constraints as described later in Section IV. Line 8 shows a soft constraint stating that the destination address should be a multiple of four. Soft constraints can be ignored by the constraint-solver if they cannot be satisfied in conjunction with the specified hard constraints. As can be seen in all the constraints, the symbolic links to the actual instances of *randv* are used. This packet will be extended step-by-step and serves as a running example for the rest of this paper. Table II shows a summary of the basic API features of *rand_obj*. Note that while constraints should be specified in a constructor for the most use cases, it is possible to add further constraints to an instantiated object using the API.

978-1-4673-2895-1/12 $31.00 © 2012 IEEE

Table I: The APIs of *randv<int>*

Description	API
Supported operators	$+, -, *, /, \%, ==, !=, >, <, >=, <=, !, \&\&, \|\|, \sim, \&, \|, <<, >>,$ ^
Add range to distribution	*addRange(left, right)*
Add weighted range	*addWeightedRange(left, right, weight)*
Generate random value	*next()* returns *true* on success
Symbolic link	*x()* with *x* being a variable of type *randv<int>*

Table II: The basic APIs of *rand_obj*

Description	API
Add hard constraint	*constraint(expression)*
Add named hard constr.	*constraint(name, expression)*
Add soft constraint	*soft_constraint(expression)*
Randomize the object	*next()* returns *true* on success

C. Constraint Inheritance

The inheritance/reuse of constraints in CRAVE is straight-forward. The user can add more fields and constraints to an existing random constrained object by using C++ class inheritance. Figure 2 shows an extension of the packet introduced in the last section. Line 2 adds a data field to the packet. The constraint for this data field is declared in line 4. The destination address of the packet is further constrained to be an odd integer on line 5. This new hard constraint contradicts the soft constraint specified on line 8 of Figure 1 and therefore renders the soft constraint useless.

The APIs of CRAVE described in this section are the basics to specify constrained random objects. Similar APIs are also available in the SCV library to a greater or lesser extent. However, CRAVE enables an enhanced usability in comparison to the SCV library as discussed in Section V. In the next section we introduce the distinctive features of CRAVE regarding dynamic constraints and data structures.

IV. DYNAMIC CONSTRAINTS AND DATA STRUCTURES

This section introduces three distinctive features of CRAVE which are not supported by the SCV library: constraints on dynamic data structures (currently only vector is supported), dynamic enabling/disabling of constraints, and the concept of references that allows the randomization to interact tightly with the verification environment.

A. Vector Constraints

The SCV library offers no direct support for the constrained randomization of dynamic data structures such as vectors, lists and trees. The user must mimic dynamic data structures by using arrays of fixed-size. This is inconvenient and not memory-efficient. Furthermore, the upper-bound on size of dynamic data structures might not be known at the time of constraint specification. CRAVE offers a template class *rand_vec<T>* for the constrained randomization of vectors.

Currently only C/C++ and SystemC built-in data types are supported as the template parameter *T*. The class *rand_vec<T>* also implements the APIs of the STL class *vector* and thus behaves as if it is an STL vector. Similar to *randv<T>*, for an instance *v* of *rand_vec<T>*, *v* refers to the actual vector, while *v()* is the symbolic vector used to specify constraints. For the symbolic vector *v()*, *v().size()* refers to the size, *v()[_i]* to a symbolic vector element, and *v()[_i - c]* to a previous element relative to *v()[_i]* (*_i* is a predefined constant in CRAVE and *c* is a positive constant). The symbolic elements *v()[_i]* and *v()[_i - c]* are used in a *foreach* constraint.

Figure 3 shows an extension of our packet with the data field, now being a constrained random vector. The constraint on the vector is declared in line 5. In the next lines, three *foreach* constraints are specified for the vector. The first two ensure that the first element is an upper case letter and the rest are lower case letters. Both are hard constraints. The third constraint (line 14) is a soft *foreach* constraint: two consecutive elements cannot be *aa*, *ab* or *ba*.

B. Dynamic Constraint Management

During the constrained random verification process, it is very useful that the user can enable/disable specific constraints of a random object. This functionality is not available in the SCV library. The user must mimic the feature by adding an auxiliary variable and constrain this variable in an implication with the constraints to be enabled/disabled. Moreover, this is inconvenient and inefficient. In the CRAVE framework, named constraints can be enabled/disabled directly via the constraint management APIs of *rand_obj*: *enable_constraint(name)* and *disable_constraint(name)*. For example, the constraint on line 7 of Figure 1 can be disabled by calling *disable_constraint("diff")*. Disabled constraints will have no effect in the randomization via *next()* until they are enabled again. Note that the vector constraint *foreach* can also be named and intentionally soft constraints cannot be enabled/disabled.

C. References

In many use cases, the randomization depends on the dynamically changing state of the verification environment. Using the SCV library, to include the state in the constraints the user must use additional variables to save the state and update them manually whenever the state is changed. For this purpose, CRAVE provides references as a convenient shortcut. References in CRAVE basically links a "real" variable with a symbolic variable which can be used during constraint

978-1-4673-2895-1/12 $31.00 © 2012 IEEE   91

```
1  struct packet2 : public packet {
2    rand_vec< char > data;
3
4    packet2() : data(this) {
5      constraint( data().size() % 4 == 0
6        && data().size() < 100 );
7
8      constraint.foreach( data, _i, IF_THEN( _i == 0,
9        'A' <= data()[_i] && data()[_i] <= 'Z') );
10
11     constraint.foreach( data, _i, IF_THEN( _i != 0,
12       'a' <= data()[_i] && data()[_i] <= 'z') );
13
14     constraint.soft_foreach( data, _i,
15       data()[_i] + data()[_i-1] > 'a' + 'b');
16   }
17 };
```

Figure 3: An inherited packet using vector constraints

```
1  packet2(int &expected_max_size) : data(this) {
2    constraint(data().size() % 4 == 0
3      && data().size() <=
4        reference(expected_max_size));
5    ...
6  }
```

Figure 4: Example of CRAVE reference

specification. Before the constraints are solved, the value of this symbolic variable is fixed to the actual value of the linked variable. Figure 4 gives an example for using references: The size of the constrained random vector *data* of the packet in Figure 3 should not exceed the value of environment variable *expected_max_size*, which is constantly changing. As can be seen, the construct *reference* links *expected_max_size* to a symbolic variable and *data().size()* is constrained to be smaller or equal to this symbolic variable.

The new features introduced in this section demonstrated different use cases where CRAVE offers clear advantages in comparison to the SCV library. The next section discusses important usability enhancements of CRAVE.

V. USABILITY

The CRAVE framework provides several usability enhancements in comparison to the SCV library. In this section we present inline constraints, incremental constraints and automatic over-constraint analysis.

A. Inline and Incremental Constraints

Constraints in CRAVE can be specified without a formal constraint class. A standalone constrained random generator can be created anywhere and used with arbitrary variables and constraints. In practice, this reduces the effort when coding non-trivial testbench environments. Here is a concrete example to demonstrate this feature:

```
1  randv<int> x,y;
2  Generator gen;
3
4  gen(x() != y());
5  for (int i =0 ; i < 1000; ++i) {
6    gen.next();
7    run_test(x,y);
8  }
9
10 gen( x()*x() == y() );
11 for (int i =0 ; i < 500; ++i) {
12   gen.next();
13   run_test(x,y) ;
14 }
15
16 gen( y()%2 == 0 );
17 for (int i =0 ; i < 500; ++i) {
18   gen.next();
19   run_test(x,y) ;
20 }
```

Figure 5: Incremental constraint modification

```
1  randv<int> x,y;
2  Generator gen;
3
4  gen( x() < y() );
5  gen( x() > 100 || y() < −100);
6
7  if (gen.next()) run_test(x,y);
```

This example declares two variables x, y (line 1) and a constrained random generator (gen, line 2). The generator can simply be called to add new constraints: the relation of x and y (line 4) is specified and that either x has to be larger than 100 or y is less that −100 (line 5) is constrained. To generate values, the next function can be called which returns false if the generator is over-constrained, i.e. the constraints are contradictory and hence no solution exists. When the constraint-solver has generated a stimulus, the *randv<int>* variables can directly be used, e.g. in run_tests.

A generator can use all features of CRAVE except for inheritance. However, incremental constraint specification is supported. This feature is very helpful in dynamic testbenches. After the generator has been executed for a certain set of constraints, new constraints can freely be added e.g. to generate more general values first and more specific ones later. We exemplify this in Figure 5.

This example first uses a general constraint (lines 4-8). Then, additional constraints are added to focus on specific behavior (lines 10-14 and 16-20). Please note in the final loop at line 20 all three constraints are respected by the constraint-solver.

B. Debugging Constraint Contradictions

For large constraint sets it can easily happen that the overall constraint contains contradiction(s). In this case the problem is over-constrained and hence the constraint-solver is unable to generate valid stimuli. Debugging the contradiction manually is very time-consuming. Therefore CRAVE can automatically identify which named constraints are part of a conflict.

As discussed earlier the soft constraint in Figure 1 (line 8) and the constraint in Figure 2 (line 5) form a conflict. If the former constraint would have been declared as a hard constraint, no constraint solution exists. For such situations CRAVE provides an analyzer, that identifies the conflicting constraints and returns their names. The analysis includes each named constraint and checks them against all unnamed constraints, which are always enabled. In a first step the analyzer determines how many and which constraints need to be disabled to resolve all contradictions. In subsequent steps each of these constraints is expanded to determine the "complete" contradiction. For the example, the first step would determine that one contradiction exists and identify the first constraint. In the next step the second constraint is identified. This is completely done on a formal level, therefore the algorithm is complete and will return all minimal subsets of the constraints that form a conflict. For the described example the run-time of the analysis was negligible.

VI. PARALLEL CONSTRAINT-SOLVING

Various alternatives to BDD-based constraint-solving have been studied, see e.g. [22]. Approaches based on *Boolean Satisfiability* (SAT) [23], [24] or *Satisfiability Modulo Theories* (SMT) [20] have shown to give very good results for constraints which are hard to solve for BDDs. However, in general it is not possible to know in advance which type of constraint-solver will show the best performance. Therefore, CRAVE uses a portfolio approach. Instead of running a specific constraint-solver, an SMT-based constraint-solver as well as a BDD-based constraint-solver are executed in parallel for the same set of constraints. This section describes the basics for constraint solving and how the portfolio approach works. The next section will compares the run-times of CRAVE and the SCV library.

To integrate the most recent reasoning engines we use *metaSMT* [25] for implementing the constraint-solving in CRAVE. Essentially, *metaSMT* allows engine independent programming by providing a unified interface to different solvers. Hence, no algorithmic changes are necessary when switching to another solver. The overall architecture is depicted in Figure 6. As can be seen CRAVE forms the top layer which implements all the features described in the previous sections. This layer connects to the *metaSMT* front-end layer using the unified input language. In the middle-end layer the transformations for optimization and the basic parallelization features (e.g. threading of different engines) is available. Finally, the backend gives access to a wide range of solvers (see e.g. SWORD [26], Z3 [27], Boolector [28], MiniSAT [29], PicoSAT [30], CUDD [31] and AIGER [32]).

Based on these constraint-solving techniques we have made the following observations for parallelization. In a simple portfolio approach each constraint would be evaluated (at least) twice using different solvers in a multi-threaded environment.

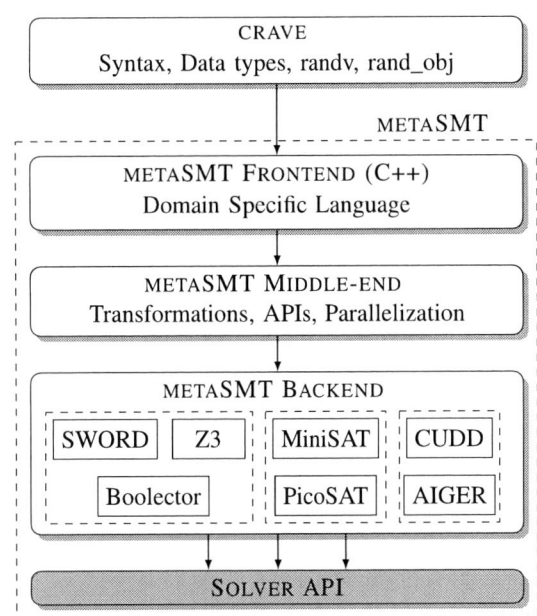

Figure 6: Constraint-Solving Architecture

Then, the result of the fastest solver would be used. However, for CRV a predictable quality of the stimuli is required. Therefore, if a BDD can be built for the overall constraint this is preferred since a uniform distribution among all solutions can be guaranteed[2]. In CRAVE this is reflected by running the SMT-solver only until the BDD is build. From this point on only the BDD is used and execution the SMT solver is stopped.

VII. EXPERIMENTAL EVALUATION

In the following examples we demonstrate the advantages of CRAVE for stimuli generation.

A. Arithmetic Constraints

Figure 7 shows a constraint object for a 16 bit ALU (later we scale the size of the ALU). The constraint specifies four operations with their respective input ranges. Table III shows the differences between the classical BDD constraints solver of the SCV and the portfolio approach of CRAVE. The first column gives the name of the library. Two rows are given for both: The first row shows the run-time needed to generate the first solution, and the second row shows the run-time in seconds for the complete execution of the constraint generator, respectively. The following columns provide the data for different bit width of the ALU constraints. As can be seen with increasing bit width of the ALU the SCV fails to solve the constraints. In contrast, in CRAVE the SMT-solver can already generate stimuli before the BDD is ready. Furthermore, note that for ALU16 the 32 bit memory restriction of the SCV

---

[2]A BDD represents all solutions and hence to select each solution with the same probability is simple. Essentially each path to the 1-terminal needs to be weighted accordingly respecting the reduction rules of reduced ordered BDDs.

```
1   struct ALU16 : public rand_obj {
2     randv< sc_bv<2> > op ;
3     randv< sc_uint<16> > a, b ;
4
5     ALU16() : op(this), a(this), b(this) {
6       constraint(IF_THEN(op() == 0,
7         65535 >= a() + b()) );
8       constraint(IF_THEN(op() == 1,
9         65535 >= a() − b())
10        && b() <= a()) );
11      constraint(IF_THEN(op() == 2,
12        65535 >= a() ∗ b()) );
13      constraint(IF_THEN(op() == 3,
14        b() != 0));
15    }
16  };
```

Figure 7: 16 bit ALU constraint

Table III: Comparison of CRAVE and the SCV

		ALU4	ALU12	ALU16	ALU24	ALU32
SCV	first	< 0.01	13.77	MO	TO	TO
	finished	0.09	19.84	MO	TO	TO
CRAVE	first	< 0.01	< 0.01	0.01	0.01	0.01
	finished	0.14	0.30	0.37	0.40	0.49

TO = time out, MO = memory out, run-time in seconds

library was hit. CRAVE can also be build on 64 bit architectures which was however not required for theses experiments.

B. Sudoku Constraints

In the second experiment we formulated the rules of the Sudoku puzzle in CRAVE and the SCV library. Figure 8 gives the respective constraints. In line 4 the 81 random variables of 4 bit size each are declared as the two-dimensional array *res_sdk*. Then, in line 6 the standard C++ two-dimensional array *given_sudoku* is declared which is filled when reading the Sudoku numbers from a file. From line 9 on five types of constraints follow. At first, the numbers of the *given_sudoku* array are assigned to the constraint variables (line 13). Note that we use here the reference feature of CRAVE. Hence, if a solution for the Sudoku-constraint is requested (by calling next), then the current values of *given_sudoku* are used as values for the constraint variables. In other words the standard C++ array *given_sudoku* can be changed anywhere and the actual values will be used in the constraint automatically. Next, the "obvious" constraints stating a valid solution range for each field (line 15), difference of rows and columns (line 20 and 26) are formulated. Finally, from line 32 on the difference per region is modeled.

For 15 puzzles the constraint-solver had to find a solution (between 16-32 numbers are set in the puzzle; 1 was unsolvable). The run-time (in seconds) and the memory consumption (in MB) were measured with a limit of 2 CPU hours or 4 GB of memory for each Sudoku instance. The evaluation showed largely homogeneous result, hence we only present the

```
1   class sudoku : public rand_obj {
2   public:
3     // variable to store solved sudoku
4     randv< sc_dt::sc_uint<4> > res_sdk[9][9];
5     // variable to hold given sudoku
6     int given_sudoku[9][9];
7
8     sudoku(rand_obj∗ parent = 0) : rand_obj(parent) {
9       // constrain given numbers
10      for (int i = 0; i < 9; i++)
11        for (int j = 0; j < 9; j++)
12          constraint( IF_THEN( reference(given_sudoku[i][j]) != 0,
13            res_sdk[i][j]() == reference(given_sudoku[i][j]) ) );
14
15      // only numbers from 1 to 9 are allowed
16      for (int i = 0; i < 9; i++)
17        for (int j = 0; j < 9; j++)
18          constraint((res_sdk[i][j]() >= 1) && (res_sdk[i][j]() <= 9));
19
20      // every number must appear exactly one time in one row
21      for (int i = 0; i < 9; i++)
22        for (int j = 0; j < 9; j++)
23          for (uint k = j + 1; k < 9; k++)
24            constraint( res_sdk[i][j]() != res_sdk[i][k]() );
25
26      // every number must appear exactly one time in one column
27      for (int j = 0; j < 9; j++)
28        for (int i = 0; i < 9; i++)
29          for (int k = i + 1; k < 9; k++)
30            constraint( res_sdk[i][j]() != res_sdk[k][j]() );
31
32      // every number must appear exactly one time in one region
33      for (int i = 0; i < 9; i++)
34        for (int j = 0; j < 9; j++)
35          constraint
36            ( res_sdk[i][j]() != res_sdk[ index(i,1) ][ j ]() )
37            ( res_sdk[i][j]() != res_sdk[ index(i,2) ][ j ]() )
38
39            ( res_sdk[i][j]() != res_sdk[ i ][ index(j,1) ]() )
40            ( res_sdk[i][j]() != res_sdk[ index(i,1) ][ index(j,1) ]() )
41            ( res_sdk[i][j]() != res_sdk[ index(i,2) ][ index(j,1) ]() )
42
43            ( res_sdk[i][j]() != res_sdk[ i ][ index(j,2) ]() )
44            ( res_sdk[i][j]() != res_sdk[ index(i,1) ][ index(j,2) ]() )
45            ( res_sdk[i][j]() != res_sdk[ index(i,2) ][ index(j,2) ]() );
46    }
47  };
48
49  int index(int x, int by ) {
50    return (x + by) % 3 + x − (x % 3);
51  }
```

Figure 8: Sudoku Constraints

Table IV: Comparison of CRAVE and the SCV

		min	max	median
SCV	time	2 636.60	4 529.07	3 165.34
	mem	MO	MO	MO
CRAVE	time	0.81	1.83	1.42
	mem	14.00	15.40	14.40

MO = memory out, run-time in seconds

aggregated results in Table IV. Although plenty of memory was provided, the SCV library could not solve a single instance[3]. In contrast, CRAVE solved all instances very fast.

[3] Note these instances are not inherently hard for BDD-based solvers. A dedicated BDD-based Sudoku solving algorithm was able to find the solution for each instance in less than a minute using no more than 216 MB of memory.

```
1   struct bubble_sort_input : public rand_obj {
2     // start address in mem
3     randv< unsigned int > start;
4     rand_vec< unsigned int > data;
5
6     bubble_sort_input() : start(this), data(this) {
7       constraint(0x70 <= start() && start() < 1024);
8       constraint(start() % 4 == 0);
9
10      constraint( 0 < data().size()
11                    && data().size() < 1024);
12      constraint(start() + 4 * data().size() <= 1024);
13
14      constraint.foreach(
15        data, _i, data()[_i] <= 0x00FFFFFF );
16      constraint.foreach(
17        data, _i, data()[_i] <= data()[_i−1] + 5);
18    }
19  };
```

Figure 9: Input data constraints for bubble sort

C. Program Input Generation for CPU Testbench

We also apply CRAVE to verify a CISC CPU with 8 registers of 32 bit data width each. The CPU implements a subset of the instructions of the IA-32 architecture including load/store, arithmetic, jump and halt instructions [33]. The CPU is available at three different levels of abstraction: an Instruction Set Architecture (ISA) model in C++, a SystemC TLM model using OSCI TLM-2.0, and a SystemC RTL model implementing a five-stage pipeline [34], [35]. We use CRAVE to generate programs (i.e. instruction sequences) as well as their inputs, which can be used as stimuli for all three models. Then, the simulation-based equivalence checking approach in [35] for models at different levels of abstraction is applied. We describe only one verification scenario: for an instruction sequence implementing the bubble sort algorithm, we randomize its input under the constraints shown in Figure 9. The first four constraints ensure that the array to be sorted fits into the CPU memory and does not collide with the loaded program. The last constraint forces the array to be nearly non-increasing (and thus challenging for bubble sort). Such a concise set of constraints would have not been possible with the SCV library due to the lack of support for dynamic data structures. The average time for CRAVE to generate the first 1000 arrays is approximately 90s (0.09s per array).

VIII. CONCLUSIONS

In this paper we have presented the advanced constrained random verification environment CRAVE. After the introduction of the API for constraint specification we have shown the advantages of CRAVE in comparison to the existing SCV library. The advantages include dynamic constraint specification and management, enhanced usability and much faster constraint-solving based on a portfolio approach. All these aspects have been demonstrated by means of examples. In summary, CRAVE improves the verification productivity for SystemC models significantly.

A possible direction for future work is to extend the support for dynamic data structures and to improve the distribution of the SAT/SMT generated stimuli.

REFERENCES

[1] B. Bailey, G. Martin, and A. Piziali, *ESL Design and Verification: A Prescription for Electronic System Level Methodology.* Morgan Kaufmann/Elsevier, 2007.
[2] Accellera Systems Initiative, "SystemC," 2012, available at http://www.systemc.org.
[3] T. Grötker, S. Liao, G. Martin, and S. Swan, *System Design with SystemC.* Kluwer Academic Publishers, 2002.
[4] D. C. Black and J. Donovan, *SystemC: From the Ground Up.* Springer-Verlag New York, Inc., 2005.
[5] D. Große and R. Drechsler, *Quality-Driven SystemC Design.* Springer, 2010.
[6] *IEEE Standard SystemC Language Reference Manual,* IEEE Std. 1666, 2005.
[7] L. Cai and D. Gajski, "Transaction level modeling: an overview," in *CODES+ISSS,* 2003, pp. 19–24.
[8] F. Ghenassia, *Transaction-Level Modeling with SystemC: TLM Concepts and Applications for Embedded Systems.* Springer, 2006.
[9] H. Foster, A. Krolnik, and D. Lacey, *Assertion-Based Design.* Kluwer Academic Publishers, 2003.
[10] S. Tasiran and K. Keutzer, "Coverage metrics for functional validation of hardware designs," *IEEE Design and Test of Computers,* vol. 18, no. 4, pp. 36–45, 2001.
[11] J. Bergeron, *Writing Testbenches Using SystemVerilog.* Springer, 2006.
[12] J. Yuan, C. Pixley, and A. Aziz, *Constraint-based Verification.* Springer, 2006.
[13] *SystemC Verification Standard Specification Version 1.0e,* SystemC Verification Working Group, http://www.systemc.org, 2003.
[14] J. Rose and S. Swan, *SCV Randomization Version 1.0,* 2003.
[15] C. N. Ip and S. Swan, "A tutorial introduction on the new SystemC verification standard," www.systemc.org, White Paper, 2003.
[16] R. E. Bryant, "Graph-based algorithms for Boolean function manipulation," *IEEE Trans. on Comp.,* vol. 35, no. 8, pp. 677–691, 1986.
[17] M. F. S. Oliveira, C. Kuznik, W. Mueller, F. Haedicke, H. M. Le, D. Große, R. Drechsler, W. Ecker, and V. Esen, "The system verification methodology for advanced TLM verification," in *CODES+ISSS,* 2012.
[18] D. Große, R. Ebendt, and R. Drechsler, "Improvements for constraint solving in the SystemC verification library," in *ACM Great Lakes Symposium on VLSI,* 2007, pp. 493–496.
[19] D. Große, R. Wille, R. Siegmund, and R. Drechsler, "Contradiction analysis for constraint-based random simulation," in *FDL,* 2008, pp. 130–135.
[20] R. Wille, D. Große, F. Haedicke, and R. Drechsler, "SMT-based stimuli generation in the SystemC verification library," in *FDL,* 2009, pp. 1–6.
[21] Accellera Systems Initiative - Universal Verification Methodology 1.1, http://www.accellera.org, 2011.
[22] N. Kitchen and A. Kuehlmann, "Stimulus generation for constrainted random simulation," in *Int'l Conf. on CAD,* 2007, pp. 258–265.
[23] S. M. Plaza, I. L. Markov, and V. Bertacco, "Random stimulus generation using entropy and XOR constraints," in *DATE,* 2008, pp. 664–669.
[24] H. Kim, H. Jin, K. Ravi, P. Spacek, J. Pierce, B. Kurshan, and F. Somenzi, "Application of formal word-level analysis to constrained random simulation," in *CAV,* 2008.
[25] F. Haedicke, S. Frehse, G. Fey, D. Große, and R. Drechsler, "metaSMT: Focus on your application not on solver integration," in *DIFTS'11: 1st International workshop on design and implementation of formal tools and systems,* 2011, pp. 22–29.
[26] R. Wille, G. Fey, D. Große, S. Eggersglüß, and R. Drechsler, "Sword: A SAT like prover using word level information," in *VLSI of System-on-Chip,* 2007, pp. 88–93.
[27] L. de Moura and N. Bjørner. "Z3: An efficient SMT solver," in *TACAS,* 2008, pp. 337–340.
[28] R. Brummayer and A. Biere, "Boolector: An efficient SMT solver for bit-vectors and arrays," in *TACAS,* 2009, pp. 174–177.
[29] N. Eén and N. Sörensson, "An extensible SAT-solver," in *SAT,* 2003, pp. 502–518.
[30] A. Biere, "Picosat essentials," *JSAT,* vol. 4, no. 2-4, pp. 75–97, 2008.
[31] F. Somenzi, *CUDD: CU Decision Diagram Package Release 2.4.1.* University of Colorado at Boulder, 2009.
[32] "Aiger," http://fmv.jku.at/aiger/.
[33] *IA-32 Architecture Software Developer's Manual,* Intel Corporation, 2003.
[34] A. Biere, D. Kroening, G. Weissenbacher, and C. Wintersteiger, *Digitaltechnik - eine praxisnahe Einführung.* Springer, 2008.
[35] D. Große, M. Groß, U. Kühne, and R. Drechsler, "Simulation-based equivalence checking between SystemC models at different levels of abstraction," in *ACM Great Lakes Symposium on VLSI,* 2011, pp. 223–228.

Asynchronous Parallel MPSoC Simulation on the Single-chip Cloud Computer

Christoph Roth, Simon Reder, Gökhan Erdogan, Oliver Sander, Gabriel M. Almeida, Harald Bucher, Jürgen Becker

Karlsruhe Institute of Technology (KIT)
Institute for Information Processing Technology (ITIV)
Karlsruhe, Germany
Email: christoph.roth, oliver.sander, gabriel.almeida, harald.bucher, becker@kit.edu

Abstract—The growing complexity of embedded applications currently causes a trend towards multi-core processors in the embedded domain. Time-consuming detailed simulations make the design of such systems increasingly sophisticated. In this work, applicability of Parallel Discrete Event Simulation (PDES) in the context of cycle-accurate Multi-Processor System-on-Chip (MPSoC) simulation is investigated on the Single-chip Cloud Computer (SCC) from Intel. The presented strategy targets asynchronous parallel model execution where only adjacent model partitions need to synchronize with each other in order to advance in simulation time. Performance of the approach is evaluated by means of a scalable cycle-accurate MPSoC model called HeMPS. For a 8x8 RTL model measurements reveal a speedup versus sequential RTL simulation of 25.3x. When exchanging RTL processing elements by cycle-accurate simulators a speedup of 56.3x versus sequential RTL simulation is obtained. These results promise good suitability of the asynchronous strategy for detailed parallel MPSoC simulation on an architecture like the SCC.

I. INTRODUCTION

Detailed system simulation is traditionally an application area that requires high computing performance. When looking at the embedded domain, architectures more and more tend towards Multi-Processor System-on-Chips (MPSoCs) which integrate evermore functionality directly into a single device. Thereby, cycle accuracy during simulation is needed to support debugging and verification by allowing to trace back errors of parallel programs. This combination strongly intensifies the demand for high performance.

The Single-chip Cloud Computer (SCC) has been developed by Intel Labs in order to provide a many-core software research platform [5]. It consists of 48 P54C cores and a network-on-chip based on fast non-cache coherent shared memory. The SCC implements a traditional message-passing cluster programming model on-chip. It allows data sharing among cooperating cores by using software maintained memory consistency. Dependence on hardware maintained cache coherency is removed. The main benefit is better scalability, helping to increase the number of cores used [13].

In this contribution a new SystemC-based runtime environment is presented that allows efficient parallel simulation of tightly-coupled MPSoC models on the SCC. The approach assumes a simulation methodology in which system models are divided into coarse-grained partitions with each being

executed on a single SCC core. Therewith, MPSoC models can be distributed and arranged on the SCC array in a natural way. The runtime environment is implemented by means of a framework targeting investigation of parallel SystemC simulation on the SCC which has first been introduced in [23]. The chosen synchronization strategy makes sure that each partition only synchronizes with its direct neighbours in a *decentralized* manner. It does not depend on global synchronization or a centralized software architecture which are both deciding reasons for limited scalability. Additionally, synchronization is relaxed from delta-cycle level to the level of timed events. In order to avoid loss of accuracy timing characteristics of a model's underlying communication infrastructure are exploited.

The synchronization strategy can be classified as *asynchronous* algorithm [12]. It is orthogonal to most of the previous works in parallel MPSoC simulation where *synchronous* algorithms that demand for global synchronization are applied (e.g. [22] [21] [25] [9]). There exist also several *asynchronous* approaches (e.g. [15] [10]). However, they either rely on a centralized implementation, assume models of higher levels of abstraction or focus on execution on distributed hosts. Instead, the proposed asynchronous protocol is tailored for an architecture like the SCC. Its application on the SCC implicates a reduction of the *number of messages* as well as communication *latency* thanks to the fast on-chip non-cache coherent shared memory. Effectiveness is demonstrated by means of a realistic cycle-accurate MPSoC model called HeMPS [6]. Results reveal good scalability and speedups in case of suitable partitionings and model sizes.

The remainder of this paper is organized as follows: In section II a selection of related work is summarized. Fundamentals, including a description of the SCC platform, the sequential OSCI SystemC kernel as well as a framework that extends the sequential SystemC kernel for parallel execution capabilities on the SCC are presented in section III. In sections IV and V the targeted simulation methodology as well as the asynchronous parallelization strategy are detailed. The simulation model that is used for performance analysis is described in section VI. Performance analysis results are presented in section VII. Finally, section VIII concludes and gives an overview on ongoing and future work.

978-1-4673-2895-1/12 $31.00 © 2012 IEEE

II. RELATED WORK

PDES denotes a technology that enables execution of discrete event simulations (DES) on parallel and/or distributed platforms. An integral part of PDES research deals with the so called *synchronization problem* which is the general problem of ensuring that events are processed in a time stamp order [12]. Various so called *conservative* and *optimistic* algorithms have been proposed for this issue in the literature. Conservative algorithms avoid violating the causality relationships between logical processes. They guarantee events to be executed in the correct time order [8] [16]. In contrast, optimistic approaches allow causality violations but provide for mechanisms to restore already past points in time [14] [17]. The literature additionally classifies between *synchronous* and *asynchronous* algorithms [12]. Synchronous algorithms regularly synchronize globally via barriers to advance in time and explicitly control when the entire simulation executes or suspends. Instead, asynchronous algorithms allow logical processes to advance in time without the necessity of global synchronization. Each logical process rather determines its possible time advancement based on the time advancement of other logical processes that may directly affect its own behaviour.

Existing approaches in the domain of parallel MPSoC simulation widely vary in applied PDES techniques and use cases. Examples targeting detailed microarchitectural analysis and debugging are [21] and [25]. The chosen synchronous and centralized simulator designs are specialized for full accurate model execution on shared-memory multi-core hosts and provide flexibility in terms of different thread scheduling policies. However, the centralized architecture together with the global barrier approach can lead to load imbalance and central-queue contention [7]. Instead of detailed microarchitectural analysis [22], [9] and [15] target virtual prototyping for software development and design space exploration. [22] targets parallel simulation of shared-memory machines. It uses a conservative synchronous algorithm based on global barriers that sacrifices simulation accuracy in order to increase parallelism. [9] is a centralized parallel simulator based on [4]. It uses global barriers but tries to alleviate the problem of global synchronization by introducing a simulation slack which allows timing violations but constrains them to remain within a specified window. [15] is an asynchronous parallel simulator that allows processor models to advance at independent rates. However, it uses a central backplane for establishing a global view. Finally [10] is a decentralized asynchronous parallel simulator that targets execution of un-timed and approximate-timed transaction level models on distributed hosts. [22], [21] and [9] are based on non-standardized libraries or tools. Instead, [25] and [15] are based on SystemC, [10] is based on SpecC. Thus [25], [15] and [10] facilitate reuse of already existing IP.

To the best of our knowledge, this work is the first one reporting about a decentralized asynchronous implementation of the SystemC kernel that targets parallel simulation of cycle-accurate MPSoC models on future many-core architectures

like the Single-chip Cloud Computer. The implementation is tailored for cases were no global cache coherent shared memory is available but a Network on Chip (NoC) architecture based on non-cache coherent shared memory for scalability reasons.

III. FUNDAMENTALS

A. Single-chip Cloud Computer Overview

The SCC is an experimental many-core CPU created by Intel Labs for the purpose of many-core software research. It consists of 24 tiles connected by a two-dimensional 6 by 4 mesh Network on Chip (NoC) with 256 GB/s bisection bandwidth (Fig. 1). Each tile contains two second generation P54C cores [11], 16 KB instruction and data L1 caches per core, a unified 256 KB L2 cache per core, a Mesh Interface Unit (MIU), two test-and-set registers and a 16 KB Message Passing Buffer (MPB). The MIU connects each tile to a router which integrates the tiles into the mesh NoC. The MIU packetizes data available within the MPB onto the NoC and de-packetizes data from the NoC to the MPB using a round-robin scheme to arbitrate between the two cores on the tile. The routers perform a fixed X,Y routing. The SCC has four DDR3 memory controllers that allow accessing external off-chip memory.

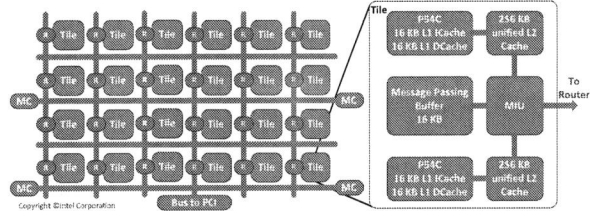

Fig. 1. SCC Architecture (source: [3])

The overall memory architecture of the SCC is composed of multiple distinct address spaces and supports both, distributed and shared memory programming models. Memory regions can be configured by the help of lookup tables (LUT) which translate the physical 32 bit address of the cores to a global system address. Thereby, the LUT configuration determines whether a physical address refers to off-chip DDR3 memory or to the on-die MPB memory. Beside that, memory regions can be configured as either private or shared [3][18].

Within this work, the default frequency settings are used (cores at 533 MHz, mesh and memory at 800 MHz). Off-chip DDR3 memory is exclusively used as private memory. Therewith, external memory corresponds to the main memory of a conventional PC. Each core has its own region. Memory access is performed via the DDR3 memory controllers, L2 and L1 caches. In contrast, the MPB space is configured as shared memory, enabling direct high-speed on-chip communication between cores in a message-based fashion. In this case, the SCC does not offer any hardware managed L2 cache coherency. Communication is performed by transferring data from private memory and L1 cache of the sending core via

978-1-4673-2895-1/12 $31.00 © 2012 IEEE

MPB and NoC to the L1 cache of the destination core. L1 caching is possible due to the featuring of a new memory type called Message Passing Buffer Type (MPBT) [18].

Basically, the off-chip DDR3 memory can be configured as shared memory too, which enables application of traditional shared memory programming models. However, due to the missing hardware managed cache coherency of shared off-chip memory, the application of programming models relying on this kind of configuration appeared to be comparatively unefficient in the context of tightly-coupled system simulation on the SCC [23].

Finally, on the software side each SCC core executes a modified Linux Kernel 2.6.16 [18]. For setting up and running parallel applications a special library called RCCE [19] is freely available. As explained later, only the administrational parts of RCCE are used. Instead, a new message-based communication pattern has been implemented within the MPB.

B. SystemC Basics

1) Sequential SystemC Kernel: In a DES like SystemC the behaviour of a system over time is modeled by processes generating discrete events which in turn can trigger the execution of the same or other processes and change the actual system state. The execution of the simulation is controlled by a simulation kernel that handles synchronization as well as process scheduling. During execution the SystemC kernel runs through different phases. Its functional principle is illustrated in Alg. 1.

Algorithm 1 SystemC Simulation Loop

1: elaboration and initialization
2: **while** timed notifications exist **do**
3:     **while** runnable processes exist **do**
4:         **while** runnable processes exist **do**
5:             select and execute a runnable process
6:             process immediate notifications
7:         **end while**
8:         process update requests
9:         **if** delta notifications exist **then**
10:             process delta notifications
11:         **end if**
12:     **end while**
13:     advance simulation time
14:     process timed notifications
15: **end while**

During *Elaboration and Initialization* the kernel is initialized. In the subsequent *Evaluation* phase, all runnable processes are executed sequentially. Processes possibly generate *immediate notifications* which cause other processes to become runnable again. If no more runnable processes exist, the kernel switches to the *Update* phase in which channel update requests are processed. These are generated when a new value is written into a channel. Channel updates possibly cause the generation of *delayed notifications* and effect a return to the *Evaluation*

phase for process execution. *Delayed notifications* prevent processes from recognizing new values immediately, but only after the next *delta cycle* [2]. If no *delayed notifications* exist simulation time advances to the next pending *timed notification* and sensitive processes are made runnable. Afterwards, the simulation loop starts over. Simulation ends if no more *timed notifications* exist.

2) Parallel SystemC Framework: The parallel SystemC framework for the SCC has been first introduced in [23]. Each SCC core taking part in simulation executes an instance of a runtime environment. Its basic structure is illustrated in Fig. 2. The foundation of the runtime environment is a hierarchical architecture consisting of three levels. Based on this, different manifestations of a parallel SystemC simulator requiring different communication and synchronization mechanisms can be built.

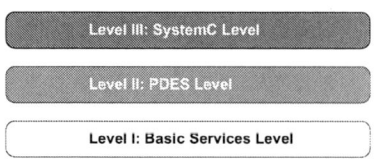

Fig. 2.   Framework Architecture

On level I (*Basic Services Level*), SCC specific routines and services for memory management and communication are made available to the upper layers. Level II (*PDES Level*) provides all fundamental functionality that is necessary to realize parallel discrete event simulation. This layer allows implementing different synchronization strategies using synchronous or asynchronous conservative time management based e.g. on a centralized or decentralized simulation architecture. Finally, on level III (*SystemC Level*) all SystemC kernel specific functionality is integrated by drawing back to the OSCI SystemC reference kernel [2]. The described framework is the basis of this contribution.

IV. Simulation Methodology

The targeted simulation methodology is illustrated in Fig. 3. It breaks down into two consecutive steps I) *Application to Simulation Model Mapping* and II) *Simulation Model to Execution Platform Mapping*. In the current and the subsequent sections the following sets are utilized in order to describe the simulation methodology as well as the functionality of the asynchronous simulation framework manifestation:

$$T = \{t \mid t = \textit{application task}\}$$
$$PE = \{pe \mid pe = \textit{processing element of the MPSoC model}\}$$
$$M = \{m \mid m = \textit{model partition}\}$$
$$A = \{a \mid a = \textit{atomic partition}\}$$
$$LP = \{lp \mid lp = \textit{logical process}\}$$
$$C = \{c \mid c = \textit{SCC core}\}$$

- *Step I: Application tasks t are mapped onto processing elements pe of the MPSoC simulation model.*

978-1-4673-2895-1/12 $31.00 © 2012 IEEE

- *Step II:* The simulation model is mapped onto the SCC. Therefore, it first needs to be clustered into distinct *model partitions* m. The granularity of the partitioning is defined by the granularity of an *atomic partition* a of a certain simulation model which is the smallest unit of work that can be mapped. E.g. the homogeneous MPSoC model described in section VI provides partitioning granularity on the level of architectural components. Each model partition $m_i \in M$ is a subset of the set of all existing atomic model partitions A. Assuming $|M| = o$ the following is true: $m_1 \cup ... \cup m_o = A$, $m_i \cap m_j = \{\}$ $\forall i, j : i \neq j$. Each of the resulting model partitions is assigned to a distinct *logical process* lp. Each of these is in turn assigned to a distinct *SCC core* c.

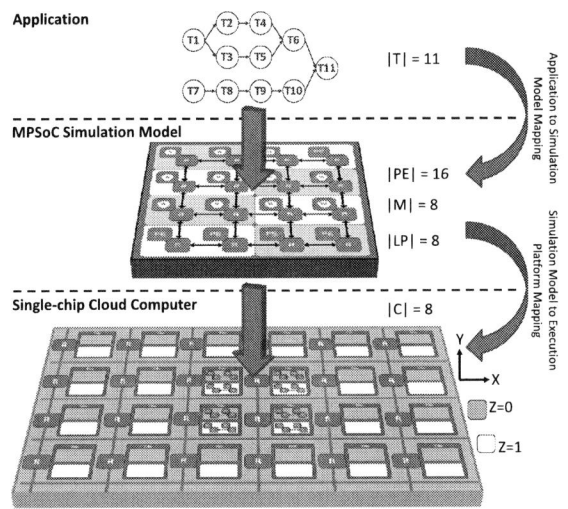

Fig. 3.   Simulation Methodology

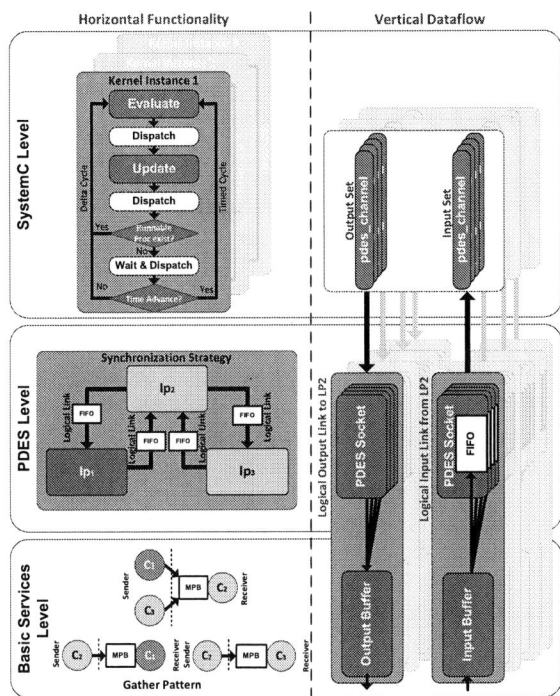

Fig. 4.   Asynchronous Framework Manifestation

V. ASYNCHRONOUS FRAMEWORK MANIFESTATION

In this section a decentralized asynchronous manifestation of the framework of section III-B2 is described. It provides a general infrastructure for executing logical processes on the SCC that were generated during step I and II of the simulation methodology. In the following, chosen design decisions and implementation aspects on all framework levels are presented and explained in detail.

A. Basic Services Level

The implementation of the MPB communication mechanism between all cores $c_{xyz} \in C$ is of particular importance for performance and applicability of the targeted synchronization method. By default, the RCCE library only provides a blocking communication mechanism where a send request does not complete until a matching receive request has been posted. This is not applicable for the targeted use case. Processors must basically be able to communicate asynchronously with each other and continue processing after having sent a message (see section V-B). Because of that, only the administrational

parts of RCCE are used for initialization and memory allocation. Based on this, a proprietary communication pattern has been developed which is adjusted to the asynchronous communication demands of the PDES Level and which completely decouples send and receive requests. The implementation corresponds to the *gather* communication pattern described in [24]. It basically allows true non-blocking FIFO-based communication between one or multiple senders and a receiver (in Fig. 4 core c_2 sends messages to c_1 and c_3 and vice versa, however, c_1 and c_3 do not exchange messages at all). The pattern exploits *receiver-based allocation* in combination with *receiver side placement* due to its performance benefits in combination with the gather pattern [24]. A circular input buffer primitive is used within the MPB of each receiver on the Basic Services Level in which incoming data from different senders can be stored. The input buffer has a size of several kilobytes (3kB in the current implementation). The size of a message is variable and depends on the data type that is transferred (see also section V-B). For a SystemC signal channel it is in the order of magnitude of several bytes. Because of that, several hundreds of messages can be buffered in the MPB concurrently without the need of calling a receive request by the receiver. From time to time (controlled by the PDES Level) the input buffer is flushed and the content is forwarded to the PDES Level. On the sender side each core is equipped with a FIFO-based output buffer in the private memory which allows buffering data before triggering transmission e.g. in the case of a full input buffer in the receiver's MPB space. Last but not least, a separate data field of 8 bytes is reserved in the MPB space of each core. The data field is necessary for storing the time stamp field (TSF) of the PDES Level that is used for the deadlock avoidance mechanism explained in section V-B. In contrast to the communication pattern previously described, *sender side placement* is applied here which allows data request by the receiver on demand.

B. PDES Level

The PDES Level implements a decentralized synchronization strategy between logical processes $lp_i \in LP$. It provides an interface to the kernel event queue of the SystemC Level for 1) dispatching messages and 2) controlled time advancement. By design, the implemented method only synchronizes on the level of timed events and not on the atomic time slice of a delta cycle. Thus, delta cycle events are generally exchanged asynchronously. However, as shown later, delta cycle level synchronization is not necessary for correct and full accurate parallel MPSoC simulation.

1) Logical Links: The decentralized strategy relies on the existence of static logical links between logical processes. Logical links transmit messages that contain time stamped data and respective events, i.e. SystemC channel updates. They guarantee data and event delivery in increasing time stamp order. Therefore, they strictly use the first-in first-out principle on the whole communication line between sender and receiver which is a necessary precondition for a decentral asynchronous algorithm to work correctly and to adhere causality relationships [12]. More formally, a logical link l_{ij} can be defined as the set of all SystemC channels that connect the model partition contained in lp_i to the one contained in lp_j. If there exists a logical link l_{ij} then lp_i is called *adjacent to* lp_j. Basically, logical links are unidirectional. In case of bidirectional SystemC channels two unidirectional logical links l_{ij} and l_{ji} of reverse orientation have to be established. In the example in Fig. 4 lp_1 and lp_3 are adjacent to lp_2 and vice versa. In contrast, lp_1 and lp_3 have no logical link connection. Hence, there must not exist a SystemC channel that interconnects the contained model partitions. Looking at their concrete implementation, logical links consist of an input and an output part which remotely connect SystemC channels via PDES sockets across all framework levels (see Fig. 4). The logical link identifies the correct input/output PDES socket for incoming/outgoing messages by unique channel IDs.

2) Message Dispatching: An outgoing message msg that is written from the SystemC Level into an output PDES socket is assigned with a time stamp $t(msg)$, directly dispatched to the Basic Services Level and hence remotely transmitted without being buffered on the PDES Level. $t(msg)$ corresponds to the current local simulation time of the sending logical process. In turn, incoming messages from the Basic Services Level are only forwarded to their corresponding input PDES sockets when initiated by the SystemC Level. Message data is then buffered within the input PDES socket's FIFO. The associated

event that effects a takeover of the buffered value to the SystemC Level is directly integrated into the event queue on the SystemC Level at simulation time $t(msg) + d$ with $t(msg)$ being the time stamp of the message and d being the delay of the channel. This is performed through the interface of the connected SystemC channel (see also section V-C). The existence of channel delays is necessary for the derivation of a value called *lookahead* of the logical link. The lookahead lo of a logical link is given by the minimum delay of all channels that are assigned to a logical link. The existence of lookahead information is fundamental for deadlock avoidance.

3) Time Advancement and Deadlock Avoidance: Beside message dispatching, the PDES Level is responsible for assuring correct time advancement while avoiding deadlocks. Therefore the PDES Level is equipped with a time stamp registry. Within this registry time stamps $t(msg_{last})$ of the messages last received on each incoming logical link are stored. If a time advancement is requested from the SystemC Level, a logical process lp_j compares the time stamp $t_j(event_{next})$ of the next timed event within the event queue on the SystemC Level to the time stamps stored within the registry. If $t_{ij}(msg_{last})$ of all logical links l_{ij} is equal or larger than $t_j(event_{next}) - lo_{ij}$ then the time advancement is granted. Otherwise the PDES Level blocks and waits for new messages until the condition is fulfilled. At this point, a deadlock can occur e.g. in case of adjacent logical processes lp_i and lp_j being connect by l_{ji} and l_{ij} and waiting on message delivery by each other. In such a case a logical process must determine a so called Lower Bound on the Time Stamps (LBTS) in order to be able to advance. Logical processes therefore have access to the time stamp field (TSF) in the MPB space of adjacent logical processes. The TSF contains the minimum time a future message can have. This is either the time stamp of the recently transmitted message or the current local simulation time if no message has been transmitted at the current time. The LBTS is given by the sum of the TSF value and the respective link lookahead. It is the lower time bound of all events that may be generated by an adjacent logical process in future and that may influence the future behaviour of the receiver. The TSF allows LBTS derivation on demand in case no messages are available. It avoids distribution of so called *null messages* through the logical links and reduces the overload probability.

C. SystemC Level

On this level an extended SystemC kernel instance is executed. SystemC modules simply can be labelled as model partitions $m_i \in M$ by means of a new template called *sc_partition<module>*. From the channel point of view the implementation of new types is needed in order to allow model partitions to communicate with each other remotely. Specific channels like signals or FIFOs need to inherit from the new abstract type *pdes_channel*. Each instance of a *pdes_channel* consists of an output and an input artifact on the sender and the receiver side. They are each assigned to a PDES sockets on the PDES Level. In the output case a *pdes_channel* forwards time stamped data to the PDES Level by directly accessing the PDES socket instance within its *write()* function without interacting with the local SystemC scheduler. In turn, data that is buffered within a PDES socket FIFO is announced to the kernel by triggering the *value_changed_event* of the appropriate *pdes_channel* instance.

From the scheduler point of view three additional phases are added to the original simulation loop of Alg. 1 (illustrated as dashed boxes on the left-hand side of the Basic Services Level in Fig. 4). In the two *dispatch* phases control is passed to the PDES Level in order to dispatch messages from Basic Services Level to SystemC Level as described in section V-B. If no more events are available within the event queue at the current simulation time, the kernel switches to the *wait & dispatch* phase and passes control to the PDES Level in order to request a time advancement. Control is returned to the SystemC Level after the next possible time advancement has been calculated.

VI. TARGET MPSOC SIMULATION MODEL

The targeted simulation model is the HeMPS system [6]. HeMPS is a homogeneous MPSoC that consists of a configurable number of processing elements $pe_i \in PE$ being interconnected by a NoC called HERMES [20] (see Fig. 5). HeMPS has a distributed memory architecture. Each processing element is assigned a dedicated NoC router $r_i \in R$. Beside that, it has its own private memory and cannot read directly from the memory of another processing element. Data transfers are implemented using a simple message passing protocol. The processing elements are based on the MIPS processor Plasma [1] which executes a multi-threaded RTOS. Applications running on HeMPS are modeled using task graphs. The NoC and its routers are RTL models whereas there exists a RTL model as well as a cycle accurate simulator (CAS) for the processing elements.

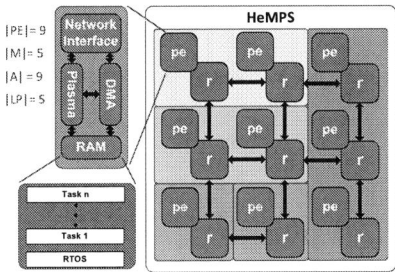

Fig. 5. Simulation Model

Step I of the simulation methodology of section IV is directly supported by the model generator tool of HeMPS. Task mapping can be performed manually. For step II, the generator tool has been extended with an option for model partitioning and generation of logical processes. A processing element with its associated router forms an atomic partition a. These can be grouped into model partitions m of arbitrary size by labelling corresponding modules using the *sc_partition<module>* template and separating model partitions through *pdes_channel*

based signals. Logical processes are assigned to SCC cores manually.

As illustrated in Fig. 6, combinational signal paths (indicated by C) between routers and processing elements of HeMPS are separated by SystemC processes that are synchronous to positive and negative clock edges (indicated by FF). This clock synchronicity is typical for such homogeneous MPSoC systems. It limits the influence of signal changes between the atomic partitions to points in time that have a minimum distance of half a clock cycle t_c. Therefore, when cutting the model through a combinational path as illustrated in Fig. 6 the clock synchronicity results in a maximum possible link lookahead of $lo < \frac{t_c}{2}$. In order to avoid the time creep problem [12] the lookahead should be set to the largest possible value.

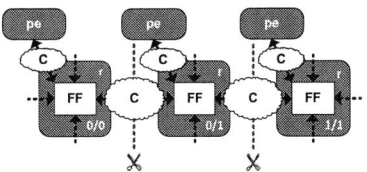

Fig. 6. HeMPS Lookahead Derivation

Synchronization on the level of timed events effects the timing of any event that is exchanged between distributed model partitions to be always relaxed by lo, even if it is a delta event. Since lo is chosen smaller than half a clock cycle remotely exchanged events are always received and executed just in time before the next clock event. However, each model partition needs to be equipped with a separate clock generation process since distribution and relaxing of clock signals would result in incorrect model behaviour.

VII. EXPERIMENTAL EVALUATION

In the following, applicability of asynchronous parallel MPSoC simulation on the SCC is demonstrated and performance is evaluated. Therefore, the parallel execution runtime was measured when varying different parameters. The first experiment focusses on the speedup versus sequential execution without explicitly paying attention to the relation of task mapping to model mapping. The second experiment explicitly focusses on the relation between task mapping and model mapping and its influence on execution performance.

A. Evaluation of Scalability

In the first experiment performance was measured when running different HeMPS models of different size on different numbers of SCC cores. On top of the models different numbers of a five stage MPEG decoder pipeline were executed. They were mapped in rows one after another onto the model. Starting from a specified number of model partitions $|M|$, atomic partitions were equally distributed across the available model partitions as far as possible. This resulted in the number of atomic partitions per model partition to differ at most by one. Furthermore, only neighbouring atomic partitions were grouped together. The model partitions were then assigned to logical processes respectively SCC cores starting from the lower left SCC core c_{000} in Fig. 3. If the number of model partitions was smaller than the current number of SCC cores then the redundant SCC cores were left idle. The simulation was executed for 10^6 cycles at a simulated clock frequency of 100 MHz. This resulted in $\frac{t_c}{2} = 5ns$ and $lo = 4.9ns$. Measured performance characteristics are shown in Fig. 7 and Fig. 8.

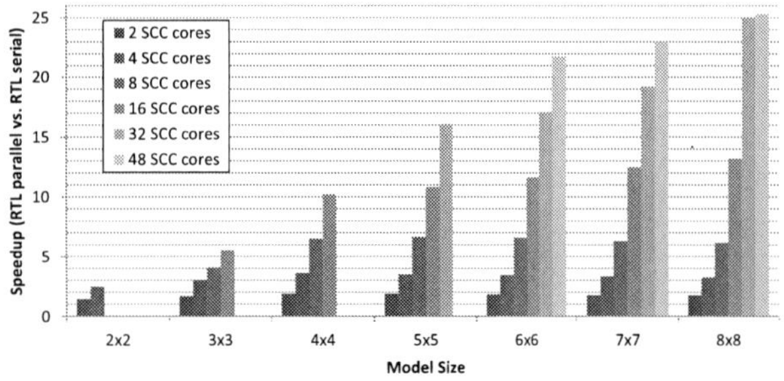

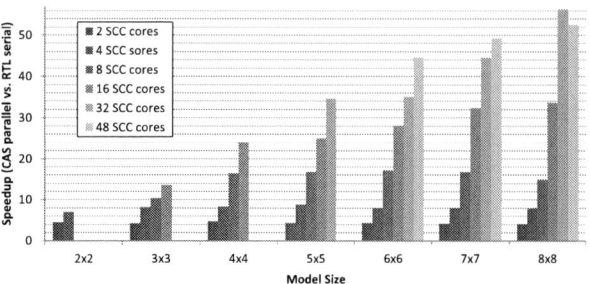

Fig. 7. Speedup RTL parallel vs. RTL serial

In Fig. 7 the speedup is illustrated that is achieved when comparing parallel execution on several SCC cores to sequential execution on a single SCC core and using the RTL model for the processing elements. The NoC size ranges from 2x2 up to 8x8. As can be observed, the speedup is always larger than one. The minimum achieved speedup is about 1.5x for a 2x2 model being executed on two SCC cores. It increases to 2.5x on four SCC cores. The speedup of small models like the 2x2 NoC is generally limited for two reasons: First, the coarse grain partitioning constrains the number of partitions that can be generated from a model (for a 2x2 NoC this number is limited to four). Second, assuming a constant number of partitions, large scale models provide a better ratio between computation and communication. E.g. when simulating a 4x4 NoC instead of a 2x2 NoC and using four partitions the speedup increases from 2.5x to 3.6x.

The largest speedup that is achieved is 25.3x in case of a 8x8 NoC being distributed across 48 SCC cores. However, this is nearly the same speedup as the one that was measured when using only 32 SCC cores for the same model size. One reason is, that the number of SCC cores only differs by a factor of 1.5x instead of two since the limit of available SCC cores is reached. Beside that, 32 SCC cores and a 8x8 model results in an ideal partitioning with each of the 32 SCC cores running two atomic partitions. Instead, the partitioning in case of 48 SCC cores is imbalanced: 32 SCC cores executed a single atomic partition whereas 16 still executed a set of two atomic partitions instead of only one.

When exchanging the RTL model of the processing elements with the CAS-based model, speedups versus serial RTL-based excution could further be increased by a factor of two to three see Fig. 7). The largest measured speedup was 56.3x in case of a 8x8 model and 32 cores. An interesting observation is that in the mentioned case the speedup is even larger than

in case of 48 SCC cores. This can be traced back to the same reasons as in the pure RTL comparison before. In the CAS case the effect is stronger due to the lower computation to communication ratio that emerged when using a more abstract model for the processing elements.

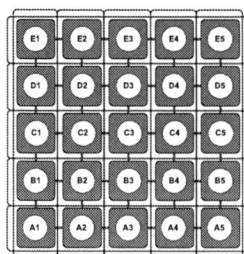

Fig. 8. Speedup CAS parallel vs. RTL serial

B. Evaluation of Task and Model Mapping

Now, the mutual influence of task mapping and model mapping is considered. Therefore, a five task dummy application has been created. Tasks of the application form a pipeline in which each task only communicates with its successor. The tasks execute an endless loop that generates as much random traffic as possible. Five instances of this application were mapped horizontally onto the rows of the HeMPS array as shown in Fig. 9. Afterwards, the model was first partitioned horizontally (identical to the orientation of the pipelines) and then vertically (contrary to the orientation of the pipelines). In the first case, communication between tasks happens exclusively inside model partitions and hence inside logical processes, due to the X,Y routing of HeMPS. In the second case, all communication between tasks always crosses logical processes. The resulting five logical processes were mapped onto the SCC side by side. The simulation was executed for 10^7 cycles.

Fig. 9. Task and Model Mapping

Tab. I shows the runtimes of different cases A to D relative to case A. They are made up of different kinds of partitionings in combination with different PE models. Switching from CAS to RTL processing elements more than doubles the runtime. However, when comparing horizontal with vertical partitioning for CAS and RTL based simulations seperately, the orientation of the partitioning does not have a great impact on performance. The measured differences in runtime are not significant for the evaluated MPSoC model (below five percent). In both cases CAS and RTL, the computational overhead obviously exceeds the communicational overhead.

TABLE I
INFLUENCE OF TASK AND MODEL MAPPING

case	Model Partitioning	PE Model	Relative Runtime
A	horizontal	CAS	1
B	vertical	CAS	1.01
C	horizontal	RTL	2.42
D	vertical	RTL	2.51

VIII. CONCLUSION AND OUTLOOK

Within this work an asynchronous strategy for parallel SystemC-based MPSoC simulation on the Single-chip Cloud Computer has been presented. In contrast to most of the previous works the strategy does not rely on global synchronization. Instead, synchronization is only done on the level of timed events which (as we believe) is sufficient for parallel execution of many MPSoC models. Application of the strategy on the SCC implicates a reduction of the number of messages in combination with a reduction of latency. Evaluations have been performed by means of a cycle-accurate MPSoC called HeMPS and revealed good scalability and large speedups. Beside that, it has been shown that the mutual impact between task mapping and model partitioning on performance is low for the used MPSoC model. Because of that, we conclude that the asynchronous method can efficiently exploit the explicit parallelism of cycle-accurate models on the SCC.

Future work comprises the development of extensions and optimizations on all framework levels. On the Basic Services Level other communication patterns will be investigated and compared. On PDES and SystemC Level the framework will be extended for concurrent support of different synchronization strategies. Based on comprehensive profiling capabilities that are currently developed an in-depth benchmarking of the presented method and a quantitative comparison to other approaches on all framework levels is planned. Beside that, different model partitioning strategies will be investigated with the goal of automization. This also includes the consideration of other simulation models described on higher levels of abstraction and the investigation of the influence on the computation to communication ratios and hence on model partitioning.

REFERENCES

[1] http://www.opencores.org.
[2] IEEE Standard System C Language Reference Manual. *IEEE Std 1666-2005)*, pages 1 –423, aug. 2006.
[3] SCC External Architecture Specification (EAS) Revision 1.1. 2010.
[4] T. Austin, E. Larson, and D. Ernst. SimpleScalar: an infrastructure for computer system modeling. *Computer*, 2002.
[5] M. Baron. The Single-Chip Cloud Computer. 2010.
[6] E. Carara, R. de Oliveira, N. Calazans, and F. Moraes. HeMPS - a framework for NoC-based MPSoC generation. In *Circuits and Systems, 2009. ISCAS 2009. IEEE International Symposium on*, 2009.

[7] R. D. Chamberlain. Parallel logic simulation of VLSI systems. In *ACM Computing Surveys*, pages 139–143, 1995.
[8] M. Chandy and J. Misra. Distributed Simulation: A Case Study in Design and Verification of distributed Programs. In *IEEE Transactions on Software Engineering SE-5, (5)*, pages 440–452, 1979.
[9] Chen, Jianwei and Annavaram, Murali and Dubois, Michel. SlackSim: a platform for parallel simulations of CMPs on CMPs. *SIGARCH Comput. Archit. News*, 2009.
[10] Chen, W. and Doemer, R. A Distributed Parallel Simulator for Transaction Level Models with Relaxed Timing. In *Center for Embedded Computer Systems, University of California, Technical Report*, 2011.
[11] T. D. Anderson. Pentium Processor System Architecture. *Addison Wesley*, 1995.
[12] R. M. Fujimoto. *Parallel and Distribution Simulation Systems*. John Wiley & Sons, Inc., New York, NY, USA, 1999.
[13] Howard, J. and Dighe, S. and Vangal, S.R. and Ruhl, G. and Borkar, N. and Jain, S. and Erraguntla, V. and Konow, M. and Riepen, M. and Gries, M. and Droege, G. and Lund-Larsen, T. and Steibl, S. and Borkar, S. and De, V.K. and Van Der Wijngaart, R. A 48-Core IA-32 Processor in 45 nm CMOS Using On-Die Message-Passing and DVFS for Performance and Power Scaling. *Solid-State Circuits, IEEE Journal of*, 2011.
[14] D. Jefferson, B. Beckman, F. Wieland, L. Blume, and M. Diloreto. Time warp operating system. In *SOSP '87*, pages 77–93, New York, NY, USA, 1987. ACM.
[15] Kim, Heekyung and Yun, Dukyoung and Ha, Soonhoi. Scalable and retargetable simulation techniquesfor multiprocessor systems. In *Proceedings of the 7th IEEE/ACM international conference on Hardware/software codesign and system synthesis*, CODES+ISSS '09, New York, NY, USA, 2009. ACM.
[16] B. D. Lubachevsky. Efficient distributed event-driven simulations of multiple-loop networks. *Commun. ACM*, January 1989.
[17] F. Mattern. Efficient algorithms for distributed snapshots and global virtual time approximation. *Journal of Parallel and Distributed Computing*, 18:423–434, 1993.
[18] T. Mattson, M. Riepen, T. Lehnig, P. Brett, W. Haas, P. Kennedy, J. Howard, S. Vangal, N. Borkar, G. Ruhl, and S. Dighe. The 48-core SCC Processor: the Programmer's View. In *Proceedings of the 2010 ACM/IEEE International Conference for High Performance Computing, Networking, Storage and Analysis*. IEEE Computer Society, 2010.
[19] T. Mattson and R. van der Wijngaart. RCCE: a Small Library for Many-Core Communication Software 1.0 - release. 2010.
[20] Moraes, Fernando and Calazans, Ney and Mello, Aline and Möller, Leandro and Ost, Luciano. HERMES: an infrastructure for low area overhead packet-switching networks on chip. *Integr. VLSI J.*, 38, 2004.
[21] Penry, D.A. and Fay, D. and Hodgdon, D. and Wells, R. and Schelle, G. and August, D.I. and Connors, D. Exploiting parallelism and structure to accelerate the simulation of chip multi-processors. In *High-Performance Computer Architecture, 2006. The Twelfth International Symposium on*, 2006.
[22] Reinhardt, Steven K. and Hill, Mark D. and Larus, James R. and Lebeck, Alvin R. and Lewis, James C. and Wood, David A. The Wisconsin Wind Tunnel: virtual prototyping of parallel computers. *SIGMETRICS Perform. Eval. Rev.*, 1993.
[23] C. Roth, O. Sander, M. Hübner, and J. Becker. A Framework for Exploration of Parallel SystemC Simulation on the Single-chip Cloud Computer. In *Proceedings of the 5th ICST Conference on Simulation Tools and Techniques*, 2012.
[24] R. Rotta. On Efficient Message Passing on the Intel SCC. In *Proceedings of the 3rd Many-core Applications Research Community (MARC) Symposium*. KIT Scientific Publishing, 2011.
[25] C. Schumacher, R. Leupers, D. Petras, and A. Hoffmann. parSC: synchronous parallel systemc simulation on multi-core host architectures. CODES/ISSS '10, pages 241–246, New York, NY, USA, 2010. ACM.

978-1-4673-2895-1/12 $31.00 © 2012 IEEE

Ultra-Low Latency NoC testing via Pseudo-Random Test Pattern Compaction

Herve' Tatenguem[†], Alessandro Strano[†], Vineeth Govind[§], Jaan Raik[§], Davide Bertozzi[†]

[†] ENDIF, University of Ferrara, 44122 Ferrara, Italy.

[§] Tallinn Institute of Technology, Estonia.

ttnhrv@unife.it[†], alessandro.strano@unife.it[†], vineeth@pld.ttu.ee[§], jaan@pld.ttu.ee[§], davide.bertozzi@unife.it[†]

Abstract— **This paper aims at devising an optimized pseudo-random test methodology for NoCs and its architectural support. The guiding principle consists of using a test pattern compaction engine for generating minimal test lengths. We show the application of this principle driven by the objective to minimize test application time, at the cost of test wrapper complexity. The achieved design point results in a reduction of test application time by two orders of magnitude with respect to state-of-the-art test architectures for NoCs exploiting pseudo-random patterns.**

I. INTRODUCTION

All systems-on-chip (SoCs) should be tested for manufacturing defects. Such testing procedure turns out to be particularly challenging for those large scale SoCs making use of a network-on-chip (NoC) as their communication backbone: the controllability/observability of NoC links and sub-blocks is relatively reduced since they are deeply embedded and spread across the chip. This issue adds up to the newer challenges of testing generic large digital designs in nanoscale technologies. For instance, pin-count limitations restrict the use of I/O pins dedicated for testing. Other concerns regard the use of external testers, which has been a mainstream testing practice so far: lack of scalability of test data volumes and high cost for full clock speed testing. Finally, wear-out mechanisms such as oxide breakdown, electro-migration and mechanical/thermal stress become more prominent in aggressively scaled technology nodes. These breakdown mechanisms occur over time, therefore the methodology and the infrastructure used for production testing should be designed for re-use during the system lifetime as well. This again urges new testing strategies.

Built-in Self-Testing (BIST) to some extent overcomes the above problems since test patterns are generated and evaluated on chip. By exploiting the precise knowledge of the architecture and of the circuits under test, a designer can come up with deterministic test patterns, thus potentially resulting into minimized test sequences and superior coverage results. Unfortunately, engineering such handcrafted deterministic patterns is a largely manual and time consuming task which should be performed again in case of technology library migrations or circuit modifications. Moreover, placement and routing of the design will certainly modify the gate-level netlist, thus making coverage expectations not fully trustworthy.

For synthesized logic, which is by far a relevant part of an embedded system, pseudo-random test patterns are frequently used because of their higher flexibility. They potentially result in a lightweight test architecture due to the simplicity of the linear feedback shift registers (LFSRs) and of the multiple input signature registers (MISRs) used to generate test patterns and signatures respectively. The work in [2] proves a 20% area saving in 45nm technology when a BIST architecture for NoCs is fed by pseudo-random patterns rather than by handcrafted deterministic ones.

Unfortunately, testing with pseudo-random patterns also typically dominates the total runtime of the BIST: in [12]

200000 cycles are reported for testing the main modules of a NoC switch with such patterns. A reduction of one order of magnitude in testing latency has been proved in [2], however authors there take a hybrid approach, where pseudo-random test patterns are combined with deterministic ones and with architecture-specific DfT optimizations.

In this paper, we view the flexibility and the reduced test latency requirements as fundamental for efficient NoC testing. On one hand, re-engineering deterministic patterns for each product evolution or technology migration is not cost-effective. On the other hand, overly long test application times are not compatible with the lifetime testing paradigm, where a testing procedure may be run at least at each system bootstrap.

As a result, the main objective of this paper is to develop an ultra low-latency testing framework for NoCs capable of achieving such unprecedented test application times (below 250 cycles regardless of the network size) without reverting to deterministic patterns. In contrast, we start from pseudo-random patterns and develop a testing methodology and its architectural support in the NoC that preserve the generic and flexible nature of the patterns. Such methodology and test architecture are therefore pretty general in scope and can be applied to any NoC architecture other than the one considered in this paper.

Test set compaction is at the core of our testing framework. In order to minimize the test application time we chose to compact test patterns for combinational logic blocks, where compaction efficiency is more likely to outperform that achievable for sequential circuits. Registers are tested with standard techniques. Combined with the concurrent testing of switch sub-blocks, this approach achieved an order of magnitude lower test application time than current literature.

The trade-off is clearly with the implementation complexity of the test wrapper, which needs to isolate the circuits that are concurrently tested and to connect them to their test pattern generators (TPGs) and response analyzers. This paper proposes also an optimization strategy to limit such overhead which consists of cascading several circuits under test (test responses of one block become test patterns for the next one) and of exploiting the synergies between them.

II. RELATED WORK

As the integration densities keep increasing, on-chip interconnection networks are becoming the reference communication backbone for multi-core computing in many embedded high performance systems [10], [7]. However, defects still continue to increase and are more prominent in scaled technology. To cope with this high defect rates, many test mechanisms have been proposed. For example, [15] and [19] propose a test mechanism for regular and modular systems like on-chip networks, but they incur a considerable area overhead. In the same way, [18], [16], [20] and [17] have proposed full-scan and boundary scan strategies, but they still have a high area

978-1-4673-2895-1/12 $31.00 © 2012 IEEE

overhead as showed in [14]. An alternative approach to reduce this overhead could be the partial scan technique, however its testing time is the main drawback (tens of thousands of clock cycles).

Another solution consists of applying test patterns from the input/output borders of the network [8]; this approach has been extended in [9] in such a way to support the diagnosis too. However, this approach is limited by the high number of necessary test pins.

[3] proposed a testing framework based on handcrafted deterministic test patterns and exploits the inherent structural redundancy of NoC switches. This deterministic approach leads to one of the fastest testing times reported in the literature for single stuck-at faults. However, this approach requires the in-depth knowledge of the architecture under test and an extensive effort to carefully engineer handcrafted test patterns for it. On the same NoC architecture of [3], the work in [2] implements a test architecture fed by pseudo-random patterns. The achieved design point cuts down on the area overhead by 20% in 45nm technology but provides comparable coverage in one order of magnitude more test application time. On a different switch architecture, [12] reports several tens of thousands of cycles for pseudo-random testing of most parts of the switch, confirming that the use of such patterns inherently plays against testing latency.

In this paper, on the same NoC architecture of [3] and [2] we provide a new design point, which achieves a high fault coverage with generic test methodology steps and architecture design techniques (i.e., no deterministic patterns). We rely on existing test set compaction tools (namely [4] and [5]) to cut down on testing latency, and elaborate on the test methodology and on the architecture support needed to achieve unprecedented values of test application times. The key challenge this paper deals with is to identify the most suitable circuits for test set compaction, so to maximize compaction efficiency and minimize test application time. At the same time, the implementation complexity of test wrappers, TPGs and response analyzers are kept under control.

III. TESTING METHODOLOGY

The philosophy behind this work is that pseudo-random test patterns are desirable for NoC testing since they can be easily reused and extended across architecture variants and technology migrations. In fact, they save the considerable effort and time to develop handcrafted deterministic patterns for the architecture under test.

On the other hand, we believe that some optimizations with respect to their naive application to NoCs are necessary to reduce test application time. This paper pushes this consideration to the limit and tackles the challenge of materializing an **ultra-low latency testing framework for NoCs starting from pseudo-random test patterns**. The applied optimizations then retain the generic and flexible nature of the testing methodology by never reverting to deterministic test patterns.

We found test set compaction a suitable architecture-agnostic step to optimize test patterns, where the specific architecture implementation comes into play only in determining the achieved compaction efficiency. In this paper we applied state-of-the-art compaction tools from the Turbo Tester [4] suite for handling test patterns for the modules of NoC switches. To note that compaction tools have more degrees of freedom for test set optimizations when they are fed by long test sequences that most likely stimulate the same error multiple times. Thus, the choice of pseudo-random test sequences is an ideal target for this case since these latter are

commonly longer than deterministic sequences generated by any ATPGs.

In order to obtain minimal test lengths, the following strategy was selected. **First, long pseudo-random test sequences were generated till the obtained coverage for single stuck-at faults was 99%. Then, an efficient static test set compaction tool (Optimize [4]) was run by requesting it to derive a compressed test set tracking the previously obtained coverage for each tested module. Finally, a verification step of the achieved coverage with the new test set was performed. The compacted vectors are then easily hardwired in hardware TPG.**

A key issue for our testing framework was to properly identify the circuit blocks the methodology should be applied to. The guiding principle in making this choice was to achieve the lowest possible testing latency for the NoC as a whole. This was pursued in two ways:

- Testing of NoC switches was engineered in such a way that all switches can be tested in parallel. The key enabler was to implement a cooperation mechanism between switches for testing their inter-switch links. In practice, each switch sends test patterns across its outgoing links and response analysis will be performed in the neighboring switches. The opposite holds for incoming links, which are analyzed locally. At the same time, switch internal blocks are tested in parallel, thus avoiding to create dependencies between testing phases and enabling the maximum testing parallelism.

- Test set compaction for combinational logic is intuitively simpler and potentially more effective than that for sequential circuits. There are a number of reasons for this. First, the compaction algorithm should deal with simple test vectors rather than with test sequences. Secondly, testing sequential logic of switch FSMs presents different requirements with respect to combinational logic testing. In fact sequential logic has fewer inputs than combinational logic but needs more clock cycles to be stimulated, thus requiring a suitable wrapper properly sequencing the outputs of the test generator.

  For these reasons, our goal of minimizing test application time motivated the choice for splitting each switch sub-block (essentially arbiters, buffers and crossbar) into their combinational logic and registers for the sake of testing. Then, registers were tested with well-known techniques (fundamentally, comparison of test responses or built-in scan-chains for self-testing), while the test set compaction methodology was applied to combinational logic.

The choice of isolating combinational logic for its efficient testing has a relevant impact on FSMs. In fact, their state registers are tested separately and their current state signals feeding the combinational logic should be made controllable to the TPG. In turn, the outputs and the next state signals from the combinational logic should be made observable to the response analyzer.

For testing the registers, we adopted the most convenient standard approach depending on the register type (state registers, configuration registers and data registers).

The above design decisions paved the way for a test architecture for NoCs aiming at competitive coverage of single stuck-at faults and unprecedented test application times with respect to current literature at the cost of test wrapper complexity. However, the choice of pseudo-random test patterns and of their compaction poses the foundation for low-overhead TPGs and response analyzers, thus partially counterbalancing the footprint of the test wrapper. The achieved trade-off and its comparison with state-of-the-art solutions will be quantified in

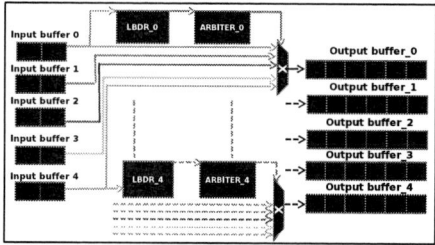

Fig. 1. Baseline Switch Architecture.

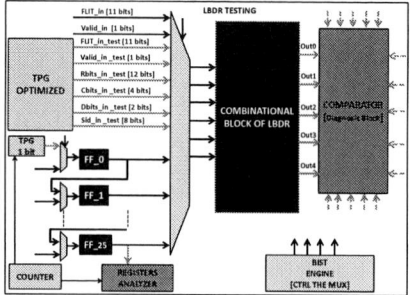

Fig. 2. Lbdr Testing Architecture.

the experimental results section.

IV. BASELINE SWITCH ARCHITECTURE

A 5x5 xpipesLite switch [6] has been used in this paper as the baseline design point without any testing support. The main blocks of this switch are illustrated in Figure 1. For each input port, a 2-slot (flit of 32 bits) input buffer and a routing module are instantiated. We use logic-based distributed routing [22], which computes target output ports by means of a simple combinational logic for each packet head. In order to provide implementation support for different routing algorithms, the logic is fed by 26 configuration bits per input port. For each output port, a port arbiter and a 6-slot output buffer are instantiated. Also, a 5x1 Multiplexer is placed in front of each output buffer, globally building up the switch crossbar. The switch implements wormhole switching and the stall/go flow control.

V. TESTING ARCHITECTURE

Next we present in detail all the changes applied to each block of the architecture. We will analyze respectively the testing scheme of the LBDR, Arbiter, Output buffer, Input buffer and Crossbar. After analyzing each block standalone, we will perform an optimization by merging some test phases together.

A. LBDR testing

Each LBDR (Logic Based Distributed Routing) routing module is split into two main blocks in such a way to be able to apply the key concept of our approach (see Fig. 2). The first block is composed of the LBDR configuration registers (FF_i blocks of fig.2), while the second one is the LBDR combinational logic ($Combinational$ block of fig.2). Combinational logic computes the information contained in the configuration registers. Configuration registers contain information about the routing restrictions of the routing algorithm (Rbits), the connectivity of the switch ports (Cbits), the local switch ID in the topology (Sid) and about deroutes to be taken in some special cases (Dbits) [22]. In test mode, these informations are randomized by the TPG, implementing the

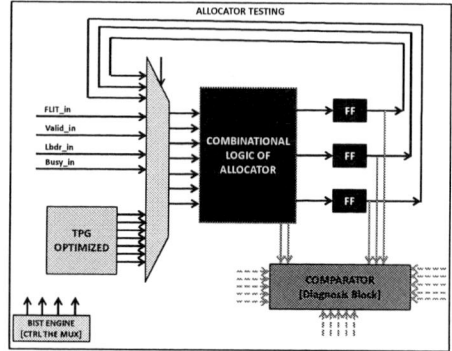

Fig. 3. Arbiter Testing Architecture.

compacted vectors. The LBDR logic reads also the destination address (11 bits) from the head flit, which needs again to be randomized in test mode. The compacted vectors codified in the $TPG\_Optimized$ block of fig.2 are used to test the combinational block of the LBDR. Before reaching the block under test, the compact test set has to cross first the wrapper placed in front of it. Response analysis is performed by comparators exploiting the output of the 5 switch ports, but nothing prevents from using MISRs. In both cases, the faulty routing module can be easily identified. All switch routing modules share the same TPG.

The scan-chain approach has been adopted for testing the registers of the LBDR. A 1 bit test pattern generator was used to inject a sequence of 0s/1s along the scan chain. A response analyzer, placed at the end of the scan chain, receives and analyzes the bit response after some amount of cycles (depending on the number of registers to cross). The configuration registers of the same LBDR modules are tested in a sequential manner. The testing of the five LBDR configuration registers occurs in parallel; they share the same TPG and counter but have different analyzers.

Finally, a BIST control engine drives the select bits of the test wrapper.

B. Arbiter testing

We distinguish between combinational logic and state registers also in the arbiter FSM. The testing diagram is depicted in Fig.3.

The arbiter state registers are tested in the same way as those of the input/output buffers (see Section V-C).

For the combinational block, besides connecting its primary inputs to the optimized TPG, we broke the feedback loop of the FSM in order to increase the controllability of the block. A comparator is again used for response analysis, thus exploiting the multiple instantiation of arbiters in the switch and their concurrent testing.

C. Output buffer testing

The output buffer belongs to the data path of the switch but it internally consists of an actual data path (data registers with selection input demux and output mux) and of a control FSM driving the read and write pointers of the mux and demux. For the sake of testing, these two internal blocks have been separated. The test architecture is illustrated in Fig.4.

The combinational block of the control-path is tested with compacted vectors generated by the Turbo Tester tool. State registers of the FSM could be tested as previously illustrated for the LBDR. However, here we present a further opportunity consisting of feeding the combinational logic outputs to the

978-1-4673-2895-1/12 $31.00 © 2012 IEEE

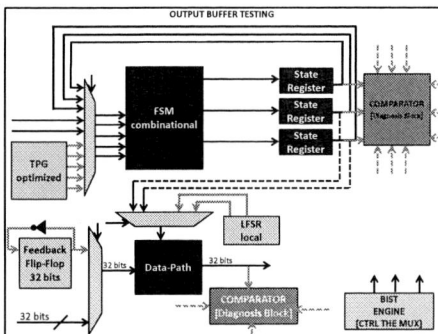

Fig. 4. Output Buffer Testing Architecture.

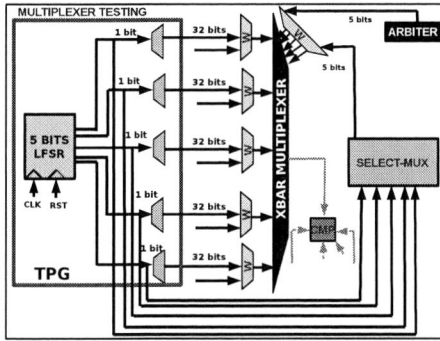

Fig. 5. Testing Architecture for Crossbar Multiplexers.

registers and analyzing their outputs by means of cycle-by-cycle comparators with the same outputs from the other switch buffers. Even in this case, full coverage is guaranteed.

In order to test the data path registers, we used a 32 bit register where the outputs are inverted and connected back to the inputs, thus generating a sequence of 0s and 1s. Due to the fact that the output signals from the control-path are not random enough, we opted for a small LFSR to drive the read/write pointers of the data path. Both pointers are connected to the same LFSR outputs, so to coherently swap all register banks.

D. Input buffer testing

The input buffer is tested exactly in the same way as the output buffer. The testing diagram of this element is identical to the testing scheme of output buffer as described in fig.4. The only difference is the number of slots (2 for the input buffer and 6 for the output buffer).

E. Testing Multiplexers of the Crossbar

The testing infrastructure adopted to test the multiplexers of the crossbar is depicted in Fig. 5. The TPG is a 5 bits maximal-length LFSR. The 5 bits LFSR generates all the patterns necessary to stimulate all the possible states of the multiplexers in less than 32 cycles. In particular, the relevant patterns for the testing of a 5x1 multiplexer are the following: 10000, 01111, 01000, 10111, 00100, 11011, 00010, 11101, 00001 & 11110.

Each pattern generated by the LFSR feeds a block called "select-mux". This latter is able to select the multiplexer input port carrying the logic value that is negated with respect to the other 4 input values. Finally, the diagnosis is performed by means of a comparator exploiting the output of the 5 multiplexers of the crossbar.

F. Testing Infrastructure Optimization

After testing each block independently, we cascaded the circuits under test exploiting the synergies between them in order to cut down on test wrapper and TPGs complexity.

The cascade is composed by the following modules:

- The Crossbar of the upstream switch

- The Output buffer of the upstream switch

- The Inter-switch link

- The Input buffer of the downstream switch

At the beginning of the cascade, we inserted a 5 bits LFSR (as described in Figure 5). Since test responses of one block become test patterns for the next one then the LFSR is responsible for the injection of test vectors for the whole cascade.

Such optimization allowed us to remove the input/output TPGs based on registers previously located in front of the input/output buffer. In the same way, we removed both the comparators located after the multiplexer and the output buffer. Finally, the comparator located after the input buffer in the downstream switch is the only preserved. The cascade testing scheme is presented in Fig.6.

To test the communication link, we synchronized the injection periods of some of the test patterns generators. In this specific case, we have three independent TPGs to synchronize:
-The TPG in front of the multiplexer input ports
-The two LFSRs of the read/write pointers of the input/output buffer

The synchronization must be performed in such a way to ensure the maximum coverage of all the blocks that belong to the cascade.

The multiplexer TPG starts to inject the test data as soon as the test mode is selected. The injection length depends on the number of cycles necessary to maximize the coverage of the cascade. Meanwhile, both the LFSRs respectively lying in the output buffer upstream switch and in the input buffer downstream switch are clocked each 31 cycles. Clearly, they must be clocked at least six times (i.e. the number of slots of the output buffer) in order to allow the data tests to cross all the data-path registers.

VI. EXPERIMENTAL RESULTS

This section presents the experimental results for a 5x5 NoC switch synthesized at 600 MHz in a 65nm industrial technology library. For the sake of comparison, two test architecture variants for the same baseline switch are available from previous work and used in this paper to assess the trade-offs of the new design point proposed by this paper. Therefore, our approach with optimized pseudo-random patterns is contrasted with the handcrafted deterministic test patterns used in [3] and with the non-compacted pseudo-random patterns used in [2], and above all with their enabling test architectures. In this comparison we do not consider ATPG generated test patterns since it has been demonstrated in [3] that these latter are not competitive with the handcrafted deterministic ones from a coverage viewpoint. This paper does not consider solutions providing less than 98% coverage on the considered NoC architecture.

978-1-4673-2895-1/12 $31.00 © 2012 IEEE

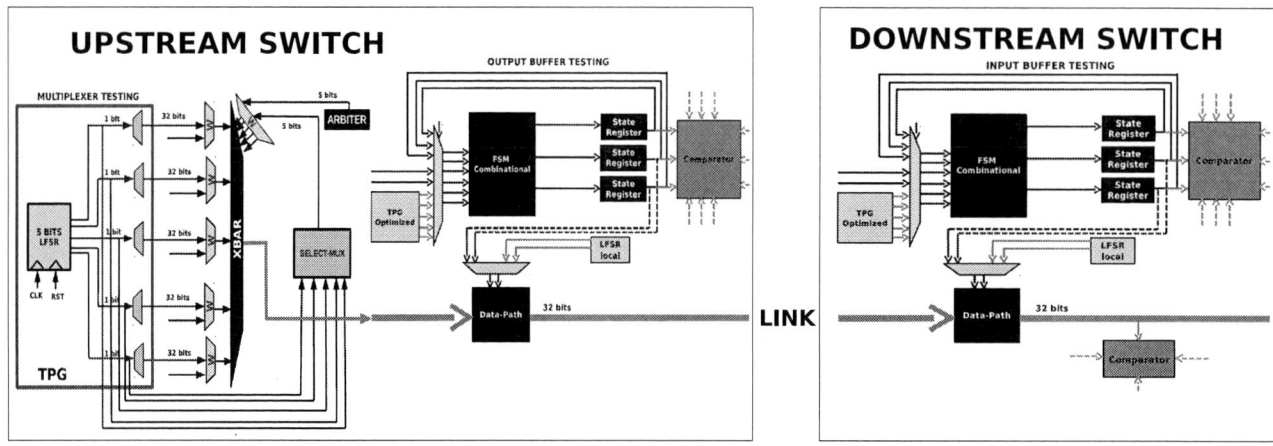

Fig. 6. Cascaded Testing Architecture.

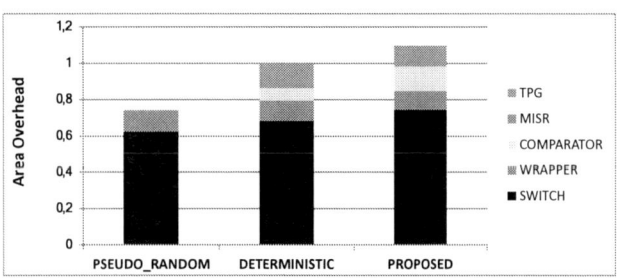

Fig. 7. Area Overhead: Deterministic [3], Pseudo-Random [2] and Proposed Compacted Pseudo-Random Approach.

A. Area Overhead

Fig.7 illustrates the area overhead of three BIST solutions for the xpipesLite switch (the one of this paper, the one with handcrafted deterministic patterns [3] and the one with non-compacted pseudo-random patterns [2]) normalized with respect to the deterministic switch. As we can see, around 11% of the area overhead of the deterministic approach comes from the wrapper, needed because of the different test phases that this approach requires, in addition to switch sub-block isolation for the sake of testing. Another 7% comes from TPGs, which encode the handcrafted test patterns, and marginally from diagnosis logic and the BIST manager. Finally, the comparator tree used for response analysis takes around 13% of the area.

When the test architecture is reconceived for non-compacted pseudo-random patterns, then the area overhead is reduced by ~26% in the considered 65nm library. Interestingly, the breakdown is completely different. MISRs are used for response analysis and account for most of the area overhead. LFSRs are extremely compact TPGs while less than 3% of the area is devoted to the test wrapper. In this architecture, block cascading was extensively used (i.e., test responses of some blocks are fed as test patterns to downstream blocks, at least until cascading does not hurt coverage too much) thus cutting down on the test wrapper overhead.

When it comes to the test architecture with test set compaction, the area overhead is 9.8% with respect to the deterministic approach. The proposed solution optimizes pseudo-random patterns thus incurring in a small area overhead, while significantly improving test application time (see section I). Most of the area overhead is due to the multiple instantiation of comparators and to the test wrapper, which needs to

provide finer-grain circuit isolation in test mode. The applied optimizations in the test infrastructure proved very effective in reducing the area overhead, to the extent that it closely tracks the overhead of the deterministic test architecture.

B. Testing time

TABLE I
TEST APPLICATION TIME PER BLOCK

TPGs	Multiplexer	Output LFSR	Input LFSR
Injection period	1 cycle	31 cycles	31 cycles
#Vectors	239	8 (7.7)	8 (7.7)
Total Cycles	239	239	239

Table I contains the injection periods of the TPGs used to optimize the cascade of Fig. 6. In fact, the TPG located at the beginning of the cascade starts to inject the data test as soon as the test mode signal driven by the BIST-engine is high. As mentioned in Table I, this injection occurs for ~239 cycles. At the same time, the multiplexer TPG controls the select inputs of the crossbar through the "select-mux" block. Meanwhile, local LFSRs of the output/input buffers in the upstream/downstream switch control the read/write pointers of data-path registers. These local LFSRs continuously inject patterns after each 31 cycles as mentioned in Table I. Overall, the test application time amounts to 239 cycles.

TABLE II
TESTING CYCLES AS FUNCTION OF THE TESTING APPROACH

Testing Technique	Testing Time (cycles)
Compressed Pseudo-Random Testing Approach	239
Deterministic Testing Approach	1104
Pseudo-Random Testing Approach	10000

Table II compares this value with those of the alternative approaches, for the same target coverage of approximately 99%. All approaches implement some form of testing cooperation between neighboring switches and share the same baseline switch architecture. However, test set compaction, proper choice of the granularity of the circuits to test (and consequent compaction efficiency) and merging of test phases make the approach of this paper the fastest. Only handcrafted deterministic patterns can somehow approach the testing time of this paper with around 1104 cycles. Although non-compacted pseudo-random patterns can achieve 96% of coverage in a comparable time with handcrafted deterministic patterns, they take around 10000 cycles to reach around 98% of coverage.

978-1-4673-2895-1/12 $31.00 © 2012 IEEE

TABLE III
TEST APPLICATION TIME AND COVERAGE OF DIFFERENT TESTING METHODS

	Test Cycle	Coverage
Our	239	98.3%
[3]	864 - 1104	99.3%
[1]	3.88×10^2 - 2.89×10^3	97.79%
[20]	4.05×10^5	95.20%
[13]	2.74×10^3	99.89%
[14]	9.45×10^3 - 3.33×10^4	98.93%
[21]	5×10^4 - 1.24×10^8	N.A.
[11]	320	99.33%
[12]	200×10^3	full (no exact numbers)

Our test application time compares favorably with previous work, as Table III shows. Only [1] and [11] are somehow competitive. However, [1] does not test the control path while [11] reports 320 cycles for a 3x3 mesh (made of a simplified switch architecture) which however grow linearly with network size. Also, this latter approach makes additional use of BIST logic for the control path not accounted for in the statistics.

We feel that area overhead is hardly comparable with previous work since whenever numbers are available, features of the testing frameworks are very different (e.g., control path not tested [1], test patterns generated externally [20], [14], diagnosis missing [20], [13], [14], [21], lack of similar test time scalability [8], [11], NoC architecture with overly costly links [13]). Moreover, the impact of synthesis constraints is never discussed.

C. Coverage

TABLE IV
COVERAGE AS FUNCTION OF THE TESTING APPROACH

Testing Technique	Coverage (%)
Compacted Pseudo-Random Testing Approach	98.3%
Deterministic Testing Approach	99.30%
Pseudo-Random Testing Approach	98.24%

The obtained coverage for single stuck-at faults is illustrated in Table IV. The technique proposed in this paper tracks the coverage of the pseudo-random approach although does far better in terms of latency. At the same time, our technique removes the burden of deriving handcrafted test patterns and enables the use of test generation and compaction tools, with one order of magnitude lower testing latency.

TABLE V
COMPACTION TABLE

Combinational Logic	Random (#vectors)	Random (coverage)	Compacted (#vectors)	Compacted (coverage)
FSM Input	1000	100.00%	10	100.00%
FSM Output	1000	100.00%	17	100.00%
LBDR	400000	93.04%	61	92.17%
ARBITER	170000	99.37%	42	99.37%

Table V shows the impact of the compaction tool on the pseudo-random test set generated for each combinational block of the switch control logic. The second and third columns of the table report the number of pseudo-random vectors together with their coverage while the last columns show the number of vectors with the respective coverage once they are compacted by the tool. It is possible to notice that the compaction operation is efficient and the compacted test set tracks the previously obtained coverage for each tested module.

VII. CONCLUSION

This paper presents a testing methodology and architecture support for NoCs that aim at the minimization of test application time, a requirement that well matches the future requirements of lifetime testing frameworks. In fact, while testing latency was not a concern for production testing, it becomes such when the testing procedure is run at system bootstrap and/or at runtime. We demonstrate NoC testing in less than 250 cycles. Above all, we do not achieve this result with handcrafted deterministic test patterns, but rather with an optimization methodology of pseudo-random patterns. The guiding principle is test set compaction, although the low-latency requirement forces a careful selection of the logic to test for best compaction efficiency. The trade-off is therefore between test application time and test wrapper overhead, although the final area footprint tracks that for a test architecture with handcrafted deterministic patterns but with one order of magnitude lower testing latency.

ACKNOWLEDGEMENTS

This work has been supported by the NaNoC European Project (FP7-ICT-248972).

REFERENCES

[1] S.Y.Lin, C.C.Hsu, A.Y.Wu, "A Scalable Built-In Self-Test/Self-Diagnosis Architecture for 2D-mesh Based Chip Multiprocessor Systems", IEEE Int. Symp. on Circuits and Systems, pp.2317 - 2320, 2009

[2] Simone Terenzi, Alessandro Strano, Davide Bertozzi, "Optimizing Built-In Pseudo-Random Self-Testing for Network-on-Chip Switches", INA-OCMC 2012.

[3] A. Strano, C. Gmez, D. Ludovici, M. Favalli, M.E. Gmez, D. Bertozzi, Exploiting Network-on-Chip Structural Redundancy for A Cooperative and Scalable Built-In Self-Test Architecture, DATE 2011.

[4] Markus, A.; Raik, J.; Ubar, R. Fast and Efficient Static Compaction of Test Sequences Using Bipartite Graph Representation, Proc. of the Second Electronic Circuits and Systems Conference (ECS'99), 1999, pp. 17 - 20.

[5] Sheng Zhang.; Sharad C seth.; Bhargab B, Bhattacharya. Efficient Test Compaction for Pseudo-Random Testing, Proc. of the 14th Asian Test Symposium (ATS '05), 2005.

[6] S.Stergiou et al., "Xpipes Lite: a Synthesis Oriented Design Library for Networks on Chips", DAC, pp.559-564, 2005.

[7] D.Wentzlaff et al., "On-Chip Interconnection Architecture of the Tile Processor", IEEE Micro, vol.27, no.5, pp.15-31, 2007.

[8] J.Raik, V.Govind, R.Ubar, "An External Test Approach for Network-on-a-Chip Switches", Proc. of the IEEE Asian Test Symposium 2006, pp.437-442, Nov. 2006.

[9] J.Raik, V.Govind, R.Ubar, "Test Configurations for Diagnosing Faulty Links in NoC Switches", Proc. ETS, 2007.

[10] D. A. Ilitzky, J. D. Hoffman, A. Chun and B. P. Esparza, "Architecture of the Scalable Communications Core's Network on Chip", IEEE MICRO, 2007, pp. 62-74.

[11] J.Raik, V.Govind, R.Ubar, "DfT-based External Test and Diagnosis of Mesh-like NoCs", IET Computers and Digital Techniques, October 2009.

[12] V.Bertacco, D.Fick, A.DeOrio, J.Hu, D.Blaauw, D.Sylvester, "VICIS: A Reliable Network for Unreliable Silicon", DAC 2009, pp.812-817.

[13] K.Peterson, J.Oberg, "Toward a Scalable Test Methodology for 2D-mesh Network-on-Chip", DATE 2007, pp.75-80, 2007.

[14] A.M. Amory, E.Briao, E.Cota, M.Lubaszewski, F.G.Moraes, "A Scalable Test Strategy for Network-on-Chip Routers", Proc. of ITC 2005.

[15] K.Arabi, "Logic BIST and Scan Test Techniques for Multiple Identical Blocks", IEEE VLSI Test Symnposium, pp.60-68, 2002.

[16] C.Grecu, P.Pande, B.Wang, A.Ivanov, R.Saleh, "Methodologies and Algorithms for Testing Switch-Based NoC Interconnects", IEEE DFT 2005, pp.238-246, 2005.

[17] R.Ubar, J.Raik, "Testing Strategies for Network on Chip", in Book: "Network on Chip", edited by A.Jantsch and H.Tenhunen, Kluwer Academic Publisher, pp.131-152, 2003.

[18] C.Aktouf, "A Complete Strategy for Testing an on-Chip Multiprocessor Architecture", IEEE Design and Test of Computers, vol.19-1, pp.18-28, 2002.

[19] S.Y.Lin, C.C.Hsu, A.Y.Wu, "A Scalable Built-In Self-Test/Self-Diagnosis Architecture for 2D-mesh Based Chip Multiprocessor Systems", IEEE Int. Symp. on Circuits and Systems, pp.2317 - 2320, 2009

[20] M.Hosseinabady, A.Banaiyan, M.N.Bojnordi, Z.Navabi, "A Concurrent Testing Method for NoC Switches", DATE, pp.1171 - 1176, 2009.

[21] C.Grecu, P.Pande, A.Ivanov, R.Saleh, "BIST for Network-on-Chip Interconnect Infrastructures", VLSI Test Symposium, page 6, 2006.

[22] S.Rodrigo, J.Flich, A.Roca, S.Medardoni, D.Bertozzi, J.Camacho, F.Silla, J.Duato, "Addressing Manufacturing Challenges with Cost-Effective Fault Tolerant Routing", NOCS 2010, pp.35-32, 2010.

A Flexible Platform Architecture for Gbps Wireless Communication

Jeroen Declerck, Prabhat Avasare, Miguel Glassee, Amir Amin,
Erik Umans, Andy Dewilde, Praveen Raghavan, Martin Palkovic
IMEC V.Z.W., Belgium.
Contact: prabhat.avasare@imec.be

Abstract—**Reprogrammable radio platforms should not only offer flexibility and low power consumption but also conform to strict throughput and latency requirements mandated by the wireless standards. To achieve these challenging goals, we introduce a platform architecture that uses a decentralized control to minimize communication and control overhead while keeping timing predictable by using state-of-the-art components and a novel interconnect. We demonstrate three main achievements in running multiple wireless standards on our platform: 1.053Gbps 4x4 80MHz WLAN 802.11ac receiver data path meeting the SIFS timing with a latency of 12.5μs, dual concurrent 173Mbps 2x2 20MHz Cat-4 3GPP-LTE receiver and platform reconfiguration from WLAN 11n to 3GPP-LTE in 52μs. The main blocks from our versatile platform architecture are currently being prepared for tape-out.**

I. INTRODUCTION

Wireless communication require continuous development of new standards to satisfy ever increasing user demands. This is true not only for cellular standards (e.g. GSM, WCDMA, HSDPA, 3GPP-LTE) but also for other wireless standards (e.g. WiMAX, WLAN, DVB-H, DVB-T2). There is a challenge to implement these standards power-efficiently with mandated latency and throughput constraints. Additionally, there are design cost and time-to-market constraints. One universally adapted solution to tackle such challenging goals is platform-based design. Such a design typically uses a mix of programmable and full custom ASICs to achieve flexibility, extensibility in design and at the same time a good control over all design costs.

In terms of implementation of wireless standards, in this paper we mainly focus on the baseband part in the receive path of the physical layer (PHY). A typical implementation architecture for such a receive path contains a Digital Front-End (DFE), BaseBand Engine (BBE) and Outer MoDem (OMD). Most of the wireless platform architectures available today emphasize on the BBE which performs the computation-intensive inner modem processing. However, we believe that equal attention should be given to the interconnect between components and overall control flow. In this paper, we introduce a platform architecture that uses decentralized control and a novel interconnect between state-of-the-art components to minimize communication and control overhead. With our architecture we have made three key achievements. First, our platform can support multiple wireless standards (WLAN, 3GPP-LTE, DVB-T2) meeting mandated latency and throughput constraints. Second, with our architecture, we can run two concurrent and independent data streams. And third, we can switch among different wireless standards within an order of 100μs. This platform is used in our overall design flow which starts with mapping different wireless standards (implemented in MATLAB) and results in a tape-out for selected components in the platform architecture.

II. RELATED WORK

One can broadly identify four types of solutions for wireless platform architectures that are proposed in academia and industry [1].

A first set of platforms uses fully custom multi-mode hardware; i.e. all standards are implemented using ASIC's. This is potentially the most power efficient, but obviously neither flexible nor extensible[2], [3]. A second set consists of the DSP-centric many-core platforms providing an end-to-end solution for a particular (or a set of) standard(s) based on a DSP architecture. It is usually composed of processing engine(s) connected to hardware-implemented accelerator blocks using a high-speed interconnect [4], [6], [7], [8], [9], [10], [11], [12]. This approach could be the fastest to implement as the platform gives an end-to-end solution but the approach comes with higher costs in terms of silicon area and power consumption. Also most of these architectures do not support multiple wireless standards running concurrently. Then the third set is the DSP-assisted architectures where the most computation intensive part is implemented on a specialized DSP [4], [5]. Such software programmability brings easier implementation and extensibility but at the cost of higher power consumption. And last there are the reconfigurable platforms where algorithmic functions and even pieces thereof (sub-functions) that are shared by multiple standards are implemented using reconfigurable data paths. The reconfiguration could either be fine-grained [28] or coarse-grained [13], [14], [15]. By hardware reuse, area can be saved but the programming model can become very complex without good mapping support.

Our platform architecture falls in the last two categories. It uses a coarse-grain reconfigurable processor as a baseband processor that implements inner modem processing. Our Digital Front End (DFE) and Outer MoDem(OMD) components are the hardware accelerators either ASICs or Domain Specific Instruction Processors (DSIPs). One can also differentiate the above-mentioned platforms in terms of interconnect architectures used for data and control flows. Most use a dedicated central bus to meet design constraints, but then contention becomes a major issue due to simultaneous accesses from various processing elements. Some platforms use a Network-on-Chip (NoC) as an interconnect [11], [12]. Even though NoC-based architectures are nicely scalable, they lack hard QoS guarantees in latency required for wireless standards [1]. Our architecture uses a segmented bus-based architecture controlled by a novel interconnect controller.

Note that the platform architecture presented in this paper is evolved from our previous platform architecture [14]: the main contribution of this architecture is the decentralized control and the separation of control and data flows. Although mapping an application on our platform is an important aspect, comparison of our mapping flow with existing flows can be found elsewhere [29] and is out of the scope of this paper. Next section describes our platform architecture in detail.

III. PLATFORM ARCHITECTURE

The platform proposed in this paper is a heterogeneous Multi-Processor System-on-Chip (MPSoC) architecture (see Figure 1) tar-

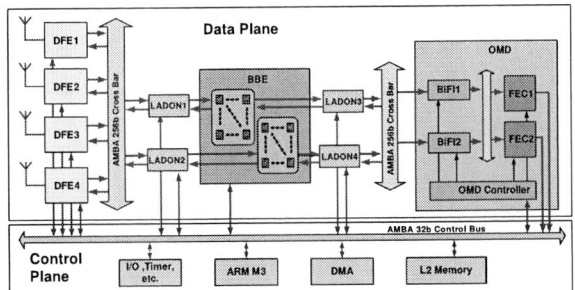

Fig. 1. The proposed platform architecture.

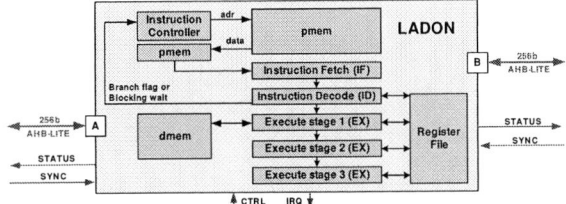

Fig. 2. The LADON processor architecture.

geted for a 40nm GP (General Purpose) process technology. This platform focuses on the receive path as it is the most complex to implement, but the platform architecture can potentially be used also for transmission. The platform design cleanly separates data and control flows into separate Data and Control Planes. The Data Plane has three types of processing blocks: the Digital Front End (DFE, see Section III-A), the BaseBand Engine (BBE, see Section III-C) and Outer MoDem (OMD, see Section III-D). All these blocks contain in-house developed DSIPs. The data is exchanged between the blocks in a flexible and programmable way by 256-bit wide AMBA AHB buses [16] and the interconnect controllers named LADON (see Section III-B). All components in the Data Plane are configured by the ARM Cortex M3 [17] control processor over a 32-bit segmented AMBA control bus. In this Control Plane, all blocks can signal events to the ARM by sending interrupts. The ARM runs a light-weight, event-based run time environment. In general the ARM is responsible for setting up a data path and reprogramming it when necessary. Reprogramming the components can be done at two levels: either by reprogramming registers in a data memory (e.g. number of symbols in a packet) or by reprogramming its complete instruction memory (e.g. changing a firmware from WLAN to 3GPP-LTE). A system RAM L2 memory is available through the control bus to store firmwares for multiple wireless standards. The receive data path starts at the DFEs: every DFE is connected to an analog Antenna Front-End interface hence we support four antenna's (Figure 1). Data then flows to the BBE for demodulation and finally to the OMD for decoding. The LADON interconnect controllers serve like Direct Memory Access (DMA) components that allow to program how the data should flow from one core to the other without intervention of the ARM control processor. The Control Plane, buses and LADONs are clocked at 400MHz, the BBE and the OMD at 800MHz. The DFE clock can be configured at run-time depending on the running standard.

A. Digital Front End (DFE)

The four DFE tiles (Figure 1) are connected to the Analog Front-End by means of an Analog to Digital Converter and Digital to Analog Converter [19]. The DFE is an instantiation of the architecture described in [20]. During reception of a packet or a stream, the DFE takes in complex time domain samples from the ADC. The DFE then detects the signal power. If the power is above a certain threshold, it will power on the rest of the DFE and if necessary, issue an interrupt to the ARM processor. The samples are then digitally filtered, down-sampled and put in FIFOs (connected to the 256-bit bus) to get transferred to the baseband processor. The DFE also features a synchronization DSIP that is capable of doing correlations and peak detection efficiently.

B. LADON interconnect controller

The LADON (Latency Aware Data Oriented Network) is a processor core with an architecture that serves as a DMA on a local 256-bit

AMBA bus on the platform. Both its architecture and connection to the platform are shown in Figure 2. The main purpose of the LADON is to transfer data from port A to port B or vice versa; for example between DFE and the BBE. Both the ports are 256-bit wide AHB-Lite master interfaces. LADON also allows synchronization between the connected components at both sides by using synchronization (SYNC) and STATUS signals e.g. synchronization between DFE and BBE. Additionally, LADON can synchronize with other LADON cores that operate in parallel on the same blocks.

The core itself has a five-stage pipeline with an Instruction Fetch (IF), Instruction Decode (ID) and three execution stages. This pipeline is fed from 16-bit wide program memory (pmem in Figure 2). For the program execution, there is a Register File (8 registers) and a separate data memory of 64 12-bit words. Specific instructions are provided to load/store data into data memory from Register File.

LADON processor is designed using processor designer tools from Target Compiler Technology [24]. It has a simple instruction set (33 instructions) with basic arithmetic instructions, transfer and synchronization instructions. The instructions allow different transfer possibilities (e.g. single or burst) between the source and destination. The instructions also allow explicit synchronization between the LADON and a connected platform block. Furthermore the burst transfer instructions are blocking allowing the LADON to write to the destination memory only when it is free (or read from the source memory only when it has data). This eases control overhead for the programmer and also for the central control processor ARM in the platform. The LADON programs can be written in a sub-set of C as the Target toolchain automatically generates a C compiler as well.

LADON processor is one of the key differentiators from our previous platform architecture. By handing over the control flow to the LADON, the ARM can avoid monitoring multiple memory transfers (and hence reduce control bus traffic significantly). This way LADON has enough flexibility to handle flow-control on its own and the traffic on the control bus is reduced significantly.

Following example illustrates the flexibility and functional description of LADON. It relates to the reception of payload data from two DFE tiles. When data is received on the two antennas, LADON1 (Figure 1) transfers samples from DFE1 and DFE2 to baseband input vector memory VMEM1 (see section III-C):

```
for( i = 1; i <= no_symbols; i++) {
    waitfor(BBE);
    for( j = 1; j <= no_dfe; j++) {
        waitfor(DFE_j);
        transfer(size,DFE_j,BBE);
    }
    ready(BBE);
}
arm_irq();
```

The first outer for-loop iterates over the number of payload symbols (*no_symbols*). LADON1 waits (*waitfor*) until the BaseBand Engine (BBE) is ready to receive data. The inner for-loop iterates

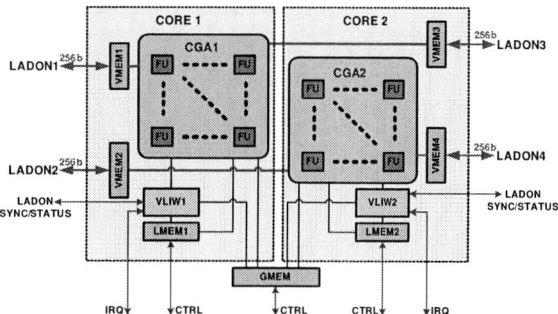

Fig. 3. The BBE architecture.

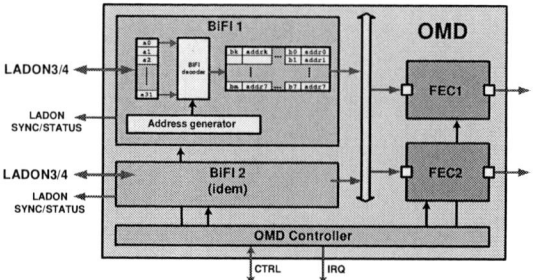

Fig. 4. The OMD architecture.

over the number of DFEs that need to be read from (DFE1 and DFE2 in this example). LADON1 does a blocking read on DFE output FIFOs. Once the FIFOs are filled with required number of samples, it transfers ($transfer$) the actual amount of data ($size$) from DFE_j to the BBE. At the end of the transfer, LADON1 signals to the BBE ($ready$). Now, the BBE has the complete payload data of a single symbol and can start the processing. At the baseband side, BBE is waiting for such a $ready$ signal before it can start processing the symbol. At the end of for-loops, the complete payload data is transferred from DFEs to BBE. The LADON notifies this to the ARM by raising an interrupt. Note that some of the loop variables are parameters (e.g. $no\_symbols, no\_dfe, size$) which are programmable by the ARM as they reside in the data memory (dmem in Figure 2). This way the LADONs can have the same firmware for one standard (IEEE 802.11ac for example) and the ARM only needs to reprogram, at run-time, the different parameters for the reception of a specific packet. This instantiation of the LADON is currently being prepared for tape-out (40nm GP technology).

C. BaseBand Engine (BBE)

The BaseBand Engine (BBE in Figure 1) performs the major part of the inner modem processing. When receiving, it takes complex time domain samples as input and generates demapped soft bits as output. It is an optimized instantiation of the coarse-grain reconfigurable processor template [18]. The instantiation used in the proposed platform is shown in Figure 3. It features a multi-core (CORE1 and CORE2) processor with two VLIW processors and two Coarse-Grained reconfigurable Array (CGA) units. The CGA contains both 256-bit and 32-bit wide Functional Units (FU). A core can either run in VLIW mode (VLIW1 and VLIW2) for control processing or in CGA mode (CGA1 and CGA2) for data processing. The multi-core architecture allows multi-threaded operation where two separate functions can run on separate parts of the processor array. This can either be done in a master-slave mode where CORE1 spawns a thread on CORE2 or in a master-master mode where the two cores run completely independent from each other (needed for multiple concurrent streams). It is also possible to claim both CGAs from a single thread. In such case, only the master VLIW1 is used for control. By using the in-house developed compiler tool chain, different baseband functionalities are mapped onto this processor: Carrier Frequency Offset (CFO) estimation, Sample Carrier Offset (SCO) estimation, Fast Fourier Transform (FFT), channel estimation, channel equalization, soft and hard demapping etc. The data plane communication between the LADONs and the BBE is done through the 256-bit wide vector memories (VMEM) that are connected to the CGA as shown in Figure 3. Every core has two of these memories to which some of the FU's have an access to. LADON1 is connected

with CORE1 at the side of the DFEs as is LADON2 with CORE2. Similarly, LADON3 and LADON4 are connected to CORE 1 and CORE2 but at the OMD side. The VLIW's can exchange status information with the LADONs e.g. VLIW can signal that it is ready to receive new input data. Each core has its own 32-bit local memory (LMEM1 and LMEM2) and share a global 32-bit memory (GMEM) between them. The ARM processor has an access to GMEM through the AMBA control bus to pass initialization parameters (e.g. number of symbols in a packet). This instantiation of the BBE is currently being prepared for tape-out (40nm GP technology).

D. Outer MoDem (OMD)

The Outer MoDem as shown in Figure 4 has two major parts; the BiFI and FEC. The BiFI (which stands for Bit Fiddler at Input) is responsible for transforming the stream of soft bits from the BBE into the appropriate format for the actual decoding in the Forward Error Correction coder (FEC). For an 802.11ac receive data path for example, this means de-interleaving, stream deparsing, removing repetition bits, depuncturing and coder parsing. All these functions can be efficiently combined by moving the data to the appropriate location at the input of the decoder (FEC). The Address generator in the BiFI calculates the destination address for every soft bit that enters the BiFI. The BiFI then transfers the appropriate bit into the input memory of the FEC. Besides simply moving data, the Address generator can add new data (for depuncturing), remove data (remove padding or repetition bits) or even invert data (for the descrambler in the LTE receive chain). The BiFI has internal data-level parallelism (it can take a full 256-bit input line at once) as well as external task-level parallelism (there are two BiFI instances). As with the LADONs and the BBE, the two BiFI instances can operate together (synchronized) as well as independent from each other. The synchronization with the LADON is done in the same way as for the BBE i.e. with the SYNC and STATUS signals. The second stage of the OMD is the decoding stage done by the FEC coder. The FEC architecture consists of a Viterbi decoder ASIC and a DSIP (generated using Target toolchain) tuned for LDPC and Turbo decoding as presented in [21].

This section detailed the heterogeneous MPSoC architecture used in our platform. Our architecture uses separate control and data planes. Further, it uses decentralized control to minimize run-time overheads. Next section shows how this platform is instantiated as a virtual platform.

IV. SIMULATION ENVIRONMENT

The platform architecture described in the previous section is simulated in Synopsys Platform Architect (PA) [23], a SystemC based TLM (Transaction-Level Model) environment for the design and development of SoC architectures. This framework allows the co-simulation of behavioral models, instruction accurate models, cycle

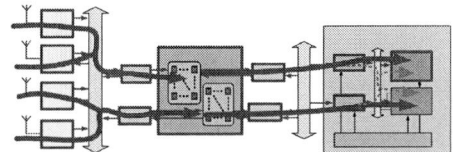

Fig. 5. Data flows on the platform: for 11ac and for multi-stream.

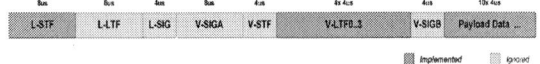

Fig. 6.   Packet structure for 4x4 80MHz WLAN 802.11ac.

accurate models and even RTL models. For the ARM Cortex M3, an instruction accurate fast model (provided by ARM Ltd) is used. The AMBA AHB buses have cycle accurate models provided by Synopsys PA. For DFE, part of it is modeled directly in SystemC. The processor part of the DFE is designed with the Target toolchain [24]. From a specified processor description, the Target toolchain generates an RTL model, a C-compiler, an Assembler and a cycle accurate SystemC model (Instruction-Set Simulator, ISS). This SystemC model is integrated inside the platform instance. The same design flow using Target toolchain is used for the LADON and the FEC. For the BBE, an in-house developed cycle accurate ISS is used. For BiFI, a loosely timed behavioral SystemC model is used. The timings of the BiFI are based on an initial hardware design of the Address Generator. All this together allows to simulate the complete platform with the actual firmwares for all processors and to obtain accurate timing information as described in the following section.

V. SIMULATION RESULTS

This section demonstrates three key simulation results from our data path implementations. First, Section V-A demonstrates an implementation of a standard-compliant 4x4 80MHz 802.11ac WLAN receive data path with a processing latency of $12.5\mu s$ (Figure 7). Second, Section V-B shows an implementation of a multi-stream data path where two completely independent and concurrent data paths are running two 2x2 20MHz Cat-4 3GPP-LTE data streams (Figure 8). And third, Section V-C explains how a reconfiguration from a WLAN data path to an LTE data path is achieved within $52\mu s$ (Figure 9).

Figure 5 show data flows on the platform during execution. Platform resource are shared or divided depending on the requirements of the application running e.g. for 4x4 80MHz 802.11ac WLAN receiver, entire platform resources are used, two cores on BBE are run in a master-slave mode and BiFIs can forward the data to one of the two FECs which is free. Whereas for multi-streaming, platform resources are distributed equally to handle two independent data streams i.e. both the cores on BBE are running in master mode.

A. WLAN Data Path

Figure 7 shows the simulation results of the platform receiving a single 4x4 80MHz WLAN 802.11ac packet. As shown in Figure 6, this packet consists of a Legacy Short Training Field (L-STF), Legacy Long Training Field (L-LTF), two Signal Fields (L-SIG and VHT-SIG), Very high throughput Short training field (VHT-STF) ,4 Very high throughput Long training field (VHT-LTFs 1, 2,3 and 4) and then ten $4\mu s$ payload symbols (PL1 till PL10) containing the actual payload data. The packet is modulated with an 80MHz bandwidth in QAM64 on four spatial streams, so four antenna's are used for reception. It is encoded with a coding rate of 3/4 using the optional LDPC coding. This results in a data rate of 1053 Mbps plus preamble overhead. The sequence of events on the platform as shown in Figure 7 shows the status (dark gray or black for heavy activity,

gray for reasonable activity and white for no activity) of the different components during the packet reception. To simplify the figure, only one LADON on each side of BBE is shown as the other one has an identical behaviour. Similarly only one DFE tile is shown as the others have identical behaviour. Following is a sequence of events:

1. The ARM programs the DFE via the control bus (CTRL BUS). The DFE starts sensing for incoming packet.
2. A packet arrives at the input of the DFE (ANTENNA1) that starts the power detection process.
3. The DFE consumes a part of the STF for detecting power and puts the rest in its output FIFOs to let LADON1 copy the required STF (L-STF) and L-LTF samples from DFE1 to the BBE for the synchronization processing. Currently, synchronization algorithm running on BBE only requires data from DFE1; therefore LADON1 flushes the same amount of samples from DFE2 FIFO. LADON2 does the same for DFE3 and DFE4. LADON1 then signals the BBE that the L-STF and L-LTF is available in its vector memory.
4. The BBE starts processing the synchronization samples and calculates the CFO compensation value and fine synchronization offset. When done, the BBE signals to LADON1-2 to receive new inputs. BBE notifies ARM about the synchronization offset.
5. The ARM reads the synchronization offset from the global memory of BBE and programs this in the DFE. From this point, all the output FIFOs of the DFEs are aligned to the beginning of the L-LTF symbol.
6. The current 11ac data path is not processing the L-LTF, L-SIG, VHT-STF and VHT-SIG symbols. Therefore all LADONs flush these samples from the FIFOs of DFEs.
7. LADON1 starts transferring VHT-LTF1 samples from DFE1 and DFE2 to BBE vector memory. LADON2 does the same for DFE3 and DFE4. This process is repeated till VHT-LTF4.
8. The BBE processes the VHT-LTF1 as part of preamble processing. This process is repeated for all four VHT-LTFs. And at the end 4x4 equalizer matrix and scaling coefficients are calculated.
9. VHT-SIG-B symbol follows the VHT-LTF. It is also not processed so LADONs 1 and 2 flush the relevant samples from the DFE FIFO.
10. During the processing of the VHT-LTF4 in the BBE, the BBE already gave a SYNC signal to LADONs 1 and 2 that its input vector memories are free. At that moment LADONs 1 and 2 start the transfer of the first payload symbol (PL1) to the BBE. At the end of the transfer, the LADONs 1 and 2 give a STATUS signal to the BBE that the transfer has finished.
11. The BBE starts processing PL1 once the data is present in the input memories. BBE signals to LADON3 and LADON4 when the demodulated PL1 data is ready.
12. The LADON3 and 4 start transferring the demodulated soft bits to OMD and signal the data availability to the OMD.
13. The BiFI in the OMD processes PL1, extracts the LDPC code words (CW1, CW2 in Figure 7) and sends them to the FEC that starts decoding them. For the 11ac decoding, 2 FEC decoders work in parallel to meet the timing requirement.
14. Steps 10 to 13 are repeated for all payload symbols (till PL10). At this point, the IQ samples for the last symbol (PL10) are sampled by the DFE and are available for transferring to the BBE vector memories.
15. This process is completed with the transfer of soft bits for the last payload symbol (PL10). After receiving last symbol, the data path pipeline is flushed and the FEC decodes the last code words.

978-1-4673-2895-1/12 $31.00 © 2012 IEEE   113

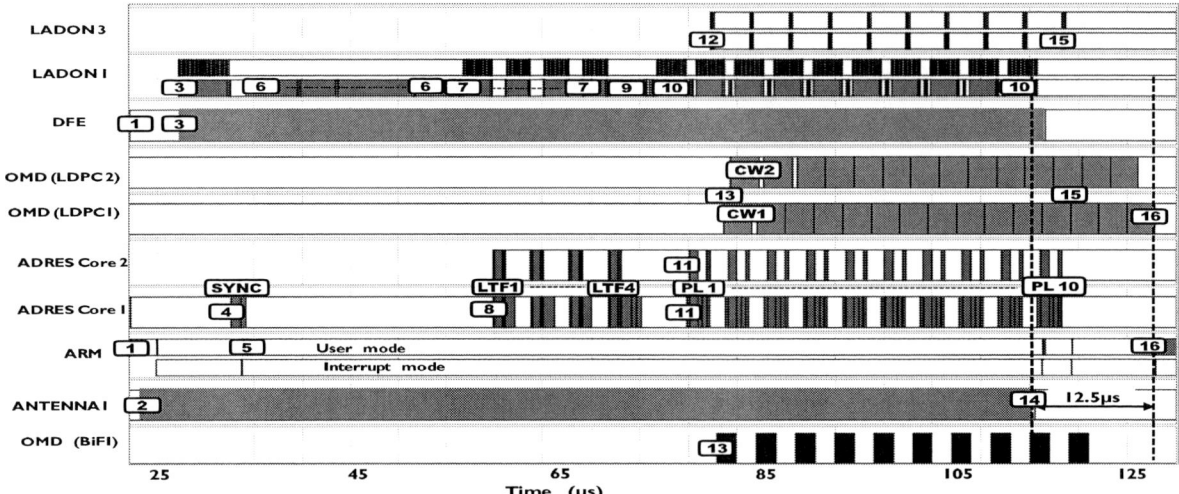

Fig. 7. Simulation result of a 4x4 80MHz WLAN 802.11ac packet reception on the proposed platform.

Fig. 8. Two concurrent LTE data streams (payload only).

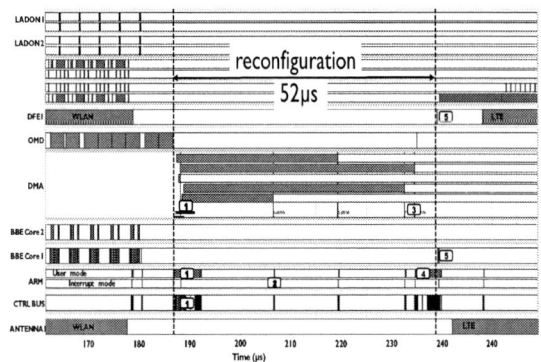

Fig. 9. Reconfiguration from a 4x4 802.11n stream to 2x2 LTE stream.

16. The OMD gives an interrupt to the ARM that the packet processing is finished.

Note that currently only the most performance critical part of the preamble processing is implemented on the platform, hence some parts of preamble processing is skipped (these samples are flushed by LADONs as explained before in steps 6 and 9). To calculate the overall latency we measured the time from the reception of the last sample at the antenna interface until the end of the decoding in the OMD. This overall latency is $12.5\mu s$ allowing sufficient time to prepare an acknowledgment (ACK) and transmit it to the sender within the $16\mu s$ SIFS (Short Inter-Frame Space) timing constraint.

B. Multi-Streaming

By multi-streaming capabilities of our platform we mean two independent streams (e.g. 2x2 LTE and 2x2 WLAN together) can be run on the platform in parallel. Figure 8 shows the simulation results when two 2x2 20MHz Cat-4 3GPP-LTE data streams run concurrently on the platform. For simplicity, only the payload BBE processing is shown in the simulation results.

At the beginning of the simulation, the first data stream starts using DFE3, DFE4, LADON2, LADON4 and BBE Core 2. Then, $100\mu s$ later, a second stream arrives at the antenna interface and is processed by DFE1, DFE2, LADON1, LADON3 and BBE Core 1. Note that each LTE stream needs only half of the platform resources for its execution. The key result of this simulation is that the two streams run completely independently and concurrently on our platform. In

other words, two streams do not interfere with each other, hence meeting latency and throughput constraints. This is achieved due to the decentralized data flow control implemented on the platform using LADONs that can operate completely independent when running concurrent streams. The remaining overhead for the central controller (i.e. the ARM processor) is very limited as can be seen by the CTRL BUS activity in Figure 8.

C. Platform Reconfiguration

Platform architecture supporting multiple wireless standards will need a way to change the datapath from one standard to another one at run-time. In our platform, the central controller ARM performs this switch by changing the firmwares running on all the platform components. We refer to this process as platform reconfiguration. Our platform reconfiguration is detailed in [26].

Figure 9 shows platform reconfiguration from a 4x4 40MHz 802.11n WLAN data stream to a 2x2 20MHz Cat-4 3GPP-LTE data stream. It shows three phases in the reconfiguration. First, the platform is at the end of running the 802.11n WLAN data stream, then it is being reconfigured to an LTE data stream and last, the LTE data stream starts running. During the reconfiguration, the important components are the ARM processor, the control bus and the DMA controller attached to the control bus (Figure 1). During reconfiguration, following sequence of events can be observed:

1) At the end of WLAN packet processing, the ARM processor programs the DMA controller (attached to the control bus, see

978-1-4673-2895-1/12 $31.00 © 2012 IEEE   114

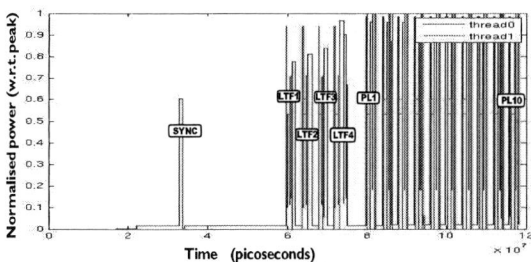

Fig. 10. Normalized BBE power profile for a WLAN 802.11ac symbol.

Figure 1) to transfer the LTE firmwares for different parts of BBE core 1: these multiple transfers are initializations for BBE data and instruction memories.

2) After every transfer, the DMA controller interrupts the ARM.

3) The DMA controller has finished all the transfers to the BBE memories and interrupts the ARM again

4) As reconfiguring some of the platform blocks to switch between the wireless streams is only configuring a few registers in their instruction memories, it is done using a `memcpy()` rather than a DMA transfer [26]. ARM does this type of reconfiguration with the DFEs, the LADONs and the OMD.

5) The ARM starts up the components in a sequence to activate the desired LTE data path. In Figure 9, it can be seen that both the DFE and BBE start.

Figure 9 shows our platform can switch from a 4x4 40MHz WLAN 11n stream to 2x2 20MHz Cat 4 3GPP-LTE stream in only $52\mu s$.

VI. PLATFORM POWER CONSUMPTION

This section discusses power consumptions of the main power-consuming platform components, i.e. the FEC, the DFE, the BBE and the LADON. In our platform, we use a FEC instance that supports both Turbo and LDPC decoding. For this instance, the peak power consumptions (excluding memories) for LDPC and Turbo Decoding are in the order of 100mW[21]. For the DFE, all details on an actual tape-out (65nm technology) and power consumption can be found in [20]. Per DFE, the synchronization for a stream consumes between 8mW and 22mW depending on the wireless standard. After the synchronization phase, the power consumption goes down as the synchronization processor is disabled. For BBE, as the instantiation used is currently under final development phase for tape-out (40nm technology), no absolute power numbers can be given. However, early gate level simulations indicate peak power consumption to be in the order of hundreds of mW. For the LADON, the gate level simulations indicate a power consumption (excluding memories) to be in the order of few mWs. Adding these numbers together, the total power consumption of our platform is in the order of 100s of mW. Note that FEC and BBE dominate our platform power consumption.

Our simulation platform can also incorporate power consumption modeling. Currently we have modeled BBE power consumption by integrating a power model into the ISS of the BBE. This model was developed based on the power reports from gate level simulation of the op-codes on different Functional Units. It is calibrated and verified with the power reports from the gate level simulation of the whole BBE. Such a technique of power modeling based on energy of individual op-codes is well-known [27]. At the end of a simulation, one can extract the BBE power consumption based on energy consumption figures spitted out during the simulation. For 802.11ac stream, a relative power profiling can be seen in Figure 10 where the normalized (w.r.t. peak) dynamic power of the BBE WLAN

11ac symbol processing (from Section V-A) is shown as a function of time. Such profiling obtained from the platform simulations gives a good overview of platform power consumption.

VII. CONCLUSIONS

In this paper, we presented a platform architecture for wireless applications that uses decentralized control and a novel interconnect to minimize communication and control overhead. We showed that our platform achieves main requirements of a reprogrammable radio platform: mandated latency and throughput constraints for Gbps traffic, fast switching between standards and concurrently running two standards. The main blocks from our versatile platform architecture are currently being prepared for tape-out.

VIII. ACKNOWLEDGMENTS

This work was supported in part by the European Commission under grants 2PARMA FP7-248716, CONSERN FP7-257542 and PHARAON FP7-288307.

REFERENCES

[1] Ramacher, "SDR prospects for multistandard mobile phones", IEEE Comp. 2007, pg 62-9.

[2] Im et al, "A Low-power and Low-complexity Baseband Processor for MIMO-OFDM WLAN Systems", Springer Journal of Signal Processing Systems, 2010, pg 62-69.

[3] Burg, "A 4-Stream 802.11n Baseband Transceiver in 0.13μm CMOS", VLSI Circuits, June'09, pg 282-3.

[4] CEVA Inc., USA http://www.ceva-dsp.com.

[5] Picochip Limited, UK http://www.picohip.com.

[6] Tensilica Inc., USA http://www.tensilica.com/.

[7] Qualcomm Inc., USA http://www.qualcomm.com/.

[8] Glossner et al, "Sandbridge SB3011 SDR platform", SympoTIC 2006.

[9] Woh, "From SODA to scotch: The evolution of a wireless baseband processor", MICRO-41, 2008, pg 152-63

[10] K. van Berkel, "Multi-core for mobile phones", DATE 2009, pg 1260-5.

[11] T. Limberg et al., "Heterogeneous MPSoC with Hardware Supported Dynamic Task Scheduling for Software Defined Radio", DAC, 2009.

[12] Clermidy et al, "A 477mW NoCBased Digital Baseband for MIMO 4G SDR", ISSCC, Feb. 2010.

[13] Montium Technology, http://www.recoresystems.com/.

[14] Jeroen Declerck, et al, "SDR Platform for 802.11n and 3-GPP LTE", Proc. of SAMOS, Greece, 2010, pg 318-23

[15] Anwar et.al., "A software defined approach for common baseband processing", J. of Systems Architecture, Elsevier, 54 (2008) pg 769-786.

[16] AMBA bus specifications, http://www.arm.com

[17] ARM cortex M3, http://www.arm.com/products/processors/cortex-m/

[18] Vander Aa T. et al, "A Multi-Threaded Coarse-Grained Array Processor for Wireless Baseband", IEEE Symp. SASP, June 2011, USA.

[19] T. Schuster et al, "Design of a low power pre-synchronization ASIP for multimode SDR terminals", Proc. of SAMOS, Greece, 2007, pg 322-32.

[20] S. Pollin et al, "Versatile sensing for mobile devices: cost, performance and hardware prototypes," in ACM Mobicom - coronet workshop, 11

[21] F. Naessens et al, "Unified C-programmable ASIP architecture for multistandard Viterbi, Turbo and LDPC decoding," IP-SOC, France, 2011.

[22] F. Naessens et al, "A 10.37 mm2 675 mW reconfigurable LDPC and Turbo encoder and decoder for 802.11n, 802.16e and 3GPP-LTE," in Symposium on VLSI Circuits, Honolulu, Hawaii, June 2010

[23] Synopsys Platform Architect http://www.synopsys.com.

[24] Target Compiler Tech., http://www.retarget.com.

[25] "802.11 Amendment 5:Enhancements for Very High Throughput for Operation in Bands below 6 GHz", IEEE P802.11ac D1.4, Nov. 2011.

[26] M. Palkovic et al, "DART - a high level software-defined radio platform model for developing the run-time controller ", Journal of Signal Processing Systems, 2012, DOI 10.1007/s11265-012-0669-3.

[27] V.Tiwari et al, "Instruction-level power analysis and optimization of software", VLSI Signal Processing, No.13, 1996, pp.223–238.

[28] Miljanic et al "The WINLAB Network Centric Cognitive Radio Hardware Platform WiNC2R", CrownCom Aug 2007, pg 155-60.

[29] M Palkovic et al, "Multicore embedded systems for future SDR platforms: architecture and mapping flow overview", IEEE Signal Process Mag., March 2010, pg 22-33.

978-1-4673-2895-1/12 $31.00 © 2012 IEEE

Efficient VLSI Architectures of QPP Interleavers for LTE Turbo Decoders

Martin Broich, Tobias G. Noll

Chair of Electrical Engineering and Computer Systems
RWTH Aachen University, D-52062 Aachen, Germany
{broich,tgn}@eecs.rwth-aachen.de

Abstract—Quadratic-permutation-polynomial (QPP) interleavers are utilized in Turbo coding of the 4G-mobile-system LTE-Advanced due to the support of parallel, contention-free memory accesses. In principle, throughput rates of 1 Gbit/s can be supported with such interleavers in today's CMOS technologies.

A systematic examination of the QPP interleaver properties has led to several design improvements concerning silicon area, energy per operation and the support of highly parallelized Turbo decoders. Regarding the interleaver network, it is proven that hardware-efficient butterfly and Beneš networks can be applied with negligible configuration overhead. With respect to the interleaver address generation, we propose and analyze a recursive address calculation method.

I. INTRODUCTION

Interleavers are basic components in digital communication systems. In Turbo codes [1], that belong to the class of parallel concatenated codes and that are used for channel coding, the information bits are interleaved before encoded once again in parallel. Therewith, the added redundancy is optimally distributed. In the Turbo decoder, the parallel received code vectors are decoded serially and iteratively. The maximum-a-posteriori (MAP) probabilities of the information bits are calculated alternately based on the non-interleaved or interleaved received code vector. Resulting reliability information, the so-called extrinsic information of the information bits, is exchanged between the iterative decoding steps thus requiring an interleaver and a deinterleaver in the decoding loop.

A schematic of a Turbo decoder is shown in Fig. 1. The interleaver is located between the MAP-decoder and the memory banks that contain the extrinsic information in the non-interleaved order. The interleaver ensures that the extrinsic information is forwarded to the MAP-decoder in the non-interleaved or interleaved version. Therewith, the interleaver is similar to a memory management unit (MMU) and performs the necessary address transformation depending on the interleaved or non-interleaved decoding phase.

Two main MAP-decoder schemes can be identified, for which the interleaver requirements are different. Both schemes are shown in Fig. 1 on the left hand side and differ in the processing schedule of the code vector symbols. A serial-MAP (SMAP) decoder processes the code vector symbols in a serial fashion. Parallelism can be easily introduced when several SMAP-decoders work in parallel on different code vector segments. In contrary, the XMAP scheme [2] supports

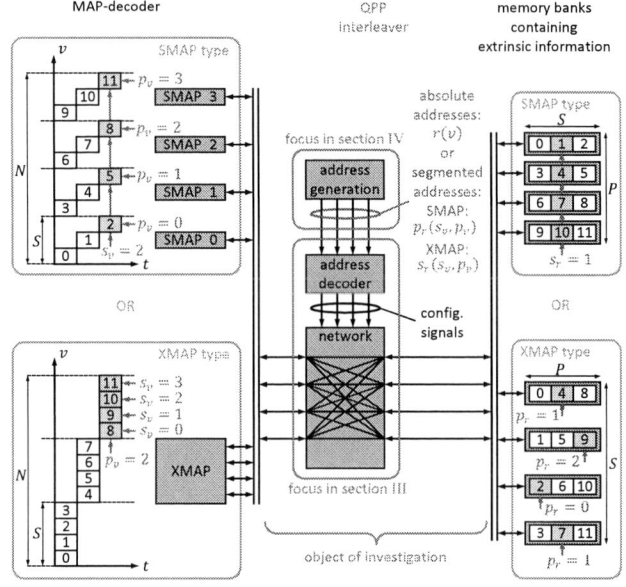

Fig. 1. SMAP vs. XMAP Turbo decoder (exemplary for $QPP(11, 6, 12)$)

just one decoder that processes all code vector symbols of a segment in parallel and all segments successively in time.

Generally, the code vector of length N is segmented into P equally-sized parts with segment length $S = \frac{N}{P}$ as shown in Fig. 1 left hand side. At one time step during decoding, the P parallel SMAP-decoders require P code symbols, one symbol of each segment at the same position within the segment. In contrary, an XMAP-decoder processes one full trellis segment in a single time step thus requiring all S code symbols of one segment at the same time.

In order to support parallel processing, the extrinsic memory has to be segmented as shown in Fig. 1 right hand side. In case of SMAP, P memory banks are needed whereas S memory banks are required for the XMAP scheme.

Let v be the (*virtual*) address of the code symbol used by the MAP decoder and r be the related *real* address of the memory where the extrinsic information of the symbol is stored. According to an interleaver function π, the interleaver maps virtual to real addresses with $r = \pi(v)$. With segmented addresses $r = s_r + p_r \cdot S$ and $v = s_v + p_v \cdot S$, $0 \leq s_r, s_v < S$, $0 \leq p_r, p_v < P$, it follows $s_r + p_r \cdot S = \pi(s_v + p_v \cdot S)$. In-

978-1-4673-2895-1/12 $31.00 © 2012 IEEE

stead of calculating the real absolute address r with $r = \pi(v)$ one can determine the segmented memory addresses via

$$s_r(s_v, p_v) = \pi(s_v + p_v \cdot S) \bmod S \qquad (1)$$

$$p_r(s_v, p_v) = \left\lfloor \frac{\pi(s_v + p_v \cdot S)}{S} \right\rfloor . \qquad (2)$$

s_v is the code vector (intra) segment address and p_v the code vector (inter) segment number. Therewith, p_v also represents the SMAP-decoder number. In the case of SMAP memory segmentation, s_r is the (intra) bank address and p_r the (inter) bank number. In contrary, for XMAP memory segmentation s_r is the (inter) bank number and p_r the (intra) bank address.

Considering the LTE standard [3], the interleaver has to be adaptable to 188 different interleaving schemes. An individually optimized parameter setup is specified for each of the 188 different interleaver functions to ensure an appropriate decoding performance. The interleaver complexity increases as the network has to be configurable on-the-fly for all 188 schemes.

In the following section II, mathematically derived properties of QPP interleavers are listed that are related to VLSI implementation aspects. In section III, the focus is on the interleaver network architecture and the address decoder for network configuration whereas section IV deals with the address generation issue.

II. Properties of QPP Interleavers

At first, the formal definition of a QPP interleaver is given.

Definition 1. (QPP interleaver) *A quadratic-permutation-polynomial interleaver $QPP(f_1, f_2, N)$ [4], [5] is defined by a polynomial index mapping function π of second degree with coefficients $f_1(odd), f_2(even) \in \mathbb{N}$ over an integer ring according to the information block length N*

$$\pi(v) := \left(f_1 \cdot v + f_2 \cdot v^2 \right) \bmod N, \qquad 0 \le v \le N-1. \quad (3)$$

The following theorems state fundamental properties of QPP interleavers with respect to implementation issues. References to proofs are denoted where required.

A. SMAP-related QPP Interleaver Properties

Theorem 1. (MV for SMAP) [6] *A QPP interleaver is maximally-vectorizable (MV) for SMAP memory segmentation and regarding intra-segment permutation, so for all segment lengths $S \mid N$ and $p_{v,x} \ne p_{v,y}$, $s_{v,x} = s_{v,y} = s_v$ it is*

$$\begin{aligned} s_{r,x} &= \pi(s_v + p_{v,x} \cdot S) \bmod S \\ &= \pi(s_v + p_{v,y} \cdot S) \bmod S = s_{r,y}. \end{aligned} \quad (4)$$

Theorem 2. (MCF for SMAP) [7], [4], [6] *A QPP interleaver is maximally-contention-free (MCF) for SMAP memory segmentation, so for $p_{v,x} \ne p_{v,y}$, $s_{v,x} = s_{v,y} = s_v$ and all segment lengths $S \mid N$ it follows*

$$p_{r,x} = \left\lfloor \frac{\pi(s_v + p_{v,x} \cdot S)}{S} \right\rfloor \ne \left\lfloor \frac{\pi(s_v + p_{v,y} \cdot S)}{S} \right\rfloor = p_{r,y}. \quad (5)$$

Theorem 3. (MD-LRS for SMAP) [8] *A QPP interleaver supports maximally-decoupled logarithmic-ring-shift (MD-LRS) for SMAP memory segmentation and regarding inter-segment permutation, if $P = 2^j$ and $P \mid N$, $j \in \mathbb{N}$. For each i in $0 \le i < j$ there are 2^i subsets of segment numbers p_v with distance 2^i, i.e. $p_{v,y} = p_{v,x} \pm 2^i$, that perform the same logarithmic ring shift according to the $(i+1)$-LSBs of the absolute shift value when mapped to the bank numbers p_r*

$$\begin{aligned} \Delta p_{rv,x} \bmod 2^{i+1} &:= (p_{r,x} - p_{v,x}) \bmod 2^{i+1} \\ &= \left(p_{r,y} - \left(p_{v,x} \pm 2^i \right) \right) \bmod 2^{i+1} = \Delta p_{rv,y} \bmod 2^{i+1}. \quad (6) \end{aligned}$$

B. XMAP-related QPP Interleaver Properties

The SMAP-related QPP interleaver properties hold for the XMAP case, too, except for the MV property. Especially, it will be proven that MD-LRS also holds for XMAP. Therewith, all benefits regarding the network design discussed in the following section apply to SMAP and XMAP schemes.

Theorem 4. (NV for XMAP) *A QPP interleaver is not-vectorizable (NV) for XMAP memory segmentation, so it exists a $s_{v,x} \ne s_{v,y}$ with $p_{v,x} = p_{v,y} = p_v$ and any segment length $S \ge 2$, $S \mid N$ for that follows*

$$p_{r,x} = \left\lfloor \frac{\pi(s_{v,x} + p_v \cdot S)}{S} \right\rfloor \ne \left\lfloor \frac{\pi(s_{v,y} + p_v \cdot S)}{S} \right\rfloor = p_{r,y}. \quad (7)$$

Theorem 5. (MCF for XMAP) [9] *A QPP interleaver is maximally-contention-free (MCF) for XMAP memory segmentation, so for $s_{v,x} \ne s_{v,y}$, $p_{v,x} = p_{v,y} = p_v$ and all segment lengths $S \mid N$ it follows*

$$\begin{aligned} s_{r,x} &= \pi(s_{v,x} + p_v \cdot S) \bmod S \\ &\ne \pi(s_{v,y} + p_v \cdot S) \bmod S = s_{r,y}. \quad (8) \end{aligned}$$

Theorem 6. (MD-LRS for XMAP) *A QPP interleaver supports maximally-decoupled logarithmic-ring-shift (MD-LRS) for XMAP memory segmentation and regarding intra-segment permutation, if $S = 2^j$ and $S \mid N$, $j \in \mathbb{N}$. For each i in $0 \le i < j$ there are 2^i subsets of segment addresses s_v with distance 2^i, i.e. $s_{v,y} = s_{v,x} \pm 2^i$, that perform the same logarithmic ring shift according to the $(i+1)$-LSBs of the absolute shift value when mapped to the bank numbers s_r*

$$\begin{aligned} \Delta s_{rv,x} \bmod 2^{i+1} &:= (s_{r,x} - s_{v,x}) \bmod 2^{i+1} \\ &= \left(s_{r,y} - \left(s_{v,x} \pm 2^i \right) \right) \bmod 2^{i+1} = \Delta s_{rv,y} \bmod 2^{i+1}. \quad (9) \end{aligned}$$

Proof:

$$\begin{aligned} \pi(s_v + p_v \cdot S) &= \left(f_1 \cdot p_v \cdot S + f_2 \cdot p_v^2 \cdot S^2 + f_1 \cdot s_v + \right. \\ &\qquad \left. 2 \cdot f_2 \cdot s_v \cdot p_v \cdot S + f_2 \cdot s_v^2 \right) \bmod N \\ &= s_r + p_r \cdot S \\ \Rightarrow s_r &= \left(f_1 \cdot s_v + f_2 \cdot s_v^2 \right) \bmod S \\ \Rightarrow \Delta s_{rv,x} &= (s_{r,x} - s_{v,x}) \bmod S \\ &= \left((f_1 - 1) \cdot s_{v,x} + f_2 \cdot s_{v,x}^2 \right) \bmod S \\ \Rightarrow \Delta s_{rv,y} &= (s_{r,y} - s_{v,y}) \bmod S \\ &= \left((f_1 - 1) \cdot s_{v,y} + f_2 \cdot s_{v,y}^2 \right) \bmod S \end{aligned}$$

$$s_{v,y} \overset{!}{=} s_{v,x} \pm 2^i$$
$$\Rightarrow \Delta s_{rv,y} = \left(s_{r,y} - \left(s_{v,x} \pm 2^i\right)\right) \bmod S$$
$$= \Big((f_1 - 1) \cdot \left(s_{v,x} \pm 2^i\right) +$$
$$+ \, f_2 \cdot \left(s_{v,x} \pm 2^i\right)^2 \Big) \bmod S$$
$$\Rightarrow \Delta s_{rv,y} \bmod 2^{i+1} = \Delta s_{rv,x} \bmod 2^{i+1}$$

C. Conclusion of the Properties

QPP interleavers are MCF for SMAP and XMAP-decoder schemes which was the reason for the adoption of QPP interleavers in the current 4G-mobile-system. Memory segmentation can easily be applied without complex scheduling techniques to support parallelism in MAP-decoders and archieve high-throughput rates.

QPP interleavers are MV only for the SMAP case. This property is used to merge the memory banks to a wide memory with just one hardware-costly address decoder. For XMAP, each memory bank has to have its own address decoder.

QPP interleavers support MD-LRS for SMAP and also for XMAP as was proven. Therewith, hardware-efficient networks can be applied to QPP interleavers with negligible configuration overhead as will be shown in the following.

III. EFFICIENT QPP INTERLEAVER NETWORKS

In this section, the focus is on the network and its address decoder that are shown in Fig. 1. The address decoder generates the configuration signals for the network using the interleaver addresses from the address generation block. The goal here is to find the most hardware-efficient network and address decoder that support all permutations defined by the QPP interleaver mapping function π.

Overviews of hardware and configuration complexities of different network approaches suitable for QPP interleavers are shown in table I and II for networks with $M = 2^i$ ports, $i \in \mathbb{N}$. The first three network approaches are known from literature. First, the network can be realized using a general crossbar (CB) approach realized with M M-to-1-multiplexer trees (MT). Second, a master-slave Batcher network (MS-BN) [10] that is based on Batcher sorting can be applied. Here, the interleaver addresses are ordered in a master network that works as an address decoder and the permutation itself is reversely performed in a slave network that is controlled by the master network. The third approach is based on a logarithmic-ring-shifter (LRS) [8] and exploits the MD-LRS property for the SMAP as well as XMAP (theorem 6) case.

Another attractive approach beeing followed here is the application of butterfly networks (BfN) or Beneš networks (BšN). A BfN features the same routing complexity as a LRS but less configuration complexity. The address decoder of a LRS needs to compute the differences between the $\frac{M}{2}$ real and virtual segmented addresses thus requiring $\frac{M}{2} \cdot \mathrm{ld}(M)$ full adders whereas the BfN can be directly configured with the real segmented addresses as will be shown. Therewith, BfNs

TABLE I
COMPARISON OF NETWORK ROUTING COMPLEXITIES

network (M ports, w bit/port)	network routing complexity (=hardware cost, number of 2-input multiplexers[a])	example: $M = 32$ $w = 6$
CB/MT	$(M \cdot (M-1)) \cdot w$	5952[a]
MS-BN	$\left(\frac{M}{2} \cdot \left(\mathrm{ld}^2(M) - \mathrm{ld}(M) + 4\right) - 2 \right) \cdot w$	2292[a]
LRS	$(M \cdot \mathrm{ld}(M)) \cdot w$	960[a]
BfN	$(M \cdot \mathrm{ld}(M)) \cdot w$	960[a]
BšN	$(2 \cdot M \cdot \mathrm{ld}(M) - M) \cdot w$	1728[a]

TABLE II
COMPARISON OF NETWORK CONFIGURATION COMPLEXITIES

network (M ports, $w = \mathrm{ld}(M)$)	network configuration complexity for QPP interleavers (=hardware cost, number of 2-input sorters[b]/ full adders[c])	example: $M = 32$ $w = 5$
CB/MT	–	0
MS-BN	$\left(\frac{M}{2} \cdot \left(\mathrm{ld}^2(M) - \mathrm{ld}(M) + 4\right) - 2 \right) \cdot w$	1910[b]
LRS	$\frac{M}{2} \cdot w$	80[c]
BfN	–	0
BšN	–	0

are the most hardware-efficient solution for realizing the QPP interleaver network inclusive the network address decoder.

A Beneš network is almost twice as large as a LRS or a BfN but it can theoretically route *all* possible permutations conflict free [11]. Therewith, the application of a BšN enables the possibility of supporting other interleaver types than QPP in case of a system that has to support multiple standards. Similiar to BfNs, BšNs can be directly configured with the real segmented addresses for QPP interleavers which is proven in the following section. MS-BNs provide similar flexibility for different standards like BšNs but they are more than twice as large for $M \geq 16$ when taking the routing and the configuration hardware costs into account.

In the following, it is proven that the Beneš and the butterfly networks are applicable to QPP interleavers with zero configuration overhead since they can be reconfigured based on the segmented addresses due to the theorems 3 and 6. So both SMAP and XMAP schemes are supported. For clarity, just the SMAP scheme is considered in the following.

A. Beneš network

A Beneš network is shown in Fig. 2. On the left hand side of the network, the P SMAP-decoders with their numbers p_v are assigned to each port of the network respectively. For each SMAP-decoder p_v a bank number $p_r(p_v)$ is determined according to (2) for a given s_v. The regarded memory bank port p_r is located on the right hand side in Fig. 2, from which the data is to be fetched respectively written back during decoding. The routing issue is to find the switching configuration so that the calculated bank numbers $p_r(p_v)$ on the left side for each p_v are routed to the ordered target ports p_r on the right. Therefore, the Beneš network is divided into

978-1-4673-2895-1/12 $31.00 © 2012 IEEE

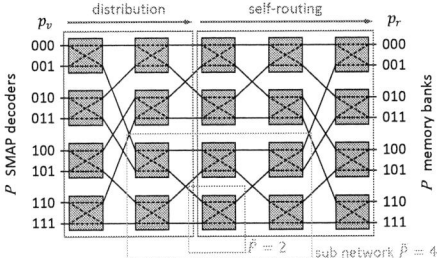

Fig. 2. Beneš network ($P = 8$) [11] for the SMAP scheme

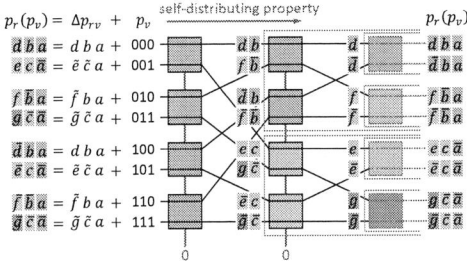

Fig. 3. Left part of the Beneš network ($P = 8$) with self-distributing property for QPP interleavers (SMAP case)

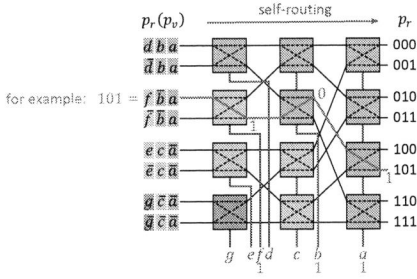

Fig. 4. Right part of the Beneš network ($P = 8$) with self-routing (SMAP case)

two parts with different independent routing algorithms [12].

The left part of the Beneš network (Fig. 3) has to ensure a proper distribution of the bank numbers $p_r(p_v)$ so that the right part of the Beneš network is able to perform the final routing. A proper configuration for distribution is found when each Beneš sub network with port width $\tilde{P} := \frac{P}{2^i}$ (see Fig. 2) has all possible different target sub port numbers at its inputs. A target sub port number is meant to be the current $p_r(p_v)$ with the i-th least significant bits (LSBs) not relevant. In case of a QPP interleaver, the distribution is trivial. Since the $p_r(p_v)$ values satisfy the bit-scheme shown in Fig. 3 on the left hand side due to the MD-LRS property (proof follows) the configuration solution is that all stages can always be in the default routing position with straight path through. The resulting proper distribution of the $p_r(p_v)$ with valid sub port numbers before all sub networks is shown in Fig. 3.

For the proof of the bit scheme, the symbolic computation of $p_r(p_v) = (p_v + \Delta p_{rv}) \bmod P$ is analyzed. The ports p_v that

have the same i-th LSBs build a subset according to i. They are routed to the same Beneš sub-network with port width $\tilde{P} = \frac{P}{2^i}$. The i-th LSBs of the p_v within each subset are equal whereas the MSBs differ. The relative shift $(\Delta p_{rv}) \bmod 2^{i+1}$ is equal for all p_v that belong to the same subset according to theorem 3. From these facts and performing the addition $p_r(p_v) = (p_v + \Delta p_{rv}) \bmod P$ for each subset it follows that the $p_r(p_v)$ have the same i-th LSBs and that the MSBs differ. All sub addresses are pairwise different as required.

For the right part (Fig. 4), that is actually a butterfly network, it is well-known that self-routing can be applied. Each port on the left side of the butterfly network is connected binary tree like to all ports p_r on the right. Therewith, the configuration signals can be directly obtained from the calculated bank numbers $p_r(p_v)$. For example, a possible self-routing path with $p_r(p_v) = \{101\}_{bin}$ is highlighted in Fig. 4. The most significant bit (MSB) of $p_r(p_v)$ is "1" so the first switching element conects the port to its lower output. The second bit is "0" so the following, second switching element switchs to its upper output and so on. The required configuration signals are $\{f\bar{b}a\}_{bin} = p_r(p_v) = \{101\}_{bin}$. Only $\frac{P}{2}$ bank numbers $p_r(p_v)$ are required to obtain all configuration signals a to g thus halving the hardware effort for address calculation.

B. Butterfly network

Actually, a Beneš network consists of two butterfly networks, one of them is mirrored, and they are merged together so that the inner stages melt into the final middle stage of the Beneš network. As was shown in the last section, the configuration complexity of a Beneš network reduces to that of a butterfly network for QPP interleavers. Therewith, it is proven that BfNs can be applied to QPP interleavers as well.

C. Results regarding the QPP interleaver network

In this section it was shown that for the implementation of QPP interleavers butterfly networks appear to be more attractive than LRS networks. A further result is the fact, that the configuration of a Beneš network is trivial for QPP interleavers. Multi standard systems with different interleaver types can be supported using the more hardware-efficient, while still fully-flexible Beneš network instead of a master-slave Batcher network.

IV. Efficient QPP Interleaver Address Generation

In this section, the focus is on the interleaver address generation according to Fig. 1 and especially on a hardware-efficient method for directly calculating the *segmented* addresses that are required for the network configuration.

Recently, Sun [13] introduced a hardware-efficient method to calculate the *absolute* QPP interleaver addresses based on a recursive address generation principle. For the XMAP-decoder, one can directly obtain the segmented addresses from the absolute addresses since the division by $S = 2^i$ in (2) is trivial in hardware.

In contrary, the calculation of segmented addresses for the SMAP case is difficult. The divisor S in (2) is generally not a power of two but can be obtained from $S = \frac{N}{P}$ with $P = 2^i$ and, therewith, depends on the interleaver scheme. In [14], a "monolithic interleaver generator" is proposed that is designed to directly calculate the bank numbers and the bank address for this SMAP case. Although the division is avoided by using a multiplier instead, the drawback of hardware- and energy-costly, multiple additions persists.

We propose to adopt the hardware-efficient method of recursive address generation to directly calculate the *segmented* addresses.

A. Recursive Address Generation of Bank Numbers and Addresses (Segmented Addresses) for parallel SMAP-decoders

For compact notations, following abbreviations are used

$$f^U(x) := (f(x)) \bmod U \tag{10}$$

$$f^{\div V}(x) := \left\lfloor \frac{f(x)}{V} \right\rfloor. \tag{11}$$

It is $f^{U \div V}(x) = \left\lfloor \frac{f(x) \bmod U}{V} \right\rfloor$, $f^{\div V, U}(x) = \left\lfloor \frac{f(x)}{V} \right\rfloor \bmod U$.

Some general function definitions used in the following are

$$f(x) := f_1 \cdot x + f_2 \cdot x^2 \tag{12}$$

$$g(x, d) := d \cdot f_1 + \left(2 \cdot d \cdot x + d^2\right) \cdot f_2 \tag{13}$$

$$c_g(d) := 2 \cdot d^2 \cdot f_2 \tag{14}$$

$$q_p(x, d) := 2 \cdot d \cdot p \cdot f_2 + g^{\div S}(x, d) \tag{15}$$

Theorem 7. (BA-FRAG) [13] *For QPP interleavers and in case of parallel SMAP decoding, forward recursive address generation (FRAG) of the bank address (BA) $a(s)$ (equation (1), $s \mathrel{\widehat{=}} s_v$ and $a(s) \mathrel{\widehat{=}} s_r(s_v)$) with d-stepwise recursion is realized with*

$$a(s) := f^S(s) = (\pi(s)) \bmod S \tag{16}$$

$$a(s + d) = \left(a(s) + g^S(s, d)\right) \bmod S \tag{17}$$

$$g^S(s + d, d) = \left(g^S(s, d) + c_g^S(d)\right) \bmod S. \tag{18}$$

Equations (17) and (18) lead to the BA-FRAG hardware architecture shown in Fig. 5. This architecture is similar to the one proposed in [13] but here with the modulo-S operation instead of modulo-N for absolute addresses. The modulo-S operation is a kind of distributed division by S over time. The control bits δ_a and δ_g that represent this distributed division are used for the proposed BN-FRAG unit.

Theorem 8. (BN-FRAG) *For QPP interleavers and in case of parallel SMAP decoding, forward recursive address generation (FRAG) of the bank number (BN) $b_p(s)$, (equation (2), $p \in [0, P-1]$, $p \mathrel{\widehat{=}} p_v$, $s \mathrel{\widehat{=}} s_v$ and $b_p(s) \mathrel{\widehat{=}} p_r(s_v, p_v)$) with d-stepwise recursion is realized with*

$$b_p(s) := f^{N \div S}(s + p \cdot S) = \left\lfloor \frac{\pi(s + p \cdot S)}{S} \right\rfloor \tag{19}$$

$$b_p(s + d) = \left(b_p(s) + q_p^P(s, d) + \delta_a(s)\right) \bmod P \tag{20}$$

$$q_p^P(s + d, d) = \left(q_p^P(s, d) + c_g^{\div S, P}(d) + \delta_g(s)\right) \bmod P \tag{21}$$

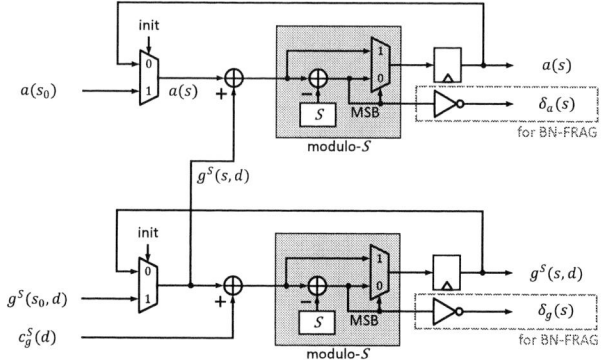

Fig. 5. Bank address forward recursive address generation (BA-FRAG)

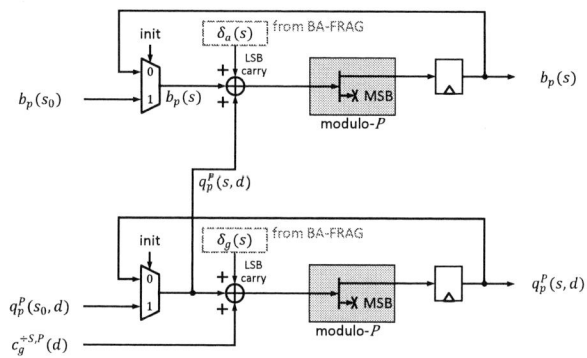

Fig. 6. Proposed bank number forward recursive address generation (BN-FRAG)

$$\delta_a(s) := \begin{cases} 0 & \text{if } a(s+d) - a(s) \geq 0 \\ 1 & \text{if } a(s+d) - a(s) < 0 \end{cases} \tag{22}$$

$$\delta_g(s) := \begin{cases} 0 & \text{if } g^S(s+d, d) - g^S(s, d) \geq 0 \\ 1 & \text{if } g^S(s+d, d) - g^S(s, d) < 0. \end{cases} \tag{23}$$

Proof:

$$b_p(s) = \left\lfloor \frac{f^N(s + p \cdot S)}{S} \right\rfloor = \frac{f^N(s + p \cdot S) - a(s)}{S}$$

$$= \frac{\left(f_1 \cdot (s + p \cdot S) + f_2 \cdot (s + p \cdot S)^2\right) \bmod (P \cdot S) - a(s)}{S}$$

$$= \left(\frac{f_1 \cdot (s + p \cdot S) + f_2 \cdot (s + p \cdot S)^2 - a(s)}{S}\right) \bmod P$$

$$b_p(s + d) =$$
$$= \left(\frac{f_1 \cdot (s + d + p \cdot S) + f_2 \cdot (s + d + p \cdot S)^2 - a(s + d)}{S}\right) \bmod P$$

$$= \left(b_p(s) + \frac{d \cdot f_1 + (2d \cdot (s + p \cdot S) + d^2) \cdot f_2 + a(s) - a(s + d)}{S}\right) \bmod P$$

$$\overset{(17)}{=} \left(b_p(s) + 2 \cdot d \cdot p \cdot f_2 + \frac{g(s, d)}{S} + \frac{-g^S(s, d) + S \cdot \delta_a(s)}{S}\right) \bmod P$$

$$= \left(b_p(s) + 2 \cdot d \cdot p \cdot f_2 + \frac{g(s, d) - g^S(s, d)}{S} + \delta_a(s)\right) \bmod P$$

$$= \left(b_p(s) + q_p^P(s, d) + \delta_a(s)\right) \bmod P$$

$$q_p^P(s+d, d) =$$

$$= \left(2 \cdot d \cdot p \cdot f_2 + \frac{g(s+d,d)}{S} - \frac{g^S(s+d,d)}{S} \right) \bmod P$$

$$= \left(2 \cdot d \cdot p \cdot f_2 + \frac{g(s,d) + c_g(d)}{S} - \frac{(g^S(s,d) + c_g^S(d)) \bmod S}{S} \right) \bmod P$$

$$= \left(q_p^P(s,d) + \frac{g^S(s,d) + c_g(d) - (g^S(s,d) + c_g^S(d)) \bmod S}{S} \right) \bmod P$$

$$= \left(q_p^P(s,d) + c_g^{\div S, P}(d) + \delta_g(s) \right) \bmod P$$

Equations (20) to (23) lead to the proposed BN-FRAG hardware architecture shown in Fig. 6. The modulo-P operation is just truncating the MSB. As can be seen, there are only two additions in total required for the calculation of each bank number $b_p(s)$. The control bits δ_a and δ_g are fed into the LSB carry inputs of the adders in the recursion loops.

Similar to the FRAG equations one can define backward RAG (BRAG) equations when substituting $s \to (s - d)$ in the FRAG equations above [13]. A BRAG unit runs reversely compared to the regarding FRAG unit. Finally, FRAG or BRAG units can be used if increasing respectively decreasing segmented addresses are required in the parallel MAP decoder.

FRAG/BRAG units can initialize other FRAG/BRAG units, so one RAG unit calculates all the initialization values for the other ones. Required initialization values for the first RAG unit can easily be precomputed using a small serial divider. The initialization ROM has just to contain N, f_1 and f_2.

B. Implementation Results

The superiority of the proposed interleaver address generation in comparison to the previously stated architecture is demonstrated using the same parametrization as [14]. The number of parallel SMAP cores is $P = 32$ and, therefore, the QPP interleaver generator consists of $\frac{P}{2} = 16$ BN-FRAG units (according to Fig. 6) for the bank numbers and a single BA-FRAG unit (according to Fig. 5) for the bank address generation. Implementation results are listed in table III. A simple technology scaling model ($A \propto s^2$, $T \propto s$, $E \propto s \cdot V_{DD}^2$) in favor of reference [14] is applied for comparison. Silicon area of the address generator is reduced by almost 65 % (A: 2.8×). The throughput is increased by a factor of two (T: 2×). The energy per operation is with 2.54 pJ only a fourth of that of the reference design (E: 3.9×).

V. Conclusion

In this paper, mainly two basic improvements regarding the implementation of QPP interleavers were proposed in the context of Turbo decoding. The improvements apply to both SMAP- and XMAP-decoder schemes.

First, the applicability of the most hardware-efficient while still fully-flexible Beneš network was proven for QPP interleavers because of zero configuration overhead. For dedicated building blocks, that do not need the full flexibility of Beneš networks, even less costly butterfly networks can be applied with even less hardware cost compared to the already field-tested logarithmic ring shifters [8]. The configuration complexity is overcome since calculated bank numbers can directly be

used as control signals for all the switching elements in the Beneš respectively butterfly network.

Second, a hardware-efficient method based on recursive calculation is developed to determine the bank numbers that are required for the configuration of the mentioned networks. Applying a conservative technology scaling, the area-time-energy complexity (ATE-product) of the proposed architecture is reduced at least by a factor of 22×.

TABLE III
COMPARISON OF QPP INTERLEAVER ADDRESS GENERATORS

		Ref. [14]	Proposed[a]
	CMOS technology, feature size s	65 nm	40 nm
	Nominal supply V_{DD} [V]	(1)[b]	0.9
	Parallel SMAP cores (P)	32	32
A	Area A [μm^2]	13 411	1 803
T	Worst case $f_{clk} = 1/T$ [MHz]	300	1000
	Nominal power P_{el} [mW]	6	2.54
E	Energy/Op. $E_{op} = P_{el}/f_{clk}$ [pJ]	20	2.54

a. Proposed design is synthesized and placed and routed using Synopsys Design Compiler respectively Encounter. Power is estimated based on an extracted netlist with parasitics and Spice simulations. b. Assumption.

References

[1] C. Berrou, A. Glavieux, and P. Thitimajshima, "Near Shannon limit error-correcting coding and decoding: Turbo-codes (1)," in *IEEE Int. Conf. on Commun. (ICC'93)*, May 1993, pp. 1064–1070.

[2] M. May, C. Neeb, and N. Wehn, "Evaluation of high throughput Turbo-decoder architectures," in *IEEE Int. Symp. on Circuits and Systems (ISCAS'07)*, 2007, pp. 2770–2773.

[3] *LTE; Evolved universal terrestrial radio access (E-UTRA); Multiplexing and channel coding*, 3GPP Std. TS 36.212, Rev. V10.1.0, Apr. 2011.

[4] O. Y. Takeshita, "On maximum contention-free interleavers and permutation polynomials over integer rings," *IEEE Trans. Inform. Theory*, vol. 52, no. 3, pp. 1249–1253, Mar. 2006.

[5] E. Rosnes and O. Y. Takeshita, "Optimum distance quadratic permutation polynomial-based interleavers for Turbo codes," in *IEEE Int. Symp. on Inform. Theory (ISIT'06)*, July 2006, pp. 1988–1992.

[6] A. Nimbalker, Y. Blankenship, B. Classon, and T. Blankenship, "ARP and QPP interleavers for LTE Turbo coding," in *IEEE Wireless Commun. and Netw. Conf. (WCNC'08)*, Mar. 2008, pp. 1032–1037.

[7] A. Nimbalker, T. E. Fuja, D. J. Costello, T. K. Blankenship, and B. Classon, "Contention-free interleavers," in *IEEE Int. Symp. on Inform. Theory (ISIT'04)*, June 2004, p. 52.

[8] C. C. Wong, M. W. Lai, C. C. Lin, H. C. Chang, and C. Y. Lee, "Turbo decoder using contention-free interleaver and parallel architecture," *IEEE J. Solid-State Circuits*, vol. 45, no. 2, pp. 422–432, Feb. 2010.

[9] M. May, T. Ilnseher, N. Wehn, and W. Raab, "A 150Mbit/s 3GPP LTE Turbo code decoder," in *Conf. Design, Automation & Test in Europe (DATE'10)*, Mar. 2010, pp. 1420–1425.

[10] C. Studer, C. Benkeser, S. Belfanti, and Q. Huang, "Design and implementation of a parallel Turbo-decoder ASIC for 3GPP-LTE," *IEEE J. Solid-State Circuits*, vol. 46, no. 1, pp. 8–17, Jan. 2011.

[11] V. E. Beneš, "Permutation groups, complexes and rearrangeable connecting networks," *Bell. Syst. Tech. J.*, vol. 43, pp. 1619–1640, July 1964.

[12] K. Y. Lee, "A new Benes network control algorithm," *IEEE Trans. Comput.*, vol. C-36, no. 6, pp. 768–772, June 1987.

[13] Y. Sun and J. R. Cavallaro, "Efficient hardware implementation of a highly-parallel 3GPP LTE/LTE-advance Turbo decoder," *Integration, the VLSI J.*, vol. 44, pp. 305–315, 2011.

[14] T. Ilnseher, M. May, and N. Wehn, "A monolithic LTE interleaver generator for highly parallel SMAP decoders," in *Wireless Telecommun. Symp. (WTS'11)*, Apr. 2011, pp. 1–4.

Tiny Application-Specific Programmable Processor for BCH Decoding

Anthony Van Herrewege
ESAT/COSIC, KU Leuven, Belgium
Email: *anthony.vanherrewege@esat.kuleuven.be*

Ingrid Verbauwhede
ESAT/COSIC, KU Leuven, Belgium
Email: *ingrid.verbauwhede@esat.kuleuven.be*

Abstract— We present a novel design for a tiny application-specific programmable processor for BCH decoding. The design is optimized for use in a PUF key extractor, where low-area overhead is extremely important. Due to it's flexible nature, it can support a wide range of BCH codes. The complete design for a BCH(413, 296, 13) decoder requires only 1% (less than 70 slices) of the available resources of a small FPGA.

Index Terms—BCH decoding, processor design, PUF key extraction, FPGA design

I. INTRODUCTION

One of the requirements of most cryptographic systems is the ability to securely generate, store and recover high-quality secret keys. The high-quality property requires the key to be both unique and unpredictable. The fact that generating such a secure key is not trivial was recently once again made clear by Lenstra et al. [1], who showed that a large amount of public RSA keys share the same prime factors, making them instantly exploitable. Designing secure storage for keys is not trivial either and often increases system implementation overhead.

Physically Unclonable Function (PUF) [2] key extractors [3–5] aim to solve both these problems. Each physical instantiation of an extractor produces a unique, unpredictable, fixed key by design, generated from the inherent randomness of the PUF. Since the key can always be regenerated with the extractor, there is no need for expensive, secure non-volatile memory. An essential part of any PUF key extractor is an error correction block.

Contribution: We present a novel design for a tiny and application-specific programmable processor for BCH decoding, a perfect fit for use in a PUF key extractor.

Paper outline: In Section II, we introduce the notation used throughout the paper and give background information on BCH code construction and decoding algorithms. Section III describes the design of our processor. Results for synthesis and runtime are presented in Section IV. Finally, conclusions are given in Section V.

II. BACKGROUND

In this section, the notation used throughout the paper is explained. Next, we look at the mathematical background of BCH codes. A short overview of BCH code construction is given and the ideas behind BCH decoding algorithms

are shown. Since this paper focuses on the design and implementation of a processor, we do not go deeper into the mathematics behind these algorithms.

A. Notation

A binary Galois field is written as $\mathbb{F}_{2^x}$. The symbol $\oplus$ is an addition over $\mathbb{F}_{2^x}$, i.e. a XOR operation, and $\otimes$ a multiplication. An element of $\mathbb{F}_{2^x}$ is written in capitals, e.g. A. $\mathcal{C}(n, k, t)$ stands for a BCH code with code length n, data length k and number of corrigible errors t.

B. BCH code construction

A BCH code $\mathcal{C}(n, k, t)$ is defined by its generator polynomial $\mathcal{G}$, which is constructed as follows [6, 7]. First, one selects the size u of the underlying field $\mathbb{F}_{2^u}$. Let $A \in \mathbb{F}_{2^u}$ be of order ord(A). For each $A^i, i = b, \ldots, b + 2t - 1$, define $\mathcal{M}_i$ as the minimal polynomial of A^i. $\mathcal{G}$ is defined as the least common multiple of all $\mathcal{M}_i$. This gives a code of length $n = \text{ord}(A)$, with $k = n - \text{ord}(\mathcal{G})$. In this paper, we only consider codes for which $A = \alpha$, a primitive element of $\mathbb{F}_{2^u}$, and $b = 1$, i.e. primitive narrow-sense BCH codes.

Codewords C are created by padding data word $D \in \mathbb{F}_{2^k}$ to length n and adding to this the modulus of the padded D and the code's generator polynomial, i.e.:

$$C = D \cdot 2^{n-k} \oplus \left(D \cdot 2^{n-k} \mod \mathcal{G} \right). \quad (1)$$

Eq. 1 clearly shows that C is always a multiple of $\mathcal{G}$, since

$$
\begin{aligned}
C \mod \mathcal{G} &= \left(D \cdot 2^{n-k} \oplus \left(D \cdot 2^{n-k} \mod \mathcal{G} \right) \right) \mod \mathcal{G} \\
&= \left(D \cdot 2^{n-k} \mod \mathcal{G} \right) \oplus \left(D \cdot 2^{n-k} \mod \mathcal{G} \right) \quad (2) \\
&= 0.
\end{aligned}
$$

By shortening the data word by m bits, the codeword will also be reduced by m bits. E.g. from $\mathcal{C}(255, 21, 55)$, one can create $\mathcal{C}(235, 1, 55)$, which has the same generator polynomial and error correction capabilities.

C. BCH decoding

BCH decoding consists of a three step process: syndrome calculation, error polynomial calculation and error position calculation. Each of these steps is explained in more detail in the next paragraphs.

978-1-4673-2895-1/12 $31.00 © 2012 IEEE

1) Syndrome calculation: The first decoding step is calculating the so called *syndromes*. One takes a received codeword $R \in \mathbb{F}_{2^n}$, which is the sum of an error-free codeword C and an error vector E, and evaluates it as a polynomial. The syndromes are the evaluation results of $R(x)$ for $x = \alpha^i$, with $i = 1, \ldots, 2t$. A syndrome S_i is thus defined as

$$
\begin{aligned}
S_i &= R(\alpha^i) \\
&= C(\alpha^i) \oplus E(\alpha^i) \quad\quad (3) \\
&= E(\alpha^i).
\end{aligned}
$$

2) Error locator polynomial calculation: Suppose we have an error vector $E = x^{l_1} + x^{l_2} + \ldots + x^{l_y}$. Then the value of the first three syndromes is:

$$
\begin{aligned}
S_1 &= \alpha^{l_1} + \alpha^{l_2} + \ldots + \alpha^{l_y} \\
S_2 &= \alpha^{2l_1} + \alpha^{2l_2} + \ldots + \alpha^{2l_y} \quad\quad (4) \\
S_3 &= \alpha^{3l_1} + \alpha^{3l_2} + \ldots + \alpha^{3l_y}
\end{aligned}
$$

The Berlekamp-Massey (BM) algorithm [8, 9], when given a list of syndromes S_i, returns an error locator polynomial

$$
\begin{aligned}
\Lambda(x) &= (\alpha^{l_1}x + 1) \cdot (\alpha^{l_2}x + 1) \cdot \ldots \cdot (\alpha^{l_y}x + 1) \\
&= (x + \alpha^{-l_1}) \cdot (x + \alpha^{-l_2}) \cdot \ldots \cdot (x + \alpha^{-l_y}).
\end{aligned} \quad (5)
$$

One of the problems with the original BM algorithm is that it requires an inversion of an element $A \in \mathbb{F}_{2^u}$ in each of its $2t$ iterations. To eliminate this costly operation, Burton [10] devised an inversionless version of the algorithm. Multiple authors have suggested improvements to this algorithm in the form of space-time tradeoffs, e.g. [11–13].

3) Error location calculation: Finding the roots of $\Lambda(x)$ gives the location of the errors in R. The Chien search algorithm [14] is an efficient way of evaluating all possible values of α^i. It does this by improving multiplications in the evaluation formula to constant factor multiplications by noting that intermediate results for $\Lambda(\alpha^{i+1})$ differ a constant factor from intermediate results for $\Lambda(\alpha^i)$:

$$
\begin{aligned}
\Lambda(\alpha^i) &= \lambda_y \cdot \alpha^{it} & + \ldots + \lambda_1 \cdot \alpha^i & + \lambda_0 \\
&\equiv \lambda_{y,i} & + \ldots + \lambda_{1,i} & + \lambda_{0,i} \\
\Lambda(\alpha^{i+1}) &= \lambda_y \cdot \alpha^{(i+1)t} & + \ldots + \lambda_1 \cdot \alpha^{i+1} & + \lambda_0 \quad (6) \\
&= \lambda_{y,i} \cdot \alpha^t & + \ldots + \lambda_{1,i} \cdot \alpha & + \lambda_{0,i} \\
&\equiv \lambda_{y,i+1} & + \ldots + \lambda_{1,i+1} & + \lambda_{0,i+1}.
\end{aligned}
$$

III. DESIGN

In general, BCH decoders are designed for high throughput, since they are most often used in high-throughput communication devices. In our case, however, the BCH decoder is intended for error correction of the output of a physically uncloneable function (PUF) [2], i.e. PUF key extraction [3–5]. In this setting, throughput is only a secondary requirement, since the PUF generates relatively few data to correct and error correction has to happen only once, at startup. Furthermore, a PUF key extractor is generally part of a larger design, and thus, should be as small as possible. As such, our design approach towards the BCH

decoder is markedly different from the de-facto standard of using systolic arrays [11–13, 15, 16]. The primary goal of the design is to be as small as possible, and be flexible, since different PUF types require different BCH parameters. Secondary comes time efficiency. In the following section, the design of the BCH decoder in explained in detail.

A. Hardware

In order to execute the three algorithms necessary for BCH decoding a controller is needed. Furthermore, this controller needs to be easily adaptable to different code parameters, because the type of BCH code used in a PUF key extraction device depends on a lot of factors such as PUF error rate, PUF output width and final key length [4]. Due to these requirements, a microcontroller design seems best suited for the decoder design.

Components The BCH decoding processor consists of three main components, which are shown in Fig. 1. Each of them is described in the next paragraphs.

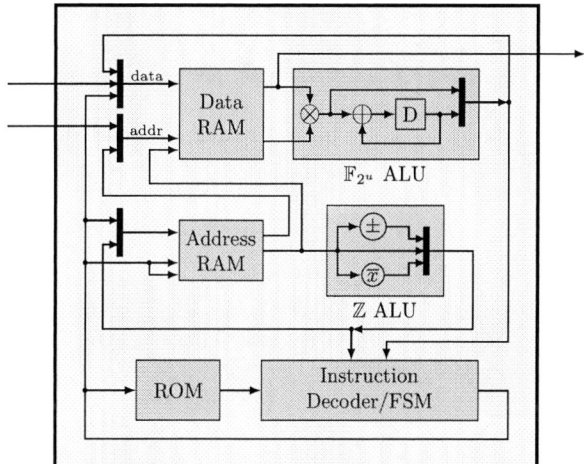

Figure 1. High-level architecture of the BCH decoder coprocessor.

1) Data block: The data block consists of a data RAM block, which stores all data necessary for the decoding as well as the corrected codeword, and an attached arithmetic unit (ALU). Since virtually all arithmetic for BCH decoding is over elements in $\mathbb{F}_{2^u}$, only a single Galois field operation is supported by the ALU: single-cycle multiply-accumulate, with the ability to execute either multiplication or addition separately. The ALU contains a single register for the accumulator and has a dual port input from the RAM.

2) Address block: Part of the novelty of our design is the use of a dedicated address block. This block consists of a tiny address RAM, of only 5 elements, and an attached ALU. The reason for including a separate address block is explained later on. The ALU works over elements in $\mathbb{Z}$ and supports increase by one, decrease by one and binary inversion, which is equal to negate and decrease by one in two's complement notation. This allows the use of the address block both for address pointer storage, for array pointer arithmetic and for keeping track of counter values.

3) Controller: The controller consists of a firmware ROM, as well as an FSM to interpret this machine code and control the microprocessor.

Communication Not only are both the data block and the address block controlled by the controller, both also have outputs connected to it. This allows the controller to compare the content of the RAMs or the result of an arithmetic operation to some fixed value. The controller can block write signals going to both RAM blocks, which allows conditional execution for all instructions.

Code analysis on the three algorithms shows that almost every arithmetic operation takes place on array elements. This lead to the development of the address block, which allows very efficient array pointer arithmetic. The address input of the data RAM is wired straight to the output of the address RAM. Therefore only indirect access of data elements is supported. Since at most five address pointers are needed at any time, the address to these pointers can be included in each instruction word. Thus, this "forced" indirect addressing actually is one of the nice aspects of the processor, driving down both firmware size and runtime. For example, an array sum can be programmed with just three instructions: accumulate, increase address pointer and conditional branch.

B. Software

The three algorithms for BCH decoding are implemented in an assembly language for the hardware described in the previous section. In this section, we list the processor's instruction set architecture (ISA) and go over some of the techniques used to achieve a time-efficient implementation.

Instruction Set Architecture Table I lists the instruction set architecture of the processor. All instructions are 10-bits wide and contain bit fields for conditional execution and (if applicable) target and destination address pointer(s). Some instructions are implemented specifically with the target algorithms in mind. E.g. the `rotr` instruction also sets a conditional execution flag depending on the LSB of the affected data word, this eliminates the need for a separate check, allowing the implementation of the *syndrome calculation* algorithm's inner loop with only two instructions.

Optimization Techniques In order to improve the runtime of our firmware a few techniques are used.

First of all, the algorithm's inner loops are all unrolled. This reduces the overhead of costly conditional jumps back to the start of the loop. Pre- and post-loop patch code is avoided by manually tuning the number of loop unrolls to the code parameters, which keeps the impact on firmware size low. This loop optimization technique improves the

Table I
INSTRUCTION SET ARCHITECTURE OF THE PROCESSOR.

Opcode	Result	Cycles
`jump`	$PC \leftarrow value$	2
`cmp_jump`	$PC \leftarrow value$ **if** $(comp = \mathbf{true})$	3
`stop`	$PC \leftarrow PC$	1
`comp`	$cond_i \leftarrow (comp = \mathbf{true})$	2
`set_cond`	$cond_i \leftarrow value$	1
`load_reg`	$reg \leftarrow data[addr_i]$	1
`load_fixed_reg`	$reg \leftarrow value$	2
`load_fixed_addr`	$addr_i \leftarrow value$	2
`mod_addr`	$addr_i \leftarrow f(addr_i)$	1
`copy_addr`	$addr_i \leftarrow addr_j$	1
`store_reg`	$data[addr_i] \leftarrow reg$	1
`store_fixed`	$data[addr_i] \leftarrow value$	2
`rotr`	$data[addr_i] \leftarrow data[addr_i] \circlearrowleft 1$	1
`shiftl_clr`	$data[addr_i] \leftarrow data[addr_i] \ll 1$	1
`shiftl_set`	$data[addr_i] \leftarrow (data[addr_i] \ll 1) \mid 1$	1
`gf2_add_mult`	$data[addr_i] \leftarrow data[addr_i] \otimes data[addr_j]$ $reg \leftarrow reg \oplus (data[addr_i] \otimes data[addr_j])$	1

runtime of our initial firmware up to 30%. The next big improvement in runtime is due to the combination of multiplication and addition in a single-cycle instruction. The merge of these two instructions results in a further 38% speedup of our error location calculation algorithm. Code duplication, in order to move conditional branches out of loops, improves the runtime of the Berlekamp-Massey implementation by another 20%. The support for conditional execution speeds up the syndrome calculation algorithm further, by 28%, due to the elimination of conditional jumps in the inner loop. Finally, the last improvement to runtime is due to improved memory management, with syndrome calculation seeing a 64% speed increase over an implementation with straightforward variable placement.

IV. IMPLEMENTATION

In the next paragraphs, we list the results for FPGA synthesis of our design and show the impact of code parameters on runtime.

Synthesis Our design is completely implemented in Verilog and was synthesized for the Xilinx© Virtex-6™ family of FPGAs using Xilinx ISE 12.2 M.63c with design strategy 'Area reduction with Physical synthesis'. As can be seen in Table II, the total size of our design is very small and changes little for different BCH codes. No separate RAM blocks are used, since our design uses RAM & ROM blocks which are implemented within LUTs. Thus, the listed slice count is the actual total size that the design requires.

Table II
SYNTHESIS RESULTS FOR IMPLEMENTATION ON A XILINX© VIRTEX-6™.

| $\mathbf{BCH}(n, k, t)$ | Area | | | F_{max} |
	[slice]	[FF]	[LUT]	[MHz]
$413, 296, 13$	65	33	244	94.4
$380, 308, 8$	66	33	244	97.8
$318, 174, 17$	68	33	251	93.6

978-1-4673-2895-1/12 $31.00 © 2012 IEEE

To the best of our knowledge, a comparison with existing BCH decoders is near impossible and makes little sense. This is due to the target application of our design: PUF key extraction. The primary goal of our design is compactness, for existing designs it is high throughput [11–13, 15–17]. Further complicating this is that the area of other implementations are either given for an ASIC implementation [11, 15, 16] or simply not stated [12, 13, 17]. Furthermore, the codes used for our target application are generally defined over $\mathbb{F}_{2^u}$ where $8 \leq u \leq 10$, with high error correcting capabilities of 3–10% [3–5], and our firmware is optimized with this in mind. We have not been able to find designs for such code parameters. Finally, most publications deal with Reed-Solomon decoding, which requires slightly different algorithms than those needed for BCH decoding, making fair comparisons even harder.

Runtime Code parameters greatly influence the runtime of each algorithm. The high-order approximate formulas for each algorithm's runtime in Table III clearly show that t has the largest influence, unless very long BCH codes are used. In this same table, formulas are given for the *ideal* runtime, which we define as: the number of cycles needed if each inner loop iteration takes one cycle, no matter how many operations are inside the loop, without parallel execution.

Comparing these *ideal* runtime formulas with the formulas for our implementation shows that the coprocessor is very efficient. Of note are the syndrome calculation and error location calculation implementations, which on average require only 2–4 times more cycles than in the *ideal* case, even with the overhead of conditional loop branches.

Table III
HIGH-ORDER APPROXIMATIONS FOR ALGORITHM RUNTIME. *Ideal*
ASSUMES SINGLE CYCLE INNER LOOPS, NO PARALLELISM.

| Algorithm | Runtime [cycles] | |
	Ideal	Actual
Syndrome calculation	$2t \cdot n$	$40t \cdot \lceil \frac{n}{u} \rceil$
Berlekamp-Massey	$3.5 \cdot (t^2 + t)$	$36t^2$
Error loc. calculation	$t \cdot n$	$3.6t \cdot n$

Table IV lists the runtime of our processor for the example BCH codes. It clearly shows that the number of corrigible errors t has the largest effect on the runtime.

Table IV
ACTUAL NUMBER OF CYCLES REQUIRED FOR BCH DECODING.

BCH(n, k, t)	Runtime [cycles]
413, 296, 13	55 379
380, 308, 8	26 165
318, 174, 17	50 320

V. CONCLUSION

We have presented the design and implementation of both hard- and software for a tiny application-specific programmable BCH decoding processor. Our design requires less than 1% (70 slices) for a BCH(413, 296, 13) decoder on a small Virtex-6 FPGA, and gets close to the *ideal* runtime for two out of three required algorithms.

Due to its extremely small size, it is the perfect match for a PUF key extraction system. Such a system will spend multiple milliseconds interfacing a PUF [4] and thus the speed of our design is well within acceptable limits.

ACKNOWLEDGMENTS

This work was supported in part by the Research Council KU Leuven: GOA TENSE (GOA/11/007) and by the European Commission through the ICT programme under contract ICT-2007-216676 ECRYPT II. In addition, it was supported by the Flemish Government, FWO G.0550.12N and by the European Commission through the ICT programme under contract FP7-ICT-2011-284833 PUFFIN and FP7-ICT-2007-238811 UNIQUE.

BIBLIOGRAPHY

[1] A. K. Lenstra, J. P. Hughes, M. Augier, J. W. Bos, T. Kleinjung, and C. Wachter, "Ron was wrong, Whit is right," Cryptology ePrint Archive, Report 2012/064, 2012, http://eprint.iacr.org/.

[2] R. Maes and I. Verbauwhede, "Physically Unclonable Functions: A Study on the State of the Art and Future Research Directions," in *Towards Hardware-Intrinsic Security*, ser. Information Security and Cryptography, A.-R. Sadeghi and D. Naccache, Eds. Springer Berlin Heidelberg, 2010, pp. 3–37.

[3] R. Maes, R. Peeters, A. Van Herrewege, C. Wachsmann, S. Katzenbeisser, A.-R. Sadeghi, and I. Verbauwhede, "Reverse Fuzzy Extractors: Enabling Lightweight Mutual Authentication for PUF-enabled RFIDs," in *Lecture Notes in Computer Science*. Springer-Verlag, Feb. 2012.

[4] R. Maes, A. Van Herrewege, and I. Verbauwhede, "PUFKY: A Fully Functional PUF-based Cryptographic Key Generator," *CHES 2012*, 2012, in press.

[5] C. Bösch, J. Guajardo, A.-R. Sadeghi, J. Shokrollahi, and P. Tuyls, "Efficient Helper Data Key Extractor on FPGAs," in *CHES*, ser. Lecture Notes in Computer Science, E. Oswald and P. Rohatgi, Eds., vol. 5154. Springer, 2008, pp. 181–197.

[6] R. C. Bose and D. K. Ray-Chaudhuri, "On a Class of Error Correcting Binary Group Codes," *Information and Control*, vol. 3, no. 1, pp. 68–79, Mar. 1960.

[7] A. Hocquenghem, "Codes Correcteurs d'Erreurs," *Chiffres*, vol. 2, pp. 147–156, Sep. 1959.

[8] E. Berlekamp, "On Decoding Binary Bose-Chadhuri-Hocquenghem Codes," *IEEE Transactions on Information Theory*, vol. 11, no. 4, pp. 577–579, Oct. 1965.

[9] J. Massey, "Shift-Register Synthesis and BCH Decoding," *IEEE Transactions on Information Theory*, vol. 15, no. 1, pp. 122–127, Jan. 1969.

[10] H. Burton, "Inversionless Decoding of Binary BCH codes," *IEEE Transactions on Information Theory*, vol. 17, no. 4, pp. 464–466, Jul. 1971.

[11] J.-I. Park, K. Lee, C.-S. Choi, and H. Lee, "High-Speed Low-Complexity Reed-Solomon Decoder using Pipelined Berlekamp-Massey Algorithm," in *2009 International SoC Design Conference (ISOCC)*, Nov. 2009, pp. 452–455.

[12] I. Reed and M. Shih, "VLSI Design of Inverse-Free Berlekamp-Massey Algorithm," *IEEE Proceedings on Computers and Digital Techniques*, vol. 138, no. 5, pp. 295–298, Sep. 1991.

[13] D. Sarwate and N. Shanbhag, "High-Speed Architectures for Reed-Solomon Decoders," *IEEE Transactions on Very Large Scale Integration (VLSI) Systems*, vol. 9, no. 5, pp. 641–655, Oct. 2001.

[14] R. Chien, "Cyclic Decoding Procedures for Bose-Chaudhuri-Hocquenghem Codes," *IEEE Transactions on Information Theory*, vol. 10, no. 4, pp. 357 – 363, Oct. 1964.

[15] W. Liu, J. Rho, and W. Sung, "Low-Power High-Throughput BCH Error Correction VLSI Design for Multi-Level Cell NAND Flash Memories," in *SiPS*. IEEE, 2006, pp. 303–308.

[16] J.-I. Park, H. Lee, and S. Lee, "An Area-Efficient Truncated Inversionless Berlekamp-Massey Architecture for Reed-Solomon Decoders," in *2011 IEEE International Symposium on Circuits and Systems (ISCAS)*, May 2011, pp. 2693–2696.

[17] H.-C. Chang and C. Shung, "New Serial Architecture for the Berlekamp-Massey Algorithm," *Communications, IEEE Transactions on*, vol. 47, no. 4, pp. 481–483, Apr. 1999.

Dataflow-Based Reconfigurable Architecture for Streaming Applications

Anja Niedermeier, Jan Kuper, Gerard Smit
University of Twente, Department of Computer Science
Enschede, The Netherlands

Abstract—**Coarse-grain reconfigurable arrays often rely on an imperative programming approach including a read/write mechanism for memory access. In this paper, we present an architecture composed of a configurable array of computing cores and memory blocks in which both the execution mechanism and configuration principle of the computing cores and the behaviour of the memory blocks are based on streaming and dataflow principles. We illustrate our ideas with the implementation of a long finite impulse response (FIR) filter where memory tiles are used to store intermediate results.**

I. MOTIVATION AND RELATED WORK

Streaming application are common in modern multimedia and wireless applications, like for example Video and Audio processing. In streaming applications, efficiency can drastically be increased if the underlying execution mechanism is based on dataflow principles, i.e. the system starts the execution as soon as the required input data is available, in contrast to conventional load/store mechanism commonly found in imperative approaches.

In embedded computing, coarse-grain reconfigurable architectures are an emerging paradigm for efficient implementations of streaming systems. Those architectures usually combine a general purpose processor (GPP) as host controller with an array of small, reconfigurable processing elements that are interconnected to form a larger, reconfigurable multicore architecture. Cores in those arrays are usually small and contain only the ALU and some local storage. Often bigger external memory is added to be able to store for example intermediate results or to provide look-up tables.

There have already been published a number of papers on coarse-grain reconfigurable architectures, an extended overview on reconfigurable architectures can be found in [1] and [2]. MorphoSys [3] is a hybrid of a host CPU and a reconfigurable array. The connection to external memory is provided via the system bus with DMA. The programming principle is based on imperative programming. ADRES [4] is a combination of a very long instruction word (VLIW) processor with a tight connection to a reconfigurable grid. The two parts are connected via a multi-port register file. Data access is performed via load/store operations. The programming is C-based. XPP [5] contains an array of 8x8 processing elements including 2 RAM-blocks per row. The XPP array can be programmed either with the low-level NML (native mapping language) or with an XPP-specific subset of C. Memory access is performed with read and write operations. DREAM [6] consists of a control unit, data path and a memory access unit. To transfer data between DREAM

and the host CPU, exchange buffers are available. DREAM is programmed using macro-instructions that are described in single-assignment C syntax. RICA [7] is a heterogeneous array of reconfigurable ICs. Dedicated control ICs are available so that RICA does not require a host control CPU. In the array, distributed memory elements are available that are accessed via special memory access ICs. The programming principle is C-based.

All the presented architectures have in common that they rely on an imperative programming approach and, thus, on a read/write method to access their memories. Whenever a certain core requires data from an external memory, it requests the data. That request can be handled via a central control unit, a memory access unit or even a dedicated IC for memory access.

Here we present a data-driven approach, where other approaches are mostly control-driven. In our architecture, the complete execution scheme is stream-based. That means, input data is expected to be available as a stream, the elements in the architecture consume and produce streams, and a stream (or multiple streams) of data is the final result. In order to achieve such an architecture, each actor was designed based on stream-based execution. In contrast to the publications mentioned above, we consider the memories as streaming actors. Our basic assumption is that a core usually does not only need a single element of data from a certain memory but a stream of data elements. A common use case for this assumption is that a certain streaming application is partitioned into several parts that are executed in sequence. The intermediate results have then to be stored in a memory and streamed back into the system at a later stage. By implementing the memories as streaming actors, the data transfer has only to be initialised once. After the initialisation, the data then automatically streams to its destination. As a consequence, dedicated load/store operations can be omitted.

In the remainder of this paper we will illustrate the proposed architecture, focusing on the streaming memory actor. Finally we will present an extended use case where different features of the architecture are exploited.

II. ARCHITECTURE

Figure 1 shows our architecture. The blocks denoted with **Cxy** represent simple reconfigurable cores with identifier **xy**, memory tiles are represented by the blocks labeled **My** where **y** is the identifier.

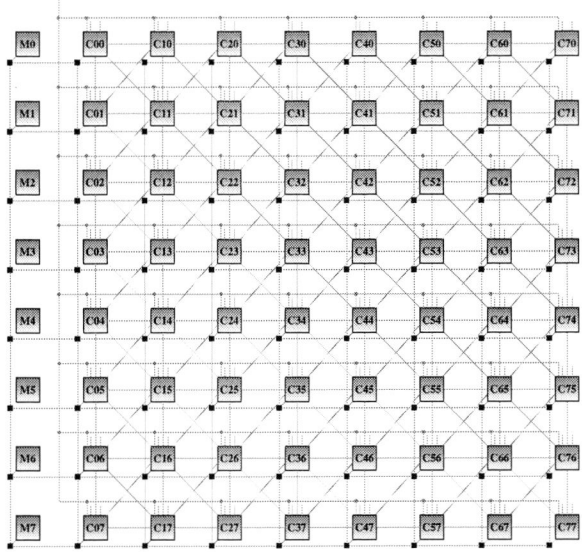

Figure 1: Architecture

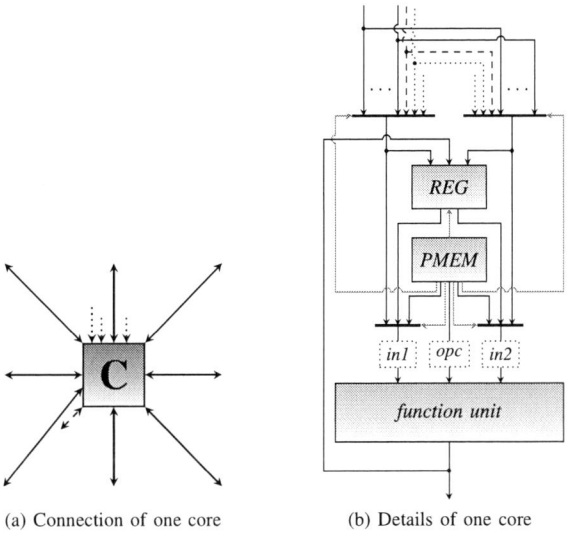

(a) Connection of one core   (b) Details of one core

Figure 2: Core

A. Cores

In Figure 2 two different views on a single core are shown. Figure 2a shows the connectivity and Figure 2b illustrates the internal composition. A core consists of:

1) Inputs: Figure 2a shows a close-up of one core of the grid shown in Figure 1 with a focus on the connectivity. The connectivity can hereby be split into three kinds: the local connection to neighbouring cores (the continuous lines), the global connection to the system via the NoC (the dashed line), and the external inputs (the dotted lines).

In Figure 2b, the same scheme is used. At the input of the core the incoming signals are connected to two multiplexers from which the correct input is selected according to the current setting in the program memory. Each multiplexer also includes one FIFO buffer to store incoming data (not shown in the graph).

2) Function unit: The function unit is responsible for the actual computations. It supports binary operations, such as numerical operations (e.g. addition, multiplication..), shifting operations and operations on the bit-level (e.g. and, or ..). Both integer and fixed-point operations are supported.

3) Local Memory: A local storage is available in form of a register file, denoted *REG* in the figure. It has three write ports and two read ports, the size of the register file can be parametrised during design-time.

4) Program Memory and Control: The program memory, labelled *PMEM*, is responsible for the actual control of the operation, the selection of the correct input, and the storage of data. It is configured with a finite-state-machine based principle which will be explained in the following section.

5) Configuration: The architecture is configured on two levels. One level represents the local view on a single core, i.e. the behaviour of one core. The other level represent the global view on the complete architecture, i.e. the flow of data in the array.

The configuration of the local view is based on dataflow principles and finite state machine (FSM) methods. The behaviour is a sequence of FSM stages, where for each stage the number of repetitions is indicated in the upper left corner of the FSM states in Figure 3. Each stage is then defined in terms of a dataflow graph with tokens on the arcs representing that either a token is required (at the inputs) or a token is produced (on the output). Furthermore, for each token it is defined where it comes from (at the inputs, *EX* represents an external input, *Rx* represents a value stored at register *R0*) or where it has to be stored (at the output).

For illustration, we use the example of a pipelined multiply-accumulate (MAC) operation on data streams. The MAC operation on the streams x and y is defined as follows:

$$mac = \sum_{i=0}^{N} x_i y_i = x_0 y_0 + x_1 y_1 + x_2 y_2 + \ldots + x_N y_N \quad (1)$$

For illustration purposes the mac operation is implemented in a pipelined fashion using separate stages for the multiplication and addition. The implementation of the complete mac operation requires three stages, of which the first one is an initial stage. In Figure 3, the configuration is shown, in Figure 4, the corresponding execution on the core is shown.

The first stage is labelled *S0*, which corresponds to Figure 4a. Here, the two external inputs (x_0 and y_0 from Equation 1) are multiplied and stored in the register file at *R0*. Following, the stage *S1*, which corresponds to Figure 4b, is executed, which represents a multiplication of x_1 and y_1 . The result of this multiplication is stored in *R1*. The final stage *S2*, shown in Figure 4c, performs an addition on the results of the *S1* and *S2* and stores the result in the register file. From here on, the core alternates between the stages *S1* and *S2*.

978-1-4673-2895-1/12 $31.00 © 2012 IEEE   127

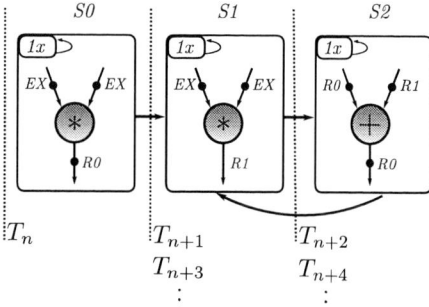

Figure 3: Configuration principle

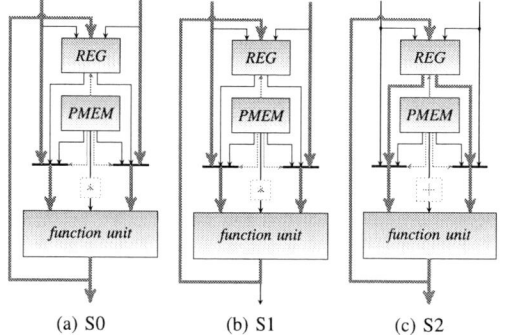

(a) S0   (b) S1   (c) S2

Figure 4: Implementation of a MAC operation on one core

B. Interconnect

In the grid, three levels of interconnects are available: local nearest neighbour connections to support locality of reference that are implemented as point-to-point links between the nodes, a global Network on Chip (NoC) to enable full connectivity of the system without the need for a fully connected network, and a broadcast network to provide an input sample simultaneously to (a subset of) all cores in the grid.

C. Memory

Each incoming packet to the memory actor is one of the following types:

1) *data*: A packet identified with D followed by the actual data, sent by a producer
2) *index*: A packet with the type identifier I followed by the identifiers of the source and final destination of the data, sent by a producer
3) *request*: A packet with the type identifier Rq followed by the identifiers of the source and final destination of the data, sent by a consumer.

A complete cycle of data transfer is illustrated in Figure 5. In this transfer, the core **C2** sends data to the memory **M0** which is at a later stage requested by core **C1**. The cycle is as follows: in the beginning, **C2** sends an index packet containing its identifier and the destination's identifier, in this case $I,(2,1)$, to **M0**. This is shown in Figure 5a. Then, **C2** sends a stream of data to **M0**, shown in Figure 5b. At a certain point in time, **C1** sends a request packet to **M0** again with an identifier-tuple

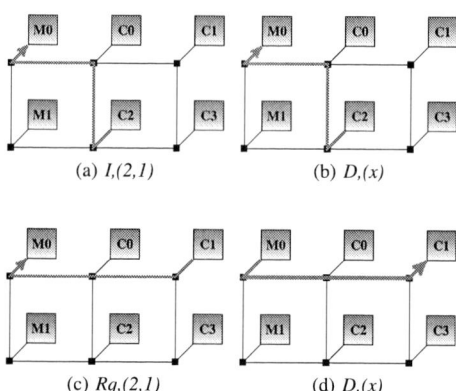

(a) $I,(2,1)$ (b) $D,(x)$

(c) $Rq,(2,1)$ (d) $D,(x)$

Figure 5: Complete data transfer

consisting of the source of its data and its own identifier, in this case $Rq,(2,1)$. This is shown in Figure 5c. **M0** will now stream data to **C1**. This is shown in Figure 5d.

III. USE CASE

The presented use case is a finite impulse response (FIR) filter [8], which is often used in the domain of digital signal processing, for example as high or low pass filter in digital audio processing.

The definition for a FIR filter is as follows:

$$y[n[= \sum_{k=0}^{M} b_k x[n - k]$$
$$= b_0 x[n] + b_1 x[n - 1] + \ldots + b_M x[n - M] \quad (2)$$

where $y[n]$ is the current output sample, $x[n]$ the current input sample, b a list of filter coefficients and $M + 1$ the length of the filter. A graphical representation is shown in Figure 6. The white rectangles represent unit delay elements, the circles represent the operations.

For long FIR filters it is often desired to partition the filter in parts which are executed in sequence. A general principle how this can be done is illustrated by the dotted rectangles in Figure 6. In this illustration, the FIR filter is partitioned into N parts of length four. The principle is quite simple: the results from one part are used as input for the next. In Figure 6, the results from the first part, denoted P_1 are streamed into the adder-row from the next part, denoted P_2. The same principle is applied to all the following parts. The final result is available at the output of P_N. If the parts are executed in sequence, the intermediate results from the parts should be stored in a memory. Consequently, the delay elements between the stages are replaced by a streaming memory actor.

A. Implementation on the proposed Architecture

On the 64 cores of our architecture, a 32-Tap FIR filter can directly be executed (each tap requires two cores, one for multiplication and one for addition). The mapping of a 32 tap filter can be seen in Figure 7. The highlighted connections

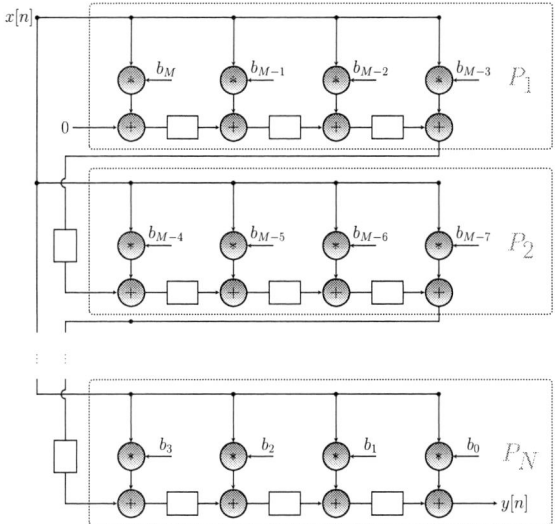

Figure 6: FIR filter - partitioned

correspond to the flow of data. Note hereby that the vast majority of the communication is implemented using the nearest neighbour network, as the FIR filter includes a high locality of reference. The delay elements are implemented by the FIFO buffers in the cores. The continuous lines are for one stage of a pipelined FIR filter, the dashed lines represent the communication between the different pipeline stages via a memory tile.

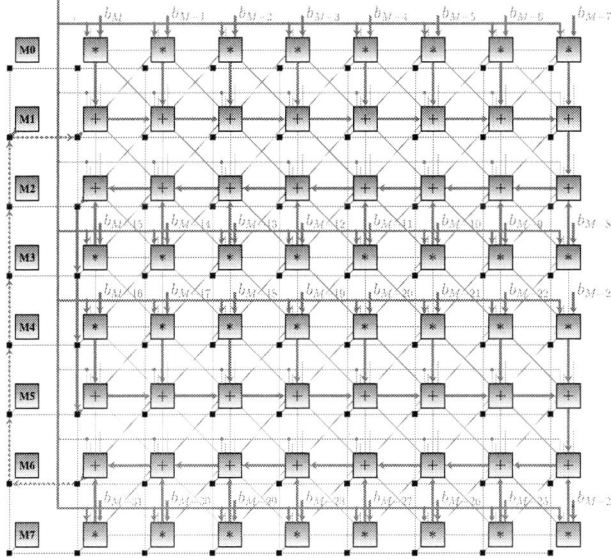

Figure 7: FIR filter

The coefficient b_M corresponds to b_{31}, coefficient b_{M-1} to b_{30} etc. The top left core (**C00** using the naming scheme of Figure 1) is the left-most multiplication node in Figure 6, i.e. it implements the multiplication of the input with coefficient b_{31}. Core **C10** implements the multiplication with b_{30} and so

on. The filter taps *M-8* to *M-15* are handled by the rows 2 and 3 (with row 0 being the top row). The next set of filter taps, *M-16* to *M-23* are handled by the rows 4 and 5. Note hereby that the connection between core **C02** (the filter tap with *M-15* and core **C05** (the filter tap *M-16* is implemented via the NoC. The remaining filter coefficients are handled by the last two rows.

For the execution of a pipelined FIR filter, the principle shown in Figure 6 is used. Furthermore, the memory transfer protocol explained in Section II-C and shown in Figure 5 is used to store intermediate values between the FIR partitions on memory **M1**. As an example, we illustrate the execution of a 128-tap FIR filter partitioned into four parts P_1 to P_4. For each part, the mapping for one 32-tap FIR filter as used in Figure 7 is used. The coefficients are adapted to each stage. First, P_1 is executed using the coefficients b_{127} to b_{96}. The resulting data is sent to the memory **M1** via the NoC, as it is shown in Figure 7. Then, P_2 with coefficients b_{95} to b_{64} is executed. For this, the results from P_1 are streamed into the grid (to core **C8**, see Figure 7). Furthermore, the results from P_2 are streamed into **M1** via the NoC, just as in the previous stage. P_3 is similar to P_2, just that the coefficients b_{63} to b_{32} are used. In P_4, the final result is produced.

IV. CONCLUSION

A coarse-grain reconfigurable architecture targeted towards streaming applications was presented, in which all blocks, including memory blocks, are streaming actors. The actors are configured by a finite state machine (FSM) logic, the flow of data in the architecture is organised in a dataflow manner.

We showed that our configuration principle gives a straightforward implementation on the presented coarse-grain reconfigurable architecture of DSP applications, as illustrated by a partitioned FIR filter.

REFERENCES

[1] C. Brunelli, F. Garzia, J. Nurmi, F. Campi, and D. Picard, "Reconfigurable hardware: The holy grail of matching performance with programming productivity," in *Field Programmable Logic and Applications, 2008. FPL 2008. International Conference on.* IEEE, 2008, pp. 409–414.

[2] B. Svensson *et al.*, "Evolution in architectures and programming methodologies of coarse-grained reconfigurable computing," *Microprocessors and microsystems*, vol. 33, no. 3, pp. 161–178, 2009.

[3] H. Singh, M. Lee, G. Lu, F. Kurdahi, N. Bagherzadeh, and E. Chaves Filho, "Morphosys: an integrated reconfigurable system for data-parallel and computation-intensive applications," *Computers, IEEE Transactions on*, vol. 49, no. 5, pp. 465–481, 2000.

[4] B. Mei, S. Vernalde, D. Verkest, H. De Man, and R. Lauwereins, "Adres: An architecture with tightly coupled vliw processor and coarse-grained reconfigurable matrix," in *Field-Programmable Logic and Applications.* Springer, 2003, pp. 61–70.

[5] V. Baumgarte, G. Ehlers, F. May, A. Nückel, M. Vorbach, and M. Weinhardt, "Pact xpp—a self-reconfigurable data processing architecture," *the Journal of Supercomputing*, vol. 26, no. 2, pp. 167–184, 2003.

[6] F. Campi, A. Deledda, M. Pizzotti, L. Ciccarelli, P. Rolandi, C. Mucci, A. Lodi, A. Vitkovski, and L. Vanzolini, "A dynamically adaptive dsp for heterogeneous reconfigurable platforms," in *Design, Automation & Test in Europe Conference & Exhibition, 2007. DATE'07.* IEEE, 2007, pp. 1–6.

[7] S. Khawam, I. Nousias, M. Milward, Y. Yi, M. Muir, and T. Arslan, "The reconfigurable instruction cell array," *Very Large Scale Integration (VLSI) Systems, IEEE Transactions on*, vol. 16, no. 1, pp. 75–85, 2008.

[8] J. McClellan, R. Schafer, and M. Yoder, *Signal processing first.* Pearson/Prentice Hall, 2003.

Enhancing Cache Coherent Architectures with Access Patterns for Embedded Manycore Systems

Jussara Marandola*, Stephane Louise[†], Loïc Cudennec[†], Jean-Thomas Acquaviva[†] and David A. Bader*

* Georgia Institute of Technology
School of Computational Science and Engineering
Email: jkofuji@usp.br

[†] CEA, LIST, CEA Saclay Nano-INNOV, Bat 862
PC 172, Gif-sur-Yvette, 91190 France
stephane.louise, loic.cudennec, jean-thomas.acquaviva @cea.fr

Abstract—One of the key challenges in advanced micro-architecture is to provide high performance hardware-components that work as application accelerators. In this paper, we present a Cache Coherent Architecture that optimizes memory accesses to patterns using both a hardware component and specialized instructions. The high performance hardware-component in our context is aimed at CMP (Chip Multi-Processing) and MPSoC (Multiprocessor System-on-Chip).

A large number of applications targeted at embedded systems are known to read and write data in memory following regular memory access patterns. In our approach, memory access patterns are fed to a specific hardware accelerator that can be used to optimize cache consistency mechanisms by prefetching data and reducing the number of transactions. In this paper, we propose to analyze this component and its associated protocol that enhance a cache coherent system to perform speculative requests when access patterns are detected. The main contributions are the description of the system architecture providing the high-level overview of a specialized hardware component and the associated transaction message model. We also provide a first evaluation of our proposal, using code instrumentation of a parallel application.

I. INTRODUCTION

Chip multi-processing (CMP) has become very popular lately, providing the power of massively parallel architectures on a single chip. One of the key challenges arising from these systems consists in designing the right programming model which would be as independent as possible of the underlying hardware. This is particularly critical in the field of data management between cache memories of many-core systems. In such architectures, each core may store a copy of a data element in its cache. Cache coherence is either directly managed by the programmer or falls under the control of a cache coherence unit (usually hardware based). This second solution makes all updates and data transfers transparent and also simplifies the development of applications. Unfortunately, it is known to have a cost in term of hardware design, refraining it from being massively adopted in embedded computing.

To make data coherence more attractive for massively-parallel embedded architectures, we think that cache coherency models and protocols should be tightly adapted to the needs of targeted applications. A large number of applications deployed on embedded devices focus on image, video, data stream and workflow processings. This class of applications tends to

This work was done in part as the first author was in CEA, LIST

access data in a regular fashion, using a given set of memory access patterns. These patterns can be used to optimize the cache coherency protocol, by prefetching data and reducing the number of memory transactions.

Using memory access patterns has already been studied in the literature but, as far as we know, our way of mixing a software and a hardware approach is unique. In this paper we describe the system architecture of a hardware component proposed to store and manage patterns, and the associated protocol which takes advantage of them for optimizing memory consistency and access time. Our main contributions provide the state of the art of directory-based cache protocols, adding an optimization for regular memory access patterns. These contributions are part of the CoCCA project standing for Co-designed Coherent Cache Architecture. We also provide in this paper a first evaluation by code instrumentation of a typical parallel program.

The remainder of this paper is organized as follows: section II presents a brief state of the art and related works; section III explains our architecture and how we optimized it to take the best advantage of regular memory access patterns that occur in applications, the principles, and the associated protocol; section IV relates an analysis of a first parallel benchmark by tracking memory traces. Finally, section V concludes and gives some perspectives about this work.

II. MEMORY CONSISTENCY AND STATE OF THE ART

A. Context: Cache Coherence for CMP Architectures

Shared Memory Chip Multi-Processor Architectures are expected to host up to hundreds of cores. These cores are connected through a scalable network (Network on Chip, NoC) usually based on a mesh topology. In this context, coherence issues occur when data are replicated on different cache memories of cores, due to concurrent read and write operations. Versions of data may differ between cores and with main memory. In order to maintain consistency, one popular approach is to use a four-state, directory-based cache coherence protocol. This protocol, called *baseline* protocol, is a derivative of the Lazy Release Consistency [1] protocol.

B. Baseline Protocol: a Directory-based Cache Protocol

In order to illustrate the behavior of the baseline protocol, we consider a CMP machine. Each core of the machine hosts a

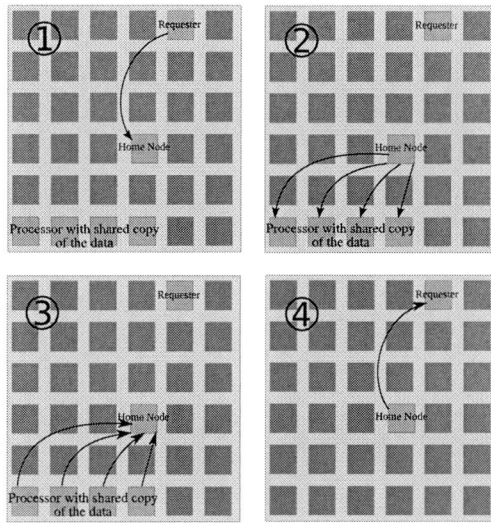

Fig. 1. Baseline Protocol: an example of a simple Write Request

L1 instructions and data caches, a L2 cache, a directory-based cache, a memory interface and a network (NoC) interface. The directory-based cache hosts coherency information of a given set of data stored in the cache (see also Figure 2 since the CoCCA approach only modifies the CMP architecture by adding the Pattern Table and modifying the protocol). The coherency information is a set of $(N+2)$ long bit-fields, sorted by memory addresses, where N is the number of cores in the system. Traditionally, the coherency information is composed by 2 bits representing the coherence state, plus a N bits-long presence vector. The coherence state field represents four states, as defined by the MESI protocol:

- M (modified): a single valid copy exists across the whole system; the core owning this copy is called *Owner* of the data and has the right to write. The value of this copy has changed since the data was cached by the owning core.
- E (exclusive): a single valid copy exists across the whole system, the core owning this copy is named the *Owner* of the data and has the right to write. The data was not modified since it was cached by the owning core.
- S (shared): multiple copies of the data exist, all copy are in read-only mode. Any associated core is named *Sharer*.
- I (invalid): the copy is currently invalid, should not be used and so will be discarded.

The length of the presence vector is equal to the number of cores in the system: 0 at the ith bit means the data is not cached in core i, and 1 at the jth bit means it is cached in core j. For each data element managed by the coherency protocol, a dedicated node, named Home-Node (HN), is in charge of managing coherency information for this particular element. In the literature, many cache coherence protocols, such as proximity-aware [4], alternative home-node [5], MESI [6] and MESIF [7] derivate from the baseline protocol.

We can illustrate the baseline protocol on a write request transaction, as shown in Figure 1: it triggers a sequence of messages transmitted between different cores. 1) The requester sends a message to the home node in charge of keeping track of the coherency information. 2) The home node checks the vector of presence and sends an exclusive access request to all the cores owning a copy of the data. 3) Then, all these cores invalidate their own copy and send an acknowledgment back to the home node. 4) Finally, the home-node grants the write permission to the requester and possibly transfers an up-to-date version of the data.

C. Optimizing the Cache Coherence Protocol

The number of messages generated by the coherency protocol is one of the most important criterion used to evaluate the overall performance. In section II-B, we have seen that a simple write request generates a four-step transaction with up to 10 messages sent over the network.

In a more sophisticated case, we can imagine an application accessing a picture column by column. This type of access cannot be handled by the baseline protocol in only one transaction. This simple example shows a case where the baseline approach falls into a worst-case scenario.

Working on columns in a picture can be achieved with the help of data access patterns. Patterns can be used to speculate on the next accesses, prefetching data where they will be most likely used in a near future. Patterns can also be used to save bandwidth, by reducing the number of protocol messages: one transaction can provide access to a whole set of data.

D. Related works

1) Exploiting Data Access Patterns: In the literature, several projects propose to optimize data consistency protocols by supporting data access patterns. This has been explored in the fields of database systems, distributed shared memories or processor cache management.

In [2], Intel uses patterns as a sequence of addresses stored in physical memory. A dedicated instruction set is provided to apply patterns, given a base memory address and an offset. A single call to these instructions can perform accesses to non-contiguous addresses in the cache. However this mechanism is limited to data stored in one cache.

In [3], IBM proposes to sort patterns by type: read-only, read-once, workflow and producer-consumer. Corresponding patterns are stored in a hardware component. Dedicated processor instructions are provided to detect and apply patterns. Here again, this mechanism is not fitted to the context of many-core computing, as it only applies patterns on a local cache.

2) MPSoC and other platforms: Our Cache Coherence Architecture was developed with MPSoC (Multiprocessor System-on-chip) in mind. This architecture is based on a multicore system with state of the art shared memory and paradigm of parallel computing. Platforms such as hybrid or heterogeneous systems (*e.g.* composed with CPU+GPU) could adopt our model of architecture.

For high performance computing, we aim at the kind of following hybrid systems: multicore processor with GPU (Graphics Processing Unit), NVIDIA Tesla (Cluster of GPU

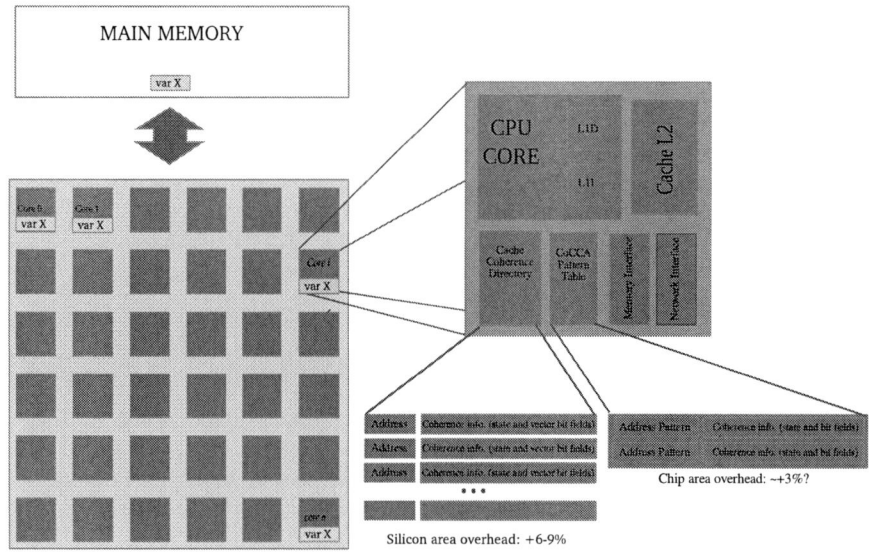

Fig. 2. CoCCA Architecture showing the addition of the Pattern Table to the cache hierarchy

processors), AMD FireStream (AMD processor + GPU), all processors presenting a new programming paradigm called HMPP - Heterogeneous Multicore Parallel Programming. This paradigm is dedicated to embedded and superscalar processors.

A related work is the TSAR project that describes a multi-core architecture with a scalable shared memory that supports cache coherency. The TSAR system [9] aims to achieve a shared memory architecture including thousands of RISC-32 bit processors.

III. CoCCA ARCHITECTURE AND PROTOCOL DESIGN

A. Principle and motivations

The main contribution is the specification of the CoCCA protocol of transaction messages that provides support for managing regular memory access patterns. The associated messages are called speculative messages. The CoCCA proto-col is a hybrid protocol designed to interleave speculative mes-sages and baseline messages through a hardware-component that has the following purposes: store patterns and control transaction messages.

The optimization of the CoCCA protocol is based on finding memory addresses of application matching a stored pattern. The requester sends the speculative message to the CoCCA Home Node (or Hybrid Home Node, HHN) if it matches a stored pattern or otherwise, the requester sends the baseline message to the ordinary Baseline Home Node (BHN). This optimized method enhance the performance of cache coherency traffic, aiming for the following advantages:

- reduction of throughput of messages,
- lower time of memory accesses.

In the next sections, we will present the design principles of our architecture, a first specification of simple patterns, the bases of our protocol and the associated data structures.

B. CoCCA Architecture principles

A typical implementation of the CoCCA Architecture would have several dozens of cores for a start. Each CPU core of the system may be involved in exchange of coherence messages, taking *four* different roles with regards to data, *i.e.* a core can play one role (or several) in transactions related to a given data, and assume different roles for different data.

- Requester, the core asking for a data.
- Home Node, the core which is in charge of tracking the coherence information of a given data in the system.
- Sharer, a core which has a copy of the data in its cache. This copy is in "*shared mode*", *i.e.* multiple copies of this data can exist at the same time for several cores.
- Owner, a core which has a copy of the data in its cache. This copy is in "*Exclusive*" or "*Modified*" mode, so one and only one instance of it can exist at this time across the whole system.

Each CPU has the following components of cache hierarchy: L1 caches, a L2 cache (shared inclusive), a directory of cache coherence and the "*CoCCA Pattern Table*", as seen on figure 2.

C. CoCCA Pattern Table

Patterns are used to summarize the spatial locality associ-ated to the access of data. The pattern table lookup process uses a signature of a pattern which is either its base address, or in a more general way a "trigger", i.e. a function that provides a specific signature of a pattern[1], as seen in figure 3. One can imagine lot of different principles for patterns, but let us illustrate it with the simplest of them: the 2D strided access, since it would cover a lot of data accesses encountered in embedded applications.

---

[1]Of course the simplest trigger, and the one we implemented in our evaluation in this paper, is the use of the base address.

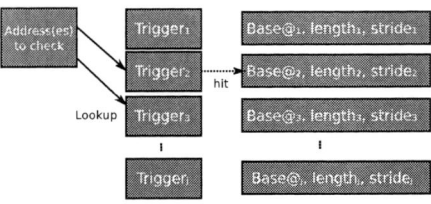

Fig. 3. Cocca Pattern Table lookup principle: in a first implementation, triggers are the base addresses of patterns

Such CoCCA patterns would be defined as triplets:

$$Pattern = (baseaddress, size, stride)$$

Where $baseaddress$ is the address of the first cache line of the pattern, $size$, the size of the pattern (number of elements), and $stride$ expresses the distance between two consecutive accesses of the pattern.

In figure 3, we present some initial concepts about the Cocca Pattern Table. The base architecture is a multi-core system, each core fitted with its memory hierarchy (L1, L2), Directory and Pattern Table, and all cores have access to a Network on Chip (NoC) that permits each core to communicate with one another and with main memory (figure 2). The CoCCA protocol optimizes the coherency protocol for the stored memory access patterns of a given application running on the system.

The pattern descriptor enables to describe the CoCCA pattern table entry:

$Desc = f_n(B_{addr}, s, \delta)$, with:

- $f_n()$ the function that builds the pattern of length n with the given characteristics,
- Desc the pattern descriptor that results from applying function $f_n()$ with the parameters B_{addr}, s and δ,
- B_{addr} represents the offset address, regarding the first address of many addresses composed by pattern access on address lookup,
- s is the size of the pattern, or number of elements,
- δ is the stride between two given accesses in the pattern

We can define an example of pattern access:

$$\{1, 4, 2\} \ following \ @1 + 1 = @2$$
$$and \ \ \{1, 4, 2\}(@1) = (@2, @5, @8, @11)$$

Applying address @1 to the pattern $\{1, 4, 2\}$ is a sequence of 4 addresses starting at @2 (@1 plus 1 offset) and with an interval of 2 addresses not belonging to the pattern between two successive addresses. This defines the base addresses of our simple patterns. In general there can be more than one address for each element of the pattern: this is given by an extra n parameter, which defines the length of each access in the pattern.

In our future work we want to simulate our principles using a transaction level model (TLM) like simSoC, but a first approach of cache coherence architecture was developed through an API that describes the hardware behavior. The implementation used the C language where a Typedef structure modelizes hardware storages and the API describes the execution of a special instruction-set to manage pattern tables.

A pattern table is similar to a hash table describing pattern ids (or triggers, as seen previously) and associated patterns.

Definition of Pattern Structure

```
typedef struct Pattern_ {
unsigned long capacity;  /* sizeof(address) */
unsigned long size;      /* address number */
unsigned long * offset;  /* pattern offset */
unsigned long * length;  /* pattern length */
unsigned long * stride;  /* pattern stride */
} Pattern_t;
```

Regarding the pattern table, we described it as composed by patterns. Our pattern structure is based on an associated stride; the key elements that compose the pattern are: offset, length and stride. To this, we added the capacity and a size. The capacity represents the memory space required to store each pattern and size is the space in memory.

In our first approach toward embedded systems, we thought that pattern table can be the result of the compilation process of an optimized application. Therefore, pattern tables can be fetched as part of a program, by using a specialized set of instructions. For our evaluation, a library is used to simulate the use of these special instructions:

- PatternNew(): function to create a pattern,
- PatternAddOffset(): function to add an offset entry,
- PatternAddLength(): function to add a length entry,
- PatternAddStride(): function to add a stride entry,
- PatternFree(): function to release the pattern after use.

D. Protocol and Home Node management

The pattern table permits the management of a hybrid protocol to improve the performance of transaction messages in the memory system. The hybrid protocol was specified for a Cache Coherent Architecture to optimize the flow of messages, interleaving baseline messages and speculative messages. We introduce the concept of granularity of messages to avoid hotspots of messages in the system.

The specification of this hybrid protocol presents the following characteristics:

- Difference between baseline and speculative messages,
- Speculative messages that permit to read all addresses of pattern through their base address,
- Requests of speculative messages by page granularity,
- Round-Robin method to choose the Home Node (HN).

The CoCCA protocol augments the baseline protocol with a dedicated protocol for managing memory access patterns. Both protocols have an important actor in message management: the Home Node (HN). Each core of the system is the Home Node of a fraction of the cached data, and the coherence information of these data are kept within an extra storage named "Coherence Directory".

When a processor accesses a data element, it needs to check the coherence state by asking to the corresponding HN. This task is handled by the coherence engine which manages all messages related to shared memory accesses. Therefore, the initial step is to determine which core in the system is the HN of the requested data. The basic, and classical, algorithm is

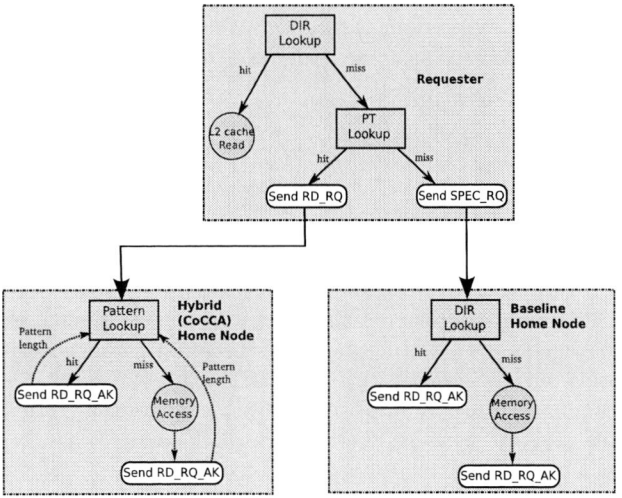

Fig. 4. Model of Transaction Messages including the Home Nodes

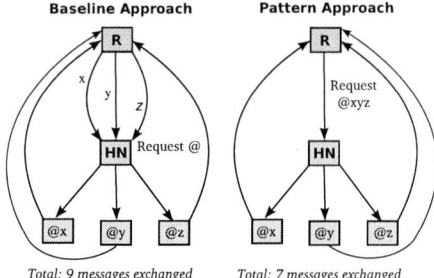

Total: 9 messages exchanged   Total: 7 messages exchanged

Fig. 5. Baseline and Pattern Approach comparison for a sequence of read requests

an allocation of HNs in a round-robin way. Round-Robin is performed by a modulo operation on the low order bits of the address of the data element.

One key question is the granularity used for the round robin algorithm. For instance, it has be shown that memory accesses are not distributed in a homogeneous way, leading to an uneven bandwidth consumption (some cores become hotspots because they are solicited often).

To reduce the hotspot issue of cache systems, the throughput of messages, and to limit the number of messages of cache coherence protocol, we chose a different granularity to determine the HN of the Baseline protocol and of the CoCCA (pattern specific) protocol: we use the line granularity and page granularity, respectively.

E. Transaction Message Model

Figure 4 shows the read transaction message model in the coherency hybrid-protocol that describes the key roles: requester, hybrid (CoCCA) HN, baseline HN. Each read access triggers the search of its address in the pattern table. If the pattern table lookup returns true, the base address is sent to the hybrid HN. Otherwise, the baseline message is sent to the baseline HN (note that table lookups can be done in parallel).

For the hybrid protocol, the roles can be divided in: core that request the data (requester), the baseline HN and the hybrid HN. The baseline HN is the core appointed by fine granularity (line) for the Round-Robin attribution, and the hybrid HN uses the coarse granularity (page) for the Round-Robin attribution. The first model of decision tree is based on the read transaction message, where the pattern table lookup is similar to a cache lookup as seen in the top box of figure 4. This schematic decision tree for read accesses describes data lookup in cache. In the case of a pattern table hit, the speculative message is sent to the Hybrid HN (determined by page granularity).

The main interest of pattern tables, is that the ranges of addresses that are defined by patterns provide a way to enhance

the baseline protocol (MESI modified protocol) by authorizing a speculative coherence traffic which is lighter (*i.e.* with less message throughput) than the baseline protocol alone. Hence, it accelerates shared memory accesses (see figure 5).

In case of a cache miss, the flow of message transaction defined by the baseline protocol is sent by requested addresses. When the pattern table and the speculative (CoCCA) protocol is added and a pattern is triggered, the flow of message transaction can be optimized in term of messages, because the pattern provides a means to use speculative messages which are in fewer number than in the baseline protocol. As a conclusion, when a pattern is discovered in our approach, the number of transaction messages is reduced by using speculation, leading to a better memory access time, less power consumption and an optimized cache coherency protocol.

In figure 5, we compared two approaches of cache coherency protocols for a cache miss case. We present two scenarios: the baseline only approach, where the requester node sends the sequential addresses x, y, z, totalizing 9 messages; and the pattern approach where the node sends speculatively the pattern with xyz, totalizing only 7 messages, and a early (speculative) prefetching of data in the cache.

IV. CODE INSTRUMENTATION AND FIRST EVALUATION

A. Choice of a first benchmark program

Our goal is to provide a first evaluation of the performance of our hybrid coherence protocol over the baseline protocol alone (but it is worth noting that the CoCCA hybrid protocol relies on the baseline protocol for the messages outside of the prefetch mechanisms).

Therefore we need a benchmark program that would be representative of algorithms found in the embedded world, easy to parallelize, and that shows an interesting variety of behaviors with regards to cache coherence and prefetch. We decided to use a cascading convolution filter: it is very typical of image processing or preprocessing, makes a good reuse of data, and is easy to parallelize. The cascading part of the convolution filter uses the destination image of one filter as the source of the new filtering, the old source image becoming the new destination image, triggering a lot of invalidation messages in the baseline protocol.

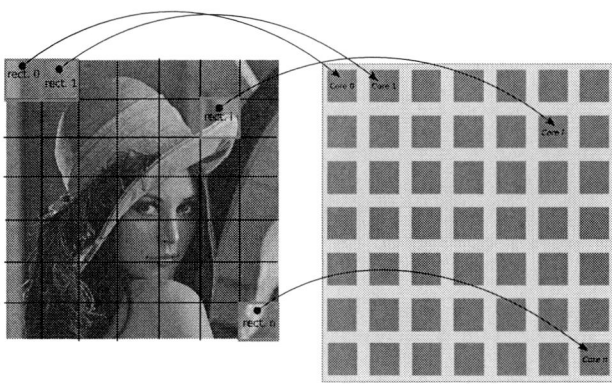

Fig. 6. Scheme of affectation of source image parts to each core in our benchmark program

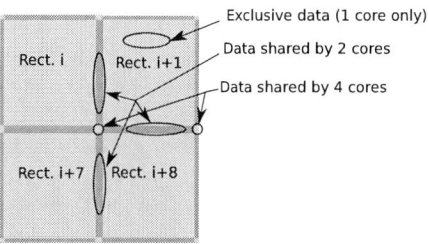

Fig. 7. Read data sharing in conterminous rectangles

The choice done was to process the algorithm by dividing the source and images in nearly equal rectangles (little variations in rectangle sizes are due to the uneven division in integers) as seen in Figure 6. Source and destination images have a resolution of 640×480 and the underlying CMP architecture is chosen as a 7×7 processor matrix, each with 256KB of L2 cache (64B by line of L2 cache). Images are defined as a set of pixels and each pixel is composed of 3 floating point values (32 bits). Both the source and the destination parts of the image managed by a given core can fit in its L2 cache. This is not the best possible implementation of a cascading filter, but this application can show lots of different behaviors regarding caches and consistency.

B. Instrumentation

In order to make a first evaluation of the hybrid protocol, we need to extract shared data read and write for each core, for this program. We decided to use the Pin/Pintools [8] software suite to that end. Pin is a framework that performs runtime binary instrumentation of programs and provides a wide variety of program analysis tools, called pintools. It uses JIT techniques to speed up instrumentation of the analyzed program. A lot of different pintools exists from the simplest (like "*inscount*") to very elaborate ones.

Let us give an example of the use of Pin, using the Simple Instruction Count (instruction instrumentation); this *inscount* pintools instruments a program to count the number of instructions executed. Here is how to run a program and display its output:

```
~/pin-2.10> ./pin -t $(PIN_OBJ_PATH)/inscount.so --
   ./cascading-convo-single-proc
~/pin-2.10/test> cat inscount.out
Count 450874769
```

This is the number of sequential instructions executed when running a mono threaded version of our program. There exists a multi-thread version of this pintool. When the multi-threaded convolution is used, we can obtain a number of instructions executed per core:

```
Number of threads ever exist = 50
Count[0]= 238062
```

```
Count[1]= 9064522
Count[2]= 9087339
...
```

An interesting pintool is *pinatrace* which is a memory reference trace (instruction instrumentation): this tool generates a trace of all memory addresses referenced by a program. We modified it to provide also the core Id on which a given memory access is done. It generates the *pinatrace.out* file:

```
0x401c5e: R    0xa0aecc   4        12
0x401d5d: W    0x6832b4   4        12
0x40119c: R    0xa02c20   4        9
0x4011c8: R    0xa02c2c   4        9
...
```

The indications of this file are, in order: the Load/Store instruction address, the Read (R) or Write (W) status, the memory access address, the size of the access in bytes, and the core Id number of the CMP architecture (the execution cores are numbered from 1 to 49 in the output). With this modification of the *pinatrace* pintool, we filtered the accesses to the shared memory accesses. The trace file has nearly 48 millions accesses for a single execution.

C. Approach to patterns

We can define three kinds of patterns on this benchmark:

- Source image prefetch and setting of old Shared values (S) to Exclusive values (E) when the source image becomes the destination (2 patterns per core),
- False concurrency of write accesses between two rectangles of the destination image. This happens because the frontiers is not alined with L2 cache lines. The associated patterns is 6 vertical lines with 0 bytes in common[2],
- Shared read data (because convolution kernels read pixels in conterminous rectangles, see figure 7). There are 6 vertical lines and 3 sets of two horizontal lines for these patterns.

As can be seen a few set of simple patterns are enough to cover all the coherence data for our benchmark program. Number of patterns is limited to 6 patterns for each core to handle all the coherency issues. This is a tiny number, showing that our approach is sustainable without having too much of impact on chip size (we can imagine to keep the pattern tables and associated components for managing pattern table lookup

---

[2]hence, the CoCCA pattern has the information that this is a false concurrency, and that the synchronization can be serialized.

and the enhanced protocol at half the size of the coherence directory, by storing only the most relevant patterns).

D. First evaluation of the protocol

The hypotheses we rely on at this point, is that the pattern tables are a given result of the compilation process, either by code instrumentation, as above, or by static analysis. They are statistically attributed for a given part of an application, and reloaded as required. This is a valid hypothesis in the embedded world where static generation is often standard because the system is tuned to a limited set of applications that are highly optimized. For pure HPC systems, an automatic dynamic generation of patterns would be preferable, but this is still future work.

In the periodic execution of our program, once initialization is passed, we have the following trend of message for the pure MESI versus the hybrid protocols:

Condition	MESI	CoCCA
Shared line invalidation	34560	17283
Exclusive line sharing (2 cores)	12768	12768
Exclusive line sharing (4 cores)	1344	772
Total throughput	48672	30723

Hence, there is a reduction of over 37% of message throughput. This does not includes the advantages of data prefetch which reduce in a large way the memory access latency. On this example, prefetch stands for about 10% of the on-chip cache sharing and nearly all main memory accesses, minus the first ones corresponding to the first access of a given pattern.

On an Intel Xeon Nehalem a single task runs in a bit less than 4490.10^3 cycles on a core, with preloaded caches (no misses, 37128 write accesses and 928200 read accesses in shared memory). This is the expected speed when the CoCCA protocol is used, since, in this case, caches are efficiently prefetched. With the baseline protocol alone, all the write accesses trigger a memory access, 17283 read misses with memory also appear and about 13000 cache sharing requests. When using 80 cycles to access main memory and a mean of 20 cycles to access on chip L2 shared data, this gives a total overhead of 3.10^6 cycles or 67% slower. For this application, using a speculative protocol like CoCCA is a huge performance boost.

V. CONCLUSION

With the growing scale of chip multi-processors, data cache consistency becomes one of the key challenge to efficiently support parallel applications. This is also true for embedded systems: a large number of embedded applications read and write data, according to regular memory access patterns. These patterns can be used to optimize cache coherence protocols and therefore, to improve application performances when sharing data among cores.

In this paper, we proposed a system architecture of Cache Coherency Architecture that make such use of memory access patterns. A regular consistency protocol has been enhanced to handle speculative requests and a new hardware component has been designed to store and retrieve patterns. We described a new hardware-component with an auxiliary memory unit that composes the cache hierarchy to implement that.

We provided a first evaluation of the benefits of our enhanced Cache Coherency Architecture. Basically, we generated the memory access traces for a benchmark program and showed the easiness of handling patterns: only a few patterns per core are sufficient to handle all the coherency traffic with the speculative protocol. On our benchmark, the evaluation shows a performance boost of over 60% thanks to the reduced access time to data (prefetched L2 caches). We showed also an optimized throughput of messages by over 35%.

As future work, we want to use an analytical model of cache coherency protocol that would permit to evaluate the effective cost of our protocol performance with regards to the standalone baseline protocol with more accuracy, and use a real simulator platform like SoClib for that. We also want to extend our protocol toward HPC friendly systems, with a dynamic (online, or at runtime) generation of pattern tables.

REFERENCES

[1] Li, K., Hudak, P.: Memory Coherence in Shared Virtual Memory Systems. ACM Transactions Computer Systems 7, 4, 321-359 (1989).

[2] Debes, E., Chen, Y-K., J.Holliman, M., M.Yeung, M.: Apparatus and Method for Performing Data Access in Accordance with Memory Access Patterns. Intel Corporation, Patent US 7,143,264 B2 (2006).

[3] Shen, X., Shafi, H.: Mechanisms and Methods for Using Data Access Patterns. International Business Machines Corporation, Patent US 7,395,407 B2 (2008).

[4] Jeffery, A., Kumar, R., Tullsen, D.: Proximity-aware Directory-based Coherence for Multicore Processor Archectures. ACM Symposium on Parallel Algorithms and Architectures SPAA'07, pp. 126-134, June 9-11, San Diego, California (2007).

[5] Zhuo H., Xudong S., Ye X., Jih-Kwon P.: Alternative Home Node: Balacing Distributed CMP Coherence Directory CMP-MSI: 2nd Workshop on Chip Multiprocessor Memory Systems and Interconnects, Beijing, China (2008).

[6] Chung, E., Hoe, J., Falsafi, B.: ProtoFlex: Co-simulation for Component-wise FPGA Emulator Development. 2nd Workshop on Architecture Research using FPGA Platforms - WARFP (2006).

[7] H.J.Hum., H., R.Goodman, J.: Forward State for Use in Cache Coherency in a Multiprocessor System. Intel Corporation, Patent US 6,922,756 B2 (2005).

[8] Moshe Bach et al.: Analyzing Parallel Programs with Pin. IEEE Computer Society 0018-9162 (2010).

[9] Tera-scale Multicore processor architecture (TSAR). Project Profile Medea+, European Project 2008-2013.

[10] Beamer, S.: Section 1: Introduction to Simics. CS152 Spring 2011, Berkeley University.

[11] K. Rupnow, J. Adriaens, W. Fu, and K. Compton. Accurately Evaluating Application Performance in Simulated Hybrid Multi-Tasking Systems. Accepted for publication at ACM/SIGDA International Symposium on Field Programmable Gate Arrays, 2010.

[12] Xudong S., Zhen Y., Jih-Kwon P., Lu P., Yen-Kuang C., Lee V., Liang B.: Coterminous locality and coterminous group data prefetching on chip-multiprocessors. In: 20th International Parallel and Distributed Processing Symposium - IPDPS'06,IEEE (2006)

[13] Stephen S., Thomas F., Anastasia A., Babak F.: Spatio-Temporal Memory Streaming. In: 36th International Symposium on Computer Architecture - ISCA'07, pp. 69-80. ACM (2009)

[14] Stephen S., Thomas F., Anastasia A., Babak F.: Spatial Memory Streaming. In: 33th International Symposium on Computer Architecture - ISCA'06, pp. 252-263. IEEE (2006)

[15] Thomas F., Stephen S., Nikolaos H., Jangwoo K., Anastasia A., Babak F.: Temporal Streaming of Shared Memory. In: 32th International Symposium on Computer Architecture - ISCA'06, pp. 222-233. IEEE (2005)

System-level Software Performance Simulation Considering Out-of-order Processor Execution

Roman Plyaskin, Thomas Wild, Andreas Herkersdorf

Institute for Integrated Systems, Technische Universität München,
Arcisstr. 21, 80290 Munich, Germany
{roman.plyaskin, thomas.wild, herkersdorf}@tum.de

Abstract—Host-compiled software simulation has become a popular method to accelerate iterative system-level design space explorations of multiprocessor systems-on-chip (MPSoCs) by abstracting the internal microarchitecture of cores. However, current approaches do not consider out-of-order processor architectures, which are emerging in the embedded system domain. Out-of-order processors exhibit complex timing behavior which is difficult to model at a high level of abstraction. In this paper, we improve the accuracy of compiled simulations for out-of-order processors. Our method is applied to binary-level compiled simulation of the target code. We discuss how to annotate timing in the target code and consider out-of-order effects at run-time of the compiled simulation. The proposed approach allows for reproducing the system-level timing behavior of an out-of-order processor observed in a cycle-accurate simulator on average 25× faster at an average error of 3%.

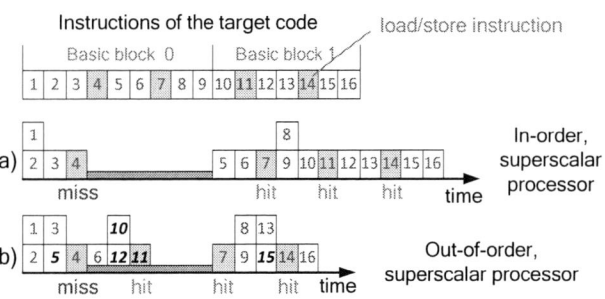

Fig. 1: Timing behavior of in-order and out-of-order processors during a data cache miss

I. INTRODUCTION

Efficient design space exploration of MPSoCs requires fast and yet accurate performance models of the processing cores. At the system level, the designer is primarily interested in allocating the software tasks on the underlying cores and the bottleneck analysis of the on-chip interconnect and shared resources. For many years, multiple instances of cycle-accurate instruction set simulators (ISS) have been employed to stimulate the system-level models of MPSoC components. However, with the increasing number of processing cores on a single chip, these simulators are becoming less applicable for system-level design space exploration (DSE). Cycle-accurate simulators, e.g. SimpleScalar [1] or GEM5 (formerly M5 [2]), explicitly model the components of processor's internal microarchitecture. Because of the high modeling complexity, the resulting simulation speed is not sufficient for fast, extensive explorations. In this paper, we are aiming at fast and yet accurate replication of the timing behavior of out-of-order processors at the system level during *iterative* DSE. Since the designer is not interested in the exploration of the core's internal microarchitecture, the core's abstraction level can be raised hiding as many microarchitectural details as possible. In this case, iterative system-level performance simulations can be done much faster compared to an ISS.

In recent years, many host-compiled techniques have been proposed to speed up the performance simulation at the system level [7], [11], [16], [17]. These methods rely on static back-annotation of computational latencies in the target code which is natively executed on the host computer. During the simulation, dynamic communication latencies are determined using performance models of caches and on-chip interconnect. However, current approaches make a number of simplifications which are not valid for out-of-order processors. For example, in the conventional way of performing host-compiled simulations, the annotated computational latencies are added to the communication latencies, i.e. the memory accesses are considered to be blocking. However, this assumption can only be made for in-order processors which stall on cache misses as shown in Fig. 1a. In contrast, out-of-order processors, which are emerging in embedded systems (e.g. Freescale e5500 or ARM Cortex-R7/Cortex-A15), are capable of executing instructions in advance. These processors exhibit more complex dynamic behavior. In case of data cache misses, an out-of-order processor continues executing independent instructions as long as the respective functional units are available (Fig. 1b). In some cases, the out-of-order execution may even result in reordering of load and store operations. Conventional compiled-simulation techniques either do not address the reordering of memory accesses [7], [11], [16], [17] or treat the latencies for computation and communication separately [14]. As a result, they overestimate the execution time.

In this paper, we improve the accuracy of compiled simulations for out-of-order processors. Compared to the conventional compiled techniques, the proposed method allows for more accurate replication of the application-specific, system-level timing behavior of a processor observed on a reference cycle-accurate ISS. Our method is not intended to replace an ISS in the design flow. Instead, we aim at supporting the designer during system-level explorations by enabling faster and yet accurate simulations. Thus, having identified a reduced set of promising design solutions, the designer may still want to perform cycle-accurate ISS-based simulation of the complete system.

978-1-4673-2895-1/12 $31.00 © 2012 IEEE

To demonstrate the necessary enhancements, we employ binary-level representation of the target code. The code is annotated with timing constants measured on the reference ISS. The paper proposes two major contributions. First, we present a new technique for measuring and annotating the timing information under consideration of out-of-order execution. Particularly, we average the context-dependent deviation of execution time of basic blocks and introduce static reordering of load/store instructions in order to increase the simulation efficiency. Second, we consider dynamic out-of-order effects of the instruction queue and non-blocking caches in order to increase the simulation accuracy. We validate the proposed enhancements using multiple benchmarks and compare the obtained results with other simulation techniques.

The following section presents related work for our paper. Background information on binary-level compiled simulations is provided in Chapter III. In Chapter IV, we present the proposed approach. Experimental results are summarized in Chapter V. Finally, Chapter VI concludes the paper.

II. RELATED WORK

Fast software performance evaluation has been studied in many research papers. In [3], the authors employ statistical simulation to speed up the design space exploration of multiprocessor architectures. In this approach, statistical data on software execution is used to generate representative synthetic traces, which, in turn, are simulated to obtain performance estimation. In SimPoint [5], the execution of a binary code is analyzed to identify representative patterns and, thus, reduce the total workload that is simulated in a cycle-accurate simulator. By simulating just a selected set of these patterns, the designer can interpolate the obtained results and predict the performance of the complete workload. Our approach can be applied complementary to SimPoint by accelerating the cycle-accurate simulation of the selected patterns. Several approaches improve performance of conventional interpretive ISSs by moving time consuming fetch and decode stages to compile time [6], [12], [13], [15]. T. Nakada et. al. [12] eliminated these stages by pre-compiling a C-code representation of the target binary code. A. Nohl et. al. [13] presented just-in-time cache compiled simulation, in which target instructions are compiled at run-time and subsequently reused using a simulation cache. D. Jones and N. Topham [6] employ dynamic binary translation to speed up the interpretive simulation of frequently used code structures consisting of multiple basic blocks. In [15], M. Reshadi et. al. employed pre-compiled instruction templates that are processed at simulation run-time by a highly optimized decoder. However, these approaches only partially abstract the internal microarchitecture of processors, and hence, achieve relatively low speedup of simulation performance compared to interpretive ISSs. In [10], X. Li et. al. propose a technique for estimating the worst case execution time (WCET) for out-of-order processors. The focus of our paper is the average software execution time which cannot be estimated with a WCET tool.

Many recent works are based on native execution of the target software on the host machine [7], [11], [16], [17]. The target code can be employed at different abstraction levels. At the source level, the major challenge is to find correspondence between compiler-optimized binary code and

```
addi r5, r3, r6  ────▶  r[5] = r[3] + r[6];
```
(a)

```
void block_function() {
    cycle += ICACHE(RD, i0);   /* call i-cache model */
    {inst.code}; cycle += c0;  /* add timing est.*/
    cycle += DCACHE(RD, a0);   /* call d-cache model */
    {inst.code}; cycle += c1;
    cycle += DCACHE(WR, a1);
    ...
    {inst.code}; cycle += cn;
}
```
(b)

Fig. 2: (a) Binary-to-C translation (b) Timing annotation in the translated code

lines in the source code. To solve this problem, S. Stattelmann et. al. [16] proposed to dynamically reconstruct binary-level control flow at simulation run-time. In [17], the authors employ intermediate-representation level of the code which is generated after compiler optimizations. However, none of these approaches addresses out-of-order processor execution. In [14], the authors use binary-level representation of the target code and propose to use atomic traces to consider reordering of memory accesses and context-dependent variations of the basic blocks' execution time. However, it does not consider dynamic out-of-order effects of the instruction queue and non-blocking caches. In addition, we introduce static reordering of memory accesses in order to speed up the simulation without significant loss in accuracy. This technique allows compensating the simulation slow-down caused by the dynamic out-of-order modeling. Thus, compared to [14], we achieve higher accuracy at the same level of simulation performance.

Trace-driven simulations employ abstract traces measured on a reference simulator or hardware prototype. The traces allow for accurate capturing of the memory access order. In [9], the authors presented a technique for run-time consideration of out-of-order effects using L2-cache traces. However, the major disadvantage of trace-based approaches is fixed control-flow of the target application. Consequently, despite the abstracted functionality, the performance of trace-driven simulations is limited by the size of generated traces. Our method follows the same principles of out-of-order modeling, however, adapted for execution-driven simulation based on the target code. In the experimental part, we will compare the accuracy of our method with the trace-based approach.

III. BACKGROUND OF BINARY-LEVEL SIMULATION

Host-compiled simulations may employ the target code at different representation levels. In this paper, we use binary-level compiled simulation (BLS). In contrast to source-level simulation, BLS allows representing the target code at the granularity of instructions, which is essential since out-of-order execution occurs at the instruction level. BLS employs the binary code obtained after optimized compilation of the source code. The target binary cannot be executed on the host machine if the host and target instruction set architectures are different. This problem is solved by static translation of the binary code into functionally equivalent C-code. During the translation, each target instruction is represented by one or

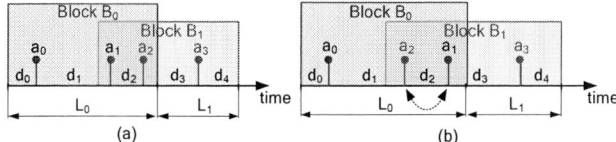

Fig. 3: Different types of basic block overlaps in out-of-order processors

multiple C-operations. The operations perform computation on variables representing the processor's registers. In the example shown in Fig. 2a, instruction `addi` is translated to the equivalent C-code that performs addition of register variables `r[3]` and `r[6]` and stores the result of the computation in variable `r[5]`.

The code of multiple target instructions representing basic blocks is encapsulated into basic block functions. BLS is organized in a while-loop, in which the basic block functions are sequentially executed according to the control-flow of the target application. To enable this, the translated code contains a variable representing the program counter (PC). The PC value is modified within the block functions, thus, determining the next function to execute. The functions are invoked using an array of pointers that are indexed by the PC variable. Further details on this method can be found in [14].

The annotation of timing information at the binary-level is similar to other host-compiled simulation techniques. The pre-estimated amount of execution cycles is inserted directly to the functional code as shown in Fig. 2b. The annotated values represent processing latencies of the processor between memory accesses. If basic block B_j contains n load/store instructions, we can represent the timing information required for the annotation as a set of time constants

$$T_{B_j} = \{c_0^j, \ c_1^j, \ \dots, \ c_n^j\}. \tag{1}$$

In turn, communication latencies of load/store operations are obtained dynamically using function calls to the performance models of caches. On a cache hit, the value of hit latency is added to the total cycle count. In case of a cache miss, the cache model initiates a transaction on the interconnect model in order to determine the value of miss penalty. The total execution time of the application is evaluated by adding static computational latencies with communication latencies determined during the simulation. However, simple addition of computational and communication latencies results in overestimation of the execution time as mentioned in Chapter I.

IV. Proposed Approach

Our method consists of static and dynamic parts. In the static part, we derive time values required for the code annotation (see Eq. 1) using measurements on a cycle-accurate simulator (CAS). The key idea behind our approach is to make use of the timing behavior of non-blocking caches and statically exchange out-of-order memory accesses in BLS. Thus, we can avoid time-consuming reordering of memory accesses at run-time of BLS [14]. We will discuss why this exchange can be accomplished without significant loss in simulation accuracy in section IV-B1. This hypothesis will be tested by the experiments in Chapter V.

In the dynamic part, we model out-of-order effects at run-time of BLS and, thus, address the capability of out-of-order

processors to hide memory access latencies during data cache misses. For this, we consider data dependencies between memory accesses and adjust the simulated time in case of independent hits in the cache. In the following sections, we explain the proposed approach in details.

A. Timing Annotation

1) Overlapping of basic blocks: The strictly sequential execution of basic block functions in conventional BLS poses a limitation in case of out-of-order execution. For the following explanation we assume that a basic block starts executing when one of its instructions (not necessarily the first instruction in the program order) is issued to the respective functional unit. Correspondingly, a basic block stops executing when its last instruction has been committed and left the pipeline. If the processor executes instructions out-of-order, the execution of adjacent basic blocks may overlap. As a result, it becomes particularly difficult to assign the time measured in the CAS to separate basic block functions.

For example, Fig. 3 shows possible executions of two adjacent basic blocks in the CAS. Within each block, the processor performs two accesses to the data cache (a_0, a_1 in block B_0 and a_2, a_3 in block B_1). Delays $d_0 \dots d_4$ represent latencies between the memory accesses measured in the CAS. The first type of block overlapping is shown in Fig. 3a. The processor performs memory accesses while the other block is still executing. In interval d_2, the processor executes instructions of both blocks. In this case, we define time constants c_k^j annotated in the respective basic block functions in BLS as

$$T_{B_0} = \{c_0^0, \ c_1^0, \ c_2^0\} = \{d_0, \ d_1, d_2\}, \tag{2}$$

$$T_{B_1} = \{c_0^1, \ c_1^1, \ c_2^1\} = \{0, \ d_3, d_4\}. \tag{3}$$

The second type of possible block overlapping is shown in Fig. 3b. In this scenario, the processor performs access a_2 out-of-order with respect to a_1. This type cannot be reproduced in BLS, because the basic block functions are executed sequentially. Access a_2 of block B_1 cannot be simulated without finishing the block function of B_0 first. In previous work [14], the authors proposed to postpone access a_1 and simulate it in the function of block B_1, i.e. perform *dynamic reordering* of memory accesses at simulation run-time. However, additional queuing of events reduces BLS performance. In this paper, we propose a simple and efficient solution. We exchange accesses a_2 and a_1, i.e. we simulate a_2 instead of a_1 and vice versa, while leaving the measured delays between them unchanged. By doing this, we intentionally introduce an error for the purpose of higher simulation performance. In Section IV-B1, we will show that this exchange will not always result in visible timing error. In the experimental results, we will provide quantitative estimation of this error in BLS. Under this assumption, this type of block overlapping can be treated as the first type depicted in Fig. 3a, and the time constants for blocks B_0 and B_1 can be defined using Eq. (2) and (3).

In the description above, we considered out-of-order execution of two adjacent memory accesses only. However, large instruction queues allow for more aggressive reordering, i.e. a memory access can be reordered with more than one access. This reordering can take place within one basic block or

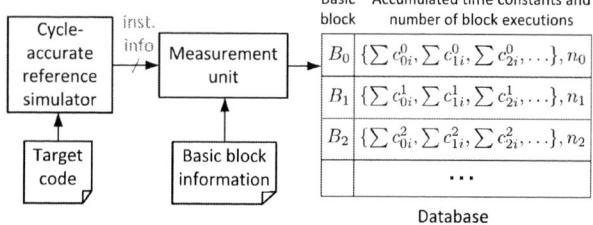

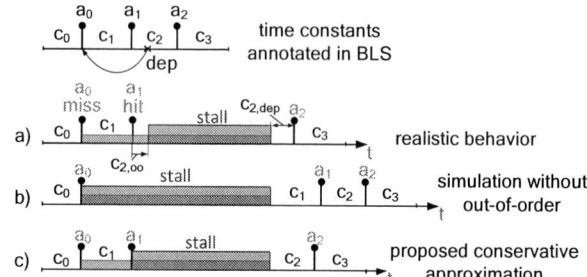

Fig. 5: Modeling of non-blocking cache behavior

Fig. 4: Measurement of basic block timing considering multiple execution contexts

among multiple blocks. Our preliminary experiments on multiple benchmarks in SimpleScalar showed that in a processor with a large instruction queue (64 entries), approximately 10% of all memory accesses were reordered along 2 or more other accesses. In case of a small queue (4 entries), this amount was less than 1%. Thus, the assumption above was valid for the most out-of-order accesses observed in the processor.

2) Timing of basic blocks: During the program execution, the processor performs dynamic scheduling of instructions on the currently available functional units. The availability of the hardware resources depends on the previously executed instructions and defines the current state of the microarchitecture or *execution context*. Therefore, the execution time of basic blocks in a superscalar processor may change depending on their context. Out-of-order processors better exploit instruction-level parallelism compared to in-order processors. As a result, the execution time of basic blocks is subject to higher context-dependent deviations, and a single measurement is not sufficient to capture the timing properties of the basic blocks correctly.

We address the context-dependent execution time of basic blocks by performing multiple measurements as shown in the workflow in Fig. 4. The reference CAS executes the binary code of the target application. During the execution, the CAS provides status information on the running instructions to the measurement unit. This information includes a stamp of the simulated time as well as the address, type and current pipeline stage of the instructions. In addition, the measurement unit requires information on the basic blocks' boundaries, i.e. addresses of the first and last instruction, as well as the number of load/store instructions in each block. The required basic block information can be obtained by disassembling the target binary code prior to the simulation. For each basic block executed in the CAS, the measurement unit determines c_k^j values according to Eq. (2) and (3) and temporarily stores them in the database. If the CAS executes a basic block multiple times, the respective time values are accumulated in the database and afterwards averaged over the number of block executions as follows:

$$c_k^j = \frac{\sum_{i=0}^{n_j} c_{ki}^j}{n_j}, \tag{4}$$

where c_k^j is a kth element in the set of time constants for block B_j (Eq. (1)), c_{ki}^j is the annotation time derived at the ith execution of block B_j, and n_j is the amount of block executions. Thus, for each basic block we average the context-

dependent variation of the execution time and annotate the mean values of the derived time intervals.

In BLS the communication latencies are obtained at run-time using system-level performance models of caches, on-chip interconnect and memory. Therefore, the measurement of basic blocks have to be performed assuming perfect (i.e. always hit) caches. In this case, it can be assured that the annotated timing is independent from the memory access latencies in the CAS environment. In contrast to caches, we do not simulate branch prediction in BLS. The time values annotated in BLS implicitly contain latencies caused by branch mispredictions during the measurements in the CAS. It should be noted that an out-of-order processor may perform memory accesses on a mispredicted execution path. When discarded, they may lead to an incorrect measurement of delays. We solve this problem by temporarily storing the time stamps of load/store instructions and postponing the measurement process. In case of a branch misprediction, respective entries in the storage queue get invalidated and do not affect the measurement. Furthermore, we assume that the target binary code is measured under typical input stimuli which will be applied in BLS as well. If a BLS simulation reveals an uncovered execution path, the database in Fig. 4 has to be correspondingly updated by performing a new CAS measurement. We performed multiple measurements of selected benchmarks and discovered that c_k^j values and the total simulation accuracy were not changing significantly among different input sets.

So far we have considered the time values that are statically annotated to the target code. Consideration of out-of-order effects at simulation run-time is discussed in the following section.

B. Dynamic modeling of out-of-order effects

1) Non-blocking cache accesses: In out-of-order processors, the instruction queue holds multiple instructions of the target code at a time. While some instructions in the queue may depend on the outcome of others, there is a portion of instructions that are independent and can start executing immediately. In case of long-latency operations, e.g. when a load or store instruction results in a cache miss, the processor continues executing independent instructions as long as there are available functional units.

Consider a segment of a basic block with three memory accesses a_0, a_1, a_2 as shown in Fig. 5. For the following explanation, we make several assumptions. Presume that the values annotated in BLS for this block equal to c_0, c_1, c_2 and c_3. Additionally, we assume that some instruction

978-1-4673-2895-1/12 $31.00 © 2012 IEEE

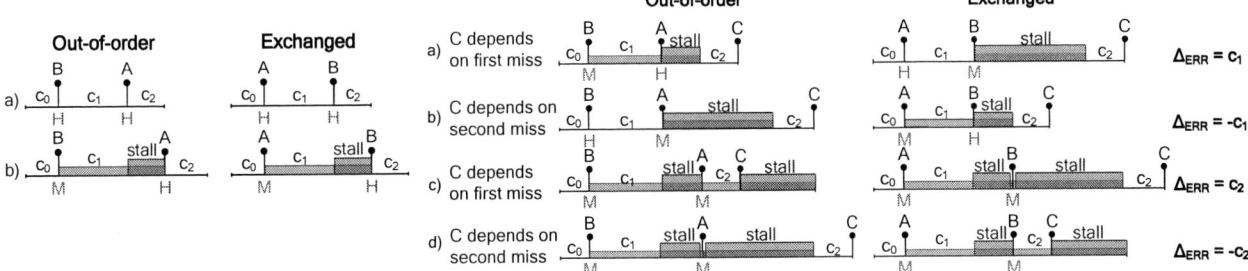

Fig. 6: Exchange of out-of-order accesses to the same cache line

Fig. 7: Exchange of out-of-order accesses to different cache lines

between a_1 and a_2 depends on the data from a_0, while all instructions before are independent from a_0. The execution of the basic block in the CAS is shown in Fig. 5a. In case of the miss caused by a_0, the processor continues executing instructions out-of-order. The non-blocking cache is capable of supplying the available data for a_1 even in the presence of the waiting miss. Therefore, the processor will perform access a_1 and stall at the dependent instruction until the data from a_0 becomes available. Note that the processor executed independent instructions during time intervals c_1 and $c_{2,oo}$. The processor continues executing from the dependent instruction and performs access a_2 after latency $c_{2,dep}$ (where $c_{2,dep} < c_2$). In BLS without out-of-order modeling (Fig. 5b), the processor stalls immediately at the cache miss. As a result, latencies c_1 and $c_{2,dep}$ are not masked by the cache miss and the execution time gets overestimated.

In order to precisely identify points at which the processor stalls (e.g. the value of $c_{2,dep}$ in Fig. 5a), we would need to perform a very large amount of cycle-accurate simulations, forcing cache misses for every load operation at each iteration. Instead, we propose to approximate the duration of out-of-order execution to the last independent memory access. Knowing that a_1 is the last access independent from a_0, we can conservatively assume that the processor stalls immediately after a_1 (Fig. 5c). Thus, in BLS the processor model will simulate full latency of c_2 before initiating access a_2. The preceding latency c_1 will not be counted for the simulated time. Note that the simulation of non-blocking caches requires knowledge on instruction dependencies. Similarly to [9], we make use of the dependency chains constructed in the cycle-accurate simulator during the program execution. Particularly, we capture the dependencies among instructions during the measurement phase. Afterwards, the observed dependency chains are traversed to determine dependency between load/store instructions.

The exchange of out-of-order memory accesses introduced in Section IV-A1 will impact the correct simulation of non-blocking cache accesses. However, not always will the exchange result in a visible error in timing. For example, assume a basic block with two independent accesses A and B to the *same* cache line. Their program order is A-B, but the CAS performed B out-of-order. In the next step, we run two binary-level simulations with different access ordering: B-A as observed in the CAS (the left part of Fig. 6) and A-B as proposed in Section IV-A1 (the right part of Fig. 6). If the required line is present in the cache, both accesses will result in a cache hit, and there will be no difference in the both simulations (Fig. 6a). If we assume that the required line is

not in the cache (Fig. 6b), the second access can be performed only after the data becomes available. From the performance perspective, there will be no difference in whether A or B caused the cache miss. In this case, the exchange will not result in a timing error as well.

The error in BLS will occur if certain conditions are met. First, the data required for A and B must reside in different cache lines. Second, at least one of the accesses must result in a miss. Finally, the memory access following A and B (access C in Fig. 7) must depend on the miss. Assume that the access order in the CAS was B-A-C. Again, we simulate two patterns in BLS: B-A-C and A-B-C. In the first scenario (Fig. 7a), access B causes a miss while A is an independent hit. In the first pattern (the left part of Fig. 7a), latency c_1 and access A will be simulated out-of-order. However, if we exchange A and B (the right part of Fig. 7a), the simulation of out-of-order execution will not take place resulting in error c_1. In the second scenario (Fig. 7b), the miss is caused by A. If we swap A and B, access B will be erroneously simulated out-of-order producing error $-c_1$. Finally, if both A and B result in a miss, the error will equal either $+c_2$ or $-c_2$ as depicted in Fig. 7(c,d).

The impact of this error on the overall time estimation will be low if the above scenarios occur infrequently during the program execution. We will quantitatively assess the resulting error in Chapter V.

2) Instruction queue: The amount of independent instructions that can be executed out-of-order during a data cache miss is limited by the size of the instruction queue. If we ignore this effect in BLS, the out-of-order execution will be simulated further, i.e. assuming the instruction queue to be infinite. For example, assume a section of the target code with three independent memory accesses shown in Fig. 8. In the CAS, the first access results in a miss (Fig. 8a). The processor continues executing the instructions available in the queue and stalls as depicted in the figure. Although there are further independent instructions that could be executed out-of-order, they cannot be fetched because the instruction queue remains full during the miss. Thus, the miss causes head-of-line (HOL) blocking for the subsequent instructions.

In BLS, the HOL blocking can be modeled by monitoring the sequential numbers of the instructions. For example, knowing that the difference of sequential numbers s_2 and s_0 is larger than the queue size, we can postpone the simulation of the third access until the data from the memory becomes available. However, it is difficult to exactly predict the time stamp at which the real processor would stall (the value

978-1-4673-2895-1/12 $31.00 © 2012 IEEE

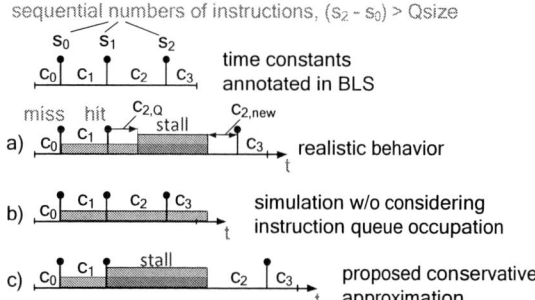

Fig. 8: Modeling of the instruction queue's occupancy

of $c_{2,Q}$ in Fig. 8a). Similarly to non-blocking caches, we conservatively assume that the processor stalls exactly at the last access which still fits in the instruction queue (Fig. 8c), i.e. c_1 will be masked by the miss penalty while c_2 not.

V. EXPERIMENTAL RESULTS

This section presents experimental results of the proposed approach. We employed *sim-outorder* tool from the SimpleScalar set [1] as a reference cycle-accurate simulator (CAS). It is a flexible and highly configurable tool for performance simulation of out-of-order processors. The simulator was extended with a unit measuring the execution time of basic blocks according to the scheme in Fig. 4. We evaluated the proposed approach on 34 different benchmarks from MiBench [4] and MediaBench [8] suites. The experiments were conducted on a PC with a 2,93 GHz Core i7 870 processor running Ubuntu OS.

Each benchmark was evaluated in two steps. First, the target code was measured in the CAS with perfect instruction and data caches. Thus, we could assure that the measured time was independent from the memory access latencies in the CAS environment. After the measurement, our tool automatically created a binary-level C-code annotated with the time constants according to Eq. (2) and (3). In the second step, the produced code was compiled on the host machine and employed for binary-level compiled simulation (BLS).

The aim of BLS was to reproduce the system-level timing behavior of the processor observed in the CAS as fast and as accurately as possible. The BLS tool incorporated performance models of caches and on-chip interconnect in order to derive the software execution time in new system-level environments. We enhanced the BLS tool with a new module (O3M) that handles the out-of-order effects of the instruction queue and non-blocking caches as described in Section IV-B. To estimate the accuracy of the proposed approach, we performed a second simulation of the target code in the CAS with the realistic models of caches and interconnect whose configuration matched the configuration in BLS. Finally, we compared the software execution time in the both tools and determined the estimation error for each benchmark as $e = (T_{BLS} - T_{CAS})/T_{CAS}$.

We compared the proposed method (*BLS+O3M*) with two simulation techniques:

1) conventional BLS employing dynamic memory access reordering which, however, does not consider non-blocking caches and out-of-order effects of the instruction queue (*old BLS*) [14];

2) trace-driven performance simulation of out-of-order processors (*TDS*). TDS employs execution traces that preserve the time constants and the order of accesses to L1-caches observed in the reference simulator.

By comparing our approach to TDS, we studied the impact of averaging the timing intervals between memory operations and static memory access reordering introduced in Sections IV-A1 and IV-A2. In turn, comparison with the old BLS method allowed for studying the impact of dynamic out-of-order effects described in Section IV-B.

The experimental results (Fig. 9) showed that the impact of out-of-order execution is application dependent. For some benchmarks (*blowfish, epic, jpeg, pegwit and rijndael*) old BLS significantly overestimated the execution time, reaching an error of 53% in *jpeg_decode*. The reason for this is the assumption of blocking memory accesses. With BLS+O3M approach, the average error was reduced to 3%. We define the average error among the benchmarks as the mean absolute error, i.e.

$$E = \frac{1}{n} \sum_{i=0}^{n} |e_i|, \tag{5}$$

where n is the number of benchmarks, and e_i is the estimation error of ith benchmark. Please note that there is almost no difference in the simulation performance of BLS+O3M and old BLS (the bottom plot in Fig. 9). This is because the slowdown due to the O3M module was compensated by the speedup due to the static exchange of out-of-order memory accesses. Both methods achieved approximately 51 MIPS (simulated instructions per second) vs. 2 MIPS in SimpleScalar, resulting in the average speedup of 25,5×.

Furthermore, the experiments showed that the static exchange of memory accesses and averaged basic block timing had relatively small impact on simulation accuracy. Compared to TDS, BLS+O3M approach showed almost the same average error (3% BLS+O3M vs. 2,3% TDS) with the exception of *crc32* and *sha* benchmarks. The error in these benchmarks originates from the static exchange of memory accesses. For these benchmarks, we could still improve the accuracy by employing dynamic memory reordering as in [14], however, at the cost of simulation performance. With the dynamic reordering, we achieved an error of 1,7% for *crc32* and -2,8% for *sha* at the reduction of simulation performance by 34% and 20% respectively. Despite the abstraction of functionality, the TDS showed the smallest simulation performance of 38 MIPS among the three methods. The reason for this is high granularity of the traces obtained at the L1-cache level. Even with the efficient representation in the binary form, the trace files were still very large. Therefore, the performance of TDS was limited by the IO bandwidth of the hard disk where the traces were stored.

The results in Fig. 9 were obtained for an out-of-order processor with a 64-entry instruction queue and 4 kB L1 instruction and data caches. Additionally, we conducted experiments for different microarchitecture configurations by taking various combinations of the queue size (4 or 64 entries) and cache sizes (4 or 32 kB). Among these configurations, our approach showed approximately the same average error of 3% achieving the average speed up of 25×.

978-1-4673-2895-1/12 $31.00 © 2012 IEEE

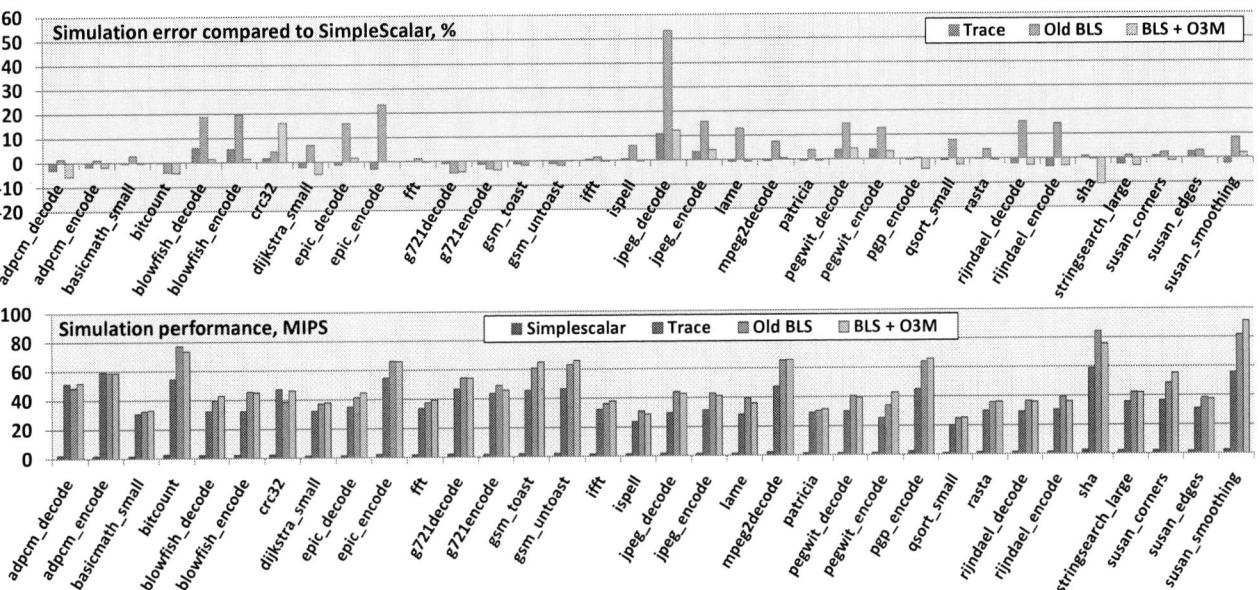

Fig. 9: Results of software simulation which reproduced the system-level timing behavior of different benchmarks on an out-of-order processor. The processor was modeled in SimpleScalar using the default configuration with the following parameters: 64-entry re-order buffer and load-store queue, 4 kB 2-way L1 instruction and data caches (1 miss status holding register, no prefetching, miss latency of 16 cycles).

VI. CONCLUSIONS AND FUTURE WORK

In this paper, we addressed fast software performance simulation at the system level considering complex timing behavior of out-of-order processors. Achieving almost the same speed up in simulation performance as in conventional BLS, the proposed approach produces closer timing estimates observed in the reference cycle-accurate simulator. One of the major drawbacks of BLS is a large size of the produced translated code which requires long compilation times. However, during design space exploration the associated overhead is spread over multiple iterations, still providing the overall gain in simulation performance [14]. We believe that the proposed methodology can be adapted at various abstraction levels of the target code, e.g. source- or intermediate-representation levels. Our approach does not yet address multi-threaded execution. We plan to accomplish this work in the future as well.

REFERENCES

[1] T. Austin, E. Larson, and D. Ernst. SimpleScalar: an infrastructure for computer system modeling. *Computer*, 35(2):59–67, 2002.

[2] N. Binkert, R. Dreslinski, L. Hsu, K. Lim, A. Saidi, and S. Reinhardt. The M5 simulator: Modeling networked systems. *Micro, IEEE*, 26(4):52 –60, Aug. 2006.

[3] D. Genbrugge and L. Eeckhout. Chip multiprocessor design space exploration through statistical simulation. *IEEE Transactions on Computers*, 58(12):1668–1681, Dec. 2009.

[4] M. Guthaus, J. Ringenberg, D. Ernst, T. Austin, T. Mudge, and R. Brown. MiBench: A free, commercially representative embedded benchmark suite. In *IEEE International Workshop on Workload Characterization*, pages 3–14, 2001.

[5] G. Hamerly, E. Perelman, J. Lau, and B. Calder. Simpoint 3.0: Faster and more flexible program analysis. In *Journal of Instruction Level Parallelism*, 2005.

[6] D. Jones and N. Topham. High speed CPU simulation using LTU dynamic binary translation. In *Proceedings of the 4th International Conference on High Performance Embedded Architectures and Compilers*, HiPEAC '09, page 5064, Berlin, Heidelberg, 2009. Springer-Verlag.

[7] T. Kempf, K. Karuri, S. Wallentowitz, G. Ascheid, R. Leupers, and H. Meyr. A SW performance estimation framework for early System-Level-Design using Fine-Grained instrumentation. In *Proceedings of the conference on Design, Automation and Test in Europe*, pages 468–473, Munich, Germany, 2006.

[8] C. Lee, M. Potkonjak, and W. Mangione-Smith. MediaBench: a tool for evaluating and synthesizing multimedia and communications systems. In *Proceedings of IEEE/ACM International Symposium on Microarchitecture*, pages 330–335, 1997.

[9] K. Lee, S. Evans, and S. Cho. Accurately approximating superscalar processor performance from traces. In *IEEE International Symposium on Performance Analysis of Systems and Software, 2009. ISPASS 2009*, pages 238–248. IEEE, Apr. 2009.

[10] X. Li, A. Roychoudhury, and T. Mitra. Modeling out-of-order processors for WCET analysis. *Real-Time Systems*, 34(3):195–227, June 2006.

[11] K. Lin, C. Lo, and R. Tsay. Source-level timing annotation for fast and accurate TLM computation model generation. In *15th Asia and South Pacific Design Automation Conference (ASP-DAC), 2010*, pages 235–240, 2010.

[12] T. Nakada, T. Tsumura, and H. Nakashima. Design and implementation of a workload specific simulator. In *Proceedings of the 39th Symposium on Simulation*, pages 230–243. IEEE Computer Society, 2006.

[13] A. Nohl, G. Braun, O. Schliebusch, R. Leupers, H. Meyr, and A. Hoffmann. A universal technique for fast and flexible instruction-set architecture simulation. In *Proceedings of the 39th Design Automation Conference, DAC '02*, page 2227, New York, NY, USA, 2002. ACM.

[14] R. Plyaskin and A. Herkersdorf. Context-aware compiled simulation of out-of-order processor behavior based on atomic traces. In *2011 IEEE/IFIP 19th International Conference on VLSI and System-on-Chip (VLSI-SoC)*, pages 386–391. IEEE, Oct. 2011.

[15] M. Reshadi, P. Mishra, and N. Dutt. Hybrid-compiled simulation: An efficient technique for instruction-set architecture simulation. *ACM Trans. Embed. Comput. Syst.*, 8(3):20:1–20:27, Apr. 2009.

[16] S. Stattelmann, O. Bringmann, and W. Rosenstiel. Fast and accurate source-level simulation of software timing considering complex code optimizations. In *Design Automation Conference (DAC)*, pages 486–491. IEEE, June 2011.

[17] Z. Wang and A. Herkersdorf. An efficient approach for system-level timing simulation of compiler-optimized embedded software. In *Proceedings of the 46th Design Automation Conference*, pages 220–225, San Francisco, California, 2009. ACM.

Coarse and Fine-Grained Monitoring and Reconfiguration for Energy-Efficient NoCs

Liang Guang
University of Turku
Finland
liagua@utu.fi

Ethiopia Nigussie
University of Turku
Finland
ethnig@utu.fi

Juha Plosila
University of Turku
Finland
juplos@utu.fi

Jouni Isoaho
University of Turku
Finland
jisoaho@utu.fi

Hannu Tenhunen
Royal Institute of Technology
Sweden
hannu@kth.se

Abstract—**Comparative evaluations of centralized, clustered and distributed architectures, for energy management in NoCs, are presented. The paper starts with the systematic examination of the monitoring, decision-making, and reconfiguration processes in building coarse and fine-grained self-adaptation architectures. With examining the physical support in modern technology, network-wide, cluster-wide and per-node energy-management architectures on NoCs are presented, utilizing either voltage regulators or multiple on-chip power delivery networks (MPNs). To identify the effectiveness and efficiency of energy-performance tradeoffs, extensive quantitative simulations are performed with various temporal and spatially changing traffics. Based on the results, we can first observe that the centralized architecture can not adapt to the traffic's spatial locality for effective energy-performance tradeoff. Second, the distributed energy management has the lowest energy-delay product mostly attributed to the fast voltage switching of MPNs, while the synchronization incurs noticeable energy overhead. The clustered architecture, last but not least, is a suitable alternative when the advanced MPN technology is not available. It has low energy and energy-delay product, with very small energy overhead from the monitoring communication.**

I. INTRODUCTION

Many-core systems with Network-on-Chip (NoC) communication architecture have become a major parallel computing platform [1]. Practical scenarios utilizing multi/many-core platforms are likely to be found in a diversity of applications. For instance, a future smartphone is expected to run various image, audio and video processing applications, whose workloads significantly differ [2].

Energy consumption is one primary design concern on NoC-based many-core systems. In particular, on-chip communication contributes a significant part of the system energy [3]. As the static configuration based on the worst-case design is no longer feasible due to the paramount variations, run-time monitoring and reconfiguration are required for adaptive power/energy management [4]. In particular, various applications may require different reconfigurations to serve the changes in traffic pattern and intensity [5].

With the system moving from monotonous to diverse applications, monitoring and reconfiguration techniques have also been evolving from coarse to fine granularity. Conventional chip-wide power management [6] has the most simple, low-overhead architecture, but cannot take into account the spatial locality of the workloads [7]. Thus, clustered and distributed

management techniques [4], [8] have been proposed to maximize the energy efficiency by providing fine-grained run-time reconfiguration.

With these previous efforts, we can still identify the need of systematic evaluation and comparison on the feasibility, effectiveness and efficiency of monitoring and reconfiguration architectures with specific physical supports. For example, per-core DVFS (dynamic voltage and frequency scaling) is a fine-grained distributed power management technique [8], which requires the advanced on-chip MPN(multiple power delivery networks) technology. Yet centralized/chip-wide DVFS can rely on conventional on-chip DC converters [7].

The paper presents a systematic study of different run-time energy management architectures for NoCs. We examine the design choices of three distinct but related processes-monitoring (M), decision-making (D) and reconfiguration (R). Then we present centralized, clustered and distributed NoC monitoring and reconfiguration architectures. We quantitatively study the influence of various parameters, including traffic patterns, synchronization, switching speed and monitoring communication, in the energy-performance tradeoff.

The rest of the paper is outlined as follows: Section II reviews existing works. Section III discusses each of the M,D and R processes. Section IV presents centralized, clustered and distributed monitoring and reconfiguration architectures. Section V provides extensive quantitative evaluations. Section VI concludes the paper.

II. RELATED WORK

In small-sized NoCs with spatially uniform traffics, centralized energy management architectures were demonstrated with lower energy consumption compared to static high voltage and frequency setting [6]. With the increase of on-chip components, fine-grained energy management and other run-time adaptive services have been widely addressed. [7] compares per-core DVFS with chip-wide DVFS in multiprocessor systems. It concludes that the energy consumption of a 4-core system with per-core DVFS is significantly lower. However, the utilization of voltage regulators on a 4-core system may not be feasible on 100s-core NoCs. [9] studies multiple voltage-frequency islands on a NoC, but the architecture is limited to static voltage/frequency assignment. [4] studies adaptively reconfigured voltage-frequency islands. But its focus lies in

978-1-4673-2895-1/12 $31.00 © 2012 IEEE

the control feedback loop design, instead of architectural comparison with other energy-management architectures, for example distributed DVFS. [8] presents distributed DVFS with a specific form of timing, while a 167-core computing platform [1] implements a general-purpose many-core system with distributed DVFS support. [10] compares clustered-based and distributed DVFS. However, the important issues of switching delay and monitoring communication were not explored, which prevents an in-depth analysis on the causes of energy benefits.

Compared to these existing efforts, this paper makes two major contributions:

- It studies and compares the M,D, and R processes in various coarse and fine-grained energy-management architectures, with the discussion on the required physical support, area overhead, effectiveness in temporal and spatial adaptation, and energy saving potential, firstly in a qualitative manner.

- It presents extensive quantitative evaluations of centralized, clustered and distributed architectures on NoCs. In particular, the communication latency, average communication energy, and energy-delay products of various traffic patterns in these architectures are analyzed, with the study on the influences of synchronization overheads, switching delays and monitoring communication.

III. MONITORING, DECISION-MAKING AND RECONFIGURATION

Run-time management architectures need to integrate three processes: monitoring (M), decision-making (D) and reconfiguration (R).

A. Monitoring

In NoCs, distributed tracing of system status has become necessary and highly feasible. In terms of necessity, the variation of workload calls for fine-grained status monitoring [9]. Distributed monitoring is also feasible due to its low cost in modern technology. For instance, distributed probes are added to network interfaces in the NoC [11], which can monitor various parameters such as throughput, timing or latency. The area overhead is only 27% of a network interface's area.

B. Decision-Making

Based on the monitored information, the decision-making process determines the reconfiguration operations to be performed. There can be distributed decision-making where the controllers are localized, or centralized decision-making where a controller decides for all nodes.

In case of the centralized decision-making (Fig. 1 (a)), the relative overhead of the decision-maker itself is small when the system has a large number of components. However, the communication from distributed monitors to the centralized decision-maker incurs timing and energy overhead. In addition, as the centralized controller needs to handle the decision-making for all system components, the approach is not scalable when the system expands.

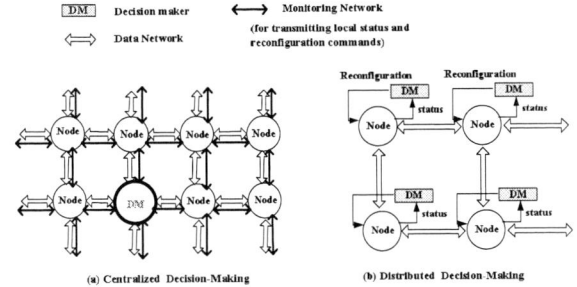

Fig. 1. Illustrating Centralized and Distributed Decision-Making on NoCs

In case of distributed decision-making (Fig. 1 (b)), the controller needs to be realizable with simple hardware so that the area overhead can be affordable. For instance, [1] presented hardware-based controllers for per-core DVFS on a 167-core computing platform. The controller is very simple, only covering 3% of each processor area. The monitoring communication overhead is minimized, as the system status and reconfiguration commands are transmitted locally. Besides, there is no scalability issue due to the distributed control.

C. Reconfiguration

After the decision-making process, reconfiguration can be performed either uniformly to the whole system (centralized reconfiguration) or individually to different components (distributed reconfiguration). For instance, the voltage and frequency can be reconfigured at once to the whole network, or individually configured for each processor or router. Distributed reconfiguration provides better optimization addressing the local workload and status in a many-core system, for instance distributed DVFS [8]. However, the realization of distributed reconfiguration is challenging, in particular due to the following two issues:

- Synchronization. Since components may run very different functions (e.g. a data processing engine vs. a controller), configuring different frequencies on various components is a common technique on NoCs. However, synchronization between different clock domains does not come for free. For instance, a bi-synchronous FIFO [12] can be added between two asynchronous routers, which incurs area, timing and energy overhead.

- Overhead of Reconfiguration Structures. For example, implementing a per-core voltage domain is very challenging due to the large overhead. Recently presented on-chip voltage regulators, despite the improvement in their conversion efficiency, still pose considerable area and power overhead [13], [7].

D. Dedicated Monitoring Network

Except for distributed monitoring with local decision-making and reconfiguration, interconnection is needed to support the communication between monitors and decision-makers. Specific properties are required for the monitoring network. Firstly, the communication has to be provided with

978-1-4673-2895-1/12 $31.00 © 2012 IEEE

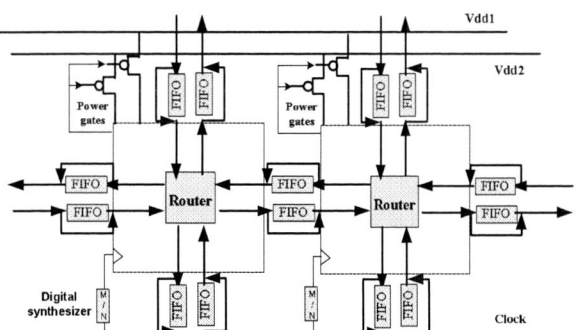

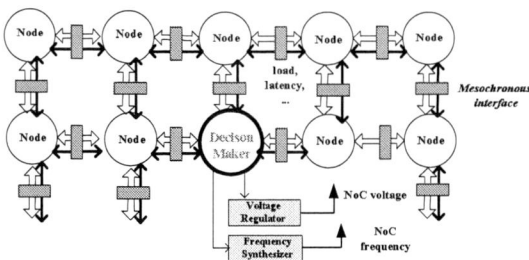

Fig. 3. Centralized Energy-Management Architecture

Fig. 2. Multiple On-Chip Power Delivery Networks with Distributed Frequency Synthesizers

guaranteed service. In particular, the flow should be isolated from the application data communication. Otherwise, if the network is congested by the application data, the monitoring information will be unpredictably delayed. Second, the latency and energy overhead of the monitoring communication should be minimized, if not with strict boundaries. Several alternative architectures can be applied, including physically separate network, TDM (time-division multiplexing) and CDM (code-division multiplexing). The simple architecture of physically separate networks reduces the design and verification efforts, which is desirable in modern parallel embedded systems. In contrast to intuitive assumption, physically separate networks are highly feasible in current technology, as multi-layer fabrication provides abundant wiring resources on-chip [14]. For instance TILE64 multi-processor integrates 5 physically separate networks, each only accounting for 1.1% of the die area [14]. This paper assumes physically separate monitoring networks (Fig. 1 (a)) without further exploration.

IV. CENTRALIZED, CLUSTERED AND DISTRIBUTED ENERGY MANAGEMENT

Run-time self-management architectures need to integrate the M,D and R processes. Given the physical support of on-chip voltage switching, we can identify centralized, clustered and distributed energy-management architectures on NoCs.

A. Physical Support for Voltage and Frequency Reconfiguration

Voltage regulators (VR), either on-chip or off-chip, are the classical techniques enabling voltage reconfiguration. Conventional off-chip regulators are usually slow, due to the large parasitics [15]. To increase the speed of voltage scaling, on-chip voltage regulators are proposed, which achieve frequencies over $10MHz$ [7]. Regardless of the improvement on speed and energy efficiency, VR-based approaches incur noticeable area overhead. For instance, [16] presents an on-chip fast voltage regulator in $130nm$ CMOS consuming $1.5mm^2$ area.

Considering the overhead of voltage regulators, a new technique, multiple on-chip power delivery networks (MPN) [1], provides a lower-cost approach (Fig. 2). Several global power delivery networks are implemented on high metal layers. The

local power networks inside each component are routed on middle metal layers. The voltage of each component can be dynamically connected to one of the global power lines via power switches. As demonstrated by [1], the voltage scaling from 0.9V to 1.3V takes less than $20ns$, which is significantly faster than voltage regulators. Power switches are implemented as parallel transistors, whose area overhead is analyzed as only 4% of a node area [1]. However, the MPN-based platform is demanding in the physical design process, including the layout of multiple power delivery networks and the design of fine-grained power switches to provide low voltage-drop and dependable voltage transition.

Compared to voltage regulators, the frequency synthesizers used for run-time frequency scaling incur much lower overhead in time and area. Even with $90nm$ technology, the clock generator in [17] only covers $0.05mm^2$ area with $4.5ns$ switching time.

B. Centralized Architecture

With VRs, the conventional architecture for energy management, centralized architecture [6], [7], is the most simple in design and implementation. As illustrated in Fig. 3, there is a centralized decision-maker in the NoC, while the run-time information is still gathered locally. The decision-maker reconfigures the network, for instance the voltage and frequency, in a unified manner. Despite that the network runs at the same frequency, it is difficult to ensure that all clocks in every router keep the same phase, thus mesochronous interface is needed [3].

The centralized architecture has low area overhead, since only one set of voltage regulator and frequency synthesizer are needed. However, chip-wide reconfiguration cannot address the spatial locality of the workloads, as will be demonstrated in Section V. It is only suitable for networks which have uniform traffics across the system. In addition, as all nodes send monitoring information to a single decision-maker, congestion may appear when there are a large number of nodes.

C. Clustered Architecture

As the centralized architecture cannot account for the workload's locality, cluster-based energy-management has been proposed [4], [10]. Instead of a centralized decision-maker, the system is divided into several clusters, each of which being supervised by a decision-maker (Fig. 4). The reconfiguration

978-1-4673-2895-1/12 $31.00 © 2012 IEEE

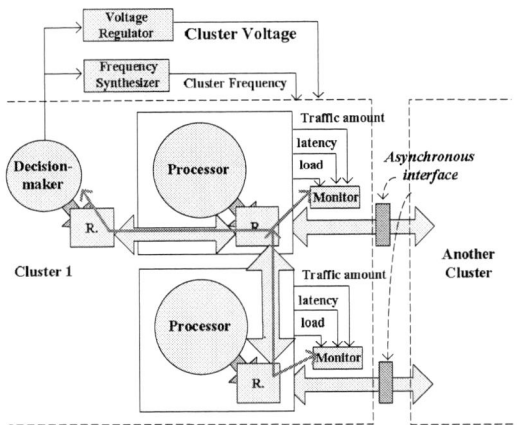

Fig. 4. Cluster-based Decision-Making and Reconfiguration (highlighting the structure in one cluster)

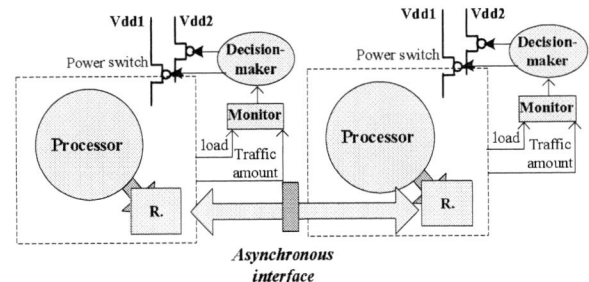

Fig. 5. Distributed Energy Management with MPN-based Physical Support

is applied to the whole cluster. Between the clusters, asynchronous interface is needed as frequencies can be different at the two ends.

The overhead of VR-based clustered architecture, due to the voltage regulators, is higher than that of the centralized architecture. But the total number of regulators only grows with the number of clusters, not with the number of processors. The major benefit of this architecture lies in its adaptation to regional traffic loads, thus it is suitable for systems with multiple applications. However, the voltage regulator still incurs considerable timing overhead (in the range of $100ns$ [7]). In case of fast-changing on-chip communication, such low voltage switching may prevent the network from adapting to the traffic variation on time, resulting in low energy efficiency (Section V will demonstrate this issue). In addition, the workload variations within a cluster cannot be addressed when all nodes in the cluster receive the same reconfiguration.

D. Distributed Architecture

In the distributed architecture, each local decision-maker supervises the reconfiguration of the corresponding node, as illustrated by Fig. 5. Thus there is no global networking between the monitors and the decision-makers. To enable such fine-grained reconfiguration, especially for voltage switching, the multiple voltage delivering networks with power switches are necessary as distributed voltage regulators incur too much overheads (Section IV-A). As each router may run on a different frequency and voltage, asynchronous interface (e.g. with FIFOs and level shifters) is needed on each link.

Compared to the clustered architecture, distributed energy-management can respond to even more fine-grained workload locality, as each router can run on a different frequency. More importantly, as MPN-based voltage switching is significantly faster than voltage regulators, the energy-management can capture the fast changes in the traffics, leading to superior energy-performance tradeoff. However, the asynchronous interface on each link incurs noticeable energy and timing overheads. In addition, local adaptation may overreact to temporary workload variations, which may lead to oscillation with worse performance.

V. QUANTITATIVE EVALUATION

This section quantitatively compares the energy efficiency of the centralized, clustered and distributed architectures. We will examine the performance and energy consumption when the system is adaptively reconfiguring itself by monitoring the run-time workload. In particular, the influences of traffic patterns, switching delay, synchronization overhead and monitoring communication will be identified.

A. Platform Setting

An in-house simulator [18] models 2-D mesh-based NoCs. Each tile in the NoC is a $2mm \times 2mm$ square. In case of centralized and clustered architectures, there are two routers in each tile for the data and monitoring communication respectively. The distributed architecture only has routers for the data communication. The data router is input buffered with 2-flit input buffers. The router utilizes wormhole-based switching and X-Y deterministic routing. Every packet is in the form of one header flit followed by 7 payload flits. Each flit is $32b$ wide, so each channel has 32 bits per direction. The monitoring network uses a similar router architecture with one-flit-deep input buffers and $8b$ channel width, as its communication volume is typically lower than that of the data traffic.

Based on the analysis of a bi-synchronous FIFO architecture [12], a 5-flit FIFO is assigned on each asynchronous interface with 3 reading cycle latencies between different frequency domains, while a 4-flit FIFO is used for each mesochronous interface with 2 reading cycle latencies. The delay on the level shifter is negligible compared to the FIFO delay [19], thus omitted. During the voltage switching, the involved nodes are paused. For VR-based switching, the delay is set as $100ns$ ($10MHz$ [7]). For the MPN-based switching, the delay is much lower as $10ns$ with clock gating (realistic based on [1]).

Energy is modeled by calculating the consumption of each packet traversing the routers and links. The choice of voltage and frequency pairs is dependent on the implementation, for instance the critical path. In this work, we adopt the values from [8] for experimental purposes, as the detailed circuit-level design is beyond the interests of our architectural comparison. Two pairs of voltage and frequency values are

(V_H=2V,F_H=1GHz),(V_L=1.05V,F_H=$\frac{1}{3}$GHz). Given the voltage and frequency, the energy consumption of routers and links can be obtained from Orion 2.0 tool [20] in $65nm$ technology. We estimate the energy consumption of synchronization interfaces also with Orion 2.0, since the FIFOs are usually designed with the same shift register structure as the input buffers of the router. In terms of energy overhead of voltage regulators, it follows $E = C \times (1 - \mu) \times \left| V_{dd2}^2 - V_{dd1}^2 \right|$ [21]. C is the filter capacitance of the power-supply regulator. μ is the conversion efficiency. V_{dd2} and V_{dd1} are the voltages before and after the transition. In the experiments, C is configured as $2.5nF$ and μ as 82.5% [13] ($90nm$; the closest figure to $65nm$ as we found). Th energy consumption of the level shifter is from [19] ($65nm$).

B. Traffic Configuration

In order to evaluate the energy-performance tradeoff for a general-purpose many-core platform, a set of synthetic traces with distinctive temporal and spatial variation patterns are devised and simulated. The synthetic traffic traces are categorized by two features: the temporal injection rate and the destination locality. In terms of injection temporal rates, uniform and b-model [22] traffics are considered. With one parameter b ($0.5 \leq b < 1$), b-model is used for modeling bursty and self-similar traffics. The closer b is to 1, the more bursty the traffic becomes. In terms of packet destination distribution, we consider uniform and hotspot traffics [23]. Hotspot traffic models the situation that some nodes are popular communication destinations. For instance 40% of all packets are destined for a specific area in the network. The combination of temporal and spatial patterns results in four traffic traces, which run simultaneously on an 8x8 mesh-based NoC (Fig. 6). The traffic patterns (e.g. (B,H) stands for b-model in the temporal injection rate, hotspot pattern in destination locality), temporal injection rate (R), hotspot regions and decision-maker locations are all labeled with experimental figures.

C. Simulation Results

The energy management, in particular energy-performance tradeoff, with different architectures is simulated with DVFS adapting to the network traffic. Buffer load, the percentage of buffers (including the FIFOs) occupied in the interested area, indicates the traffic congestion. The higher buffer load, the longer waiting time on average for the packets, leading to worse performance. Thus the decision-makers adjust the frequency and voltage of the corresponding nodes (centralized, clustered or distributed), in order to keep the buffer load within a design-specific range. In our simulation, the range is chosen as $(0.1, 0.2)$. If the buffer load is above 0.2, the corresponding nodes are reconfigured with the high frequency and voltage. Similarly, the nodes are reconfigured to the low frequency and voltage in case the load is below 0.1. The report of buffer load, in case of the clustered or centralized architecture, contains 2 flits.

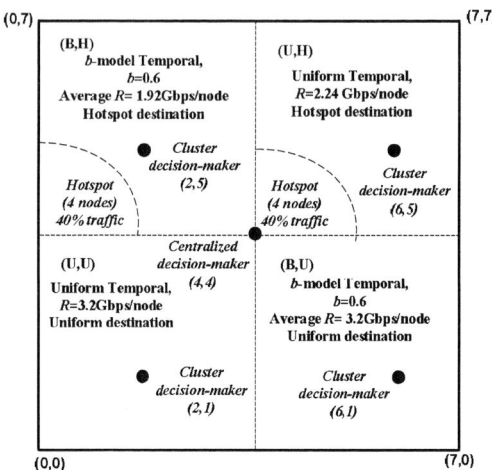

Fig. 6. Experimental Platform (8x8 NoC) Running Four Traffic Traces; each cluster (4x4) runs a different traffic pattern

Fig. 7 shows the average flit latency of the four traces running in the three architectures (CE-centralized, CL- clustered, DI- distributed). The latency is reported in every time window (200 cycles at 1GHz). For analysis purposes, an additional setting (DL) with distributed DVFS with a long switching delay ($100ns$, the same as the voltage regulator) is also simulated to identify the influence of the switching delay. We can see from Fig. 7 that distributed DVFS (DI) has much better performance than other architectures in all traffic patterns. The clustered DVFS performs much better than the centralized architecture, which has very drastic change due to the uneven traffic in the network. In addition, the centralized architecture suffers from congested monitoring network, due to one single decision-maker in a large-size network (average latency 558 cycles vs. maximal 84 cycles in the clustered architecture; not shown in the figure for brevity). From the experiment with distributed DVFS with long switching delay (DL), it is interesting to note that the performance benefit is much reduced in this scenario. The distributed decision-maker changes upon the local traffic, which leads to significant oscillation compared to the reconfiguration based on the average traffic in the cluster. Frequent switching incurs performance penalty, which is clearly seen when we configure the delay longer.

The average communication energy consumption of all four traces with different architectures and the energy-delay product (EDP) are reported in Fig. 8. The overheads contributed by the synchronization interfaces and the monitoring communication are also shown. We can observe that the energy consumption of distributed DVFS is not necessarily the lowest, since the other two architectures may reduce more energy by sacrificing the performance. When EDP is examined, the per-core DVFS has a clear advantage. The maximal saving is 50.7% compared to the centralized architecture, and 43.9% compared to the clustered architecture. The only exception is for the (B,H) traffic. Since the centralized architecture utilizes extra energy

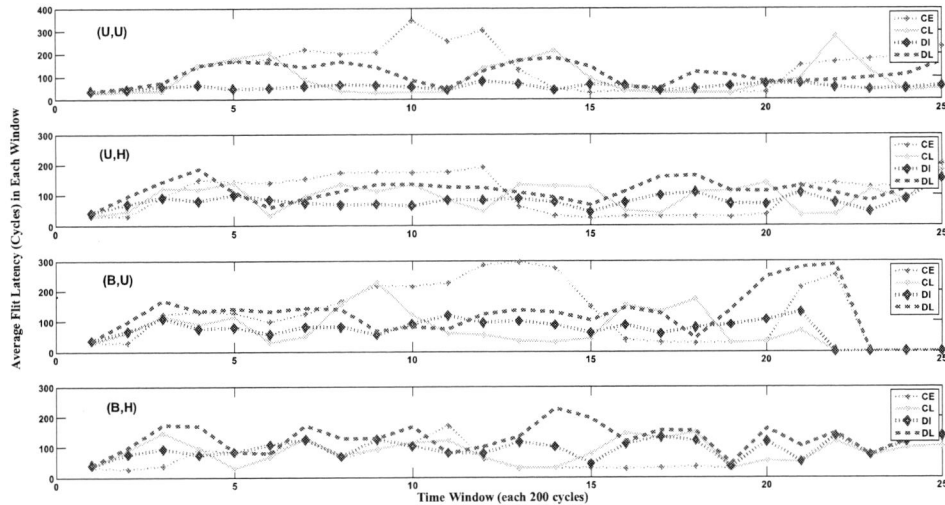

Fig. 7.   Temporal Latency of Four Traffic Traces in Different Energy-Management Architectures

for this traffic based on the average buffer load in the whole network, (B,H) traffic alone gets good performance while the other three traces suffer from long and drastically changing latencies. In addition, from Fig. 8, we can observe that the synchronizing FIFOs incur significant energy overhead. If the FIFOs were removed, the energy could be 17.8% to 25.5% lower. Besides, the monitoring communication adds little energy overhead (maximal 4.8% for the centralized architecture, 4.6% for the clustered architecture), due to its low volume compared to the data communication.

As a summary of the comparison between the centralized, clustered and distributed architectures, Table I lists the pros and cons of each architecture based on the quantitative evaluation.

VI. Conclusion

With various monitoring and reconfiguration techniques being proposed on NoCs, it is necessary to systematically examine and study the evolution from centralized to distributed, or from coarse-grained to fine-grained architectures, in particular for energy-performance tradeoff. This paper studied centralized, clustered and distributed energy-management architectures. The centralized architecture, though with the most simple design, lacks in adaptation to spatial locality. The monitoring communication also leads to congestion in this architecture. The clustered architecture, with much better scalability, is enabled with high-speed voltage regulators, and significantly improves the energy-performance tradeoff. However, the switching delay of voltage regulators is still noticeable for fast-changing on-chip traffics. Thus the distributed architecture with MPN is proposed to provide even more fine-grained adaptation. The extensive quantitative study shows that the energy-delay product of the distributed DVFS is

lower or similar to the clustered DVFS, while the performance is more stable with lower maximal temporal latency. The advantage in energy-performance tradeoff of distributed DVFS is mostly attributed to the shorter switching delay. In terms of the clustered DVFS, it already provides significant energy reduction compared to the centralized architecture, as long as the traffic's spatial locality is captured by the cluster. When the advanced MPN technology is not available for a specific design, the clustered architecture is a suitable alternative.

References

[1] D. Truong, W. Cheng, T. Mohsenin, Z. Yu, A. Jacobson, G. Landge, M. Meeuwsen, C. Watnik, A. Tran, Z. Xiao, E. Work, J. Webb, P. Mejia, and B. Baas, "A 167-processor computational platform in 65 nm cmos," *IEEE Journal of Solid State Circuits*, vol. 44, no. 4, pp. 1130–1144, 2009.

[2] C. van Berkel, "Multi-core for mobile phones," in *Proceedings of the Conference on Design, Automation and Test in Europe*, 2009, pp. 1260–1265.

[3] S. Vangal, J. Howard, G. Ruhl, S. Dighe, H. Wilson, J. Tschanz, D. Finan, A. Singh, T. Jacob, S. Jain, V. Erraguntla, C. Roberts, Y. Hoskote, N. Borkar, and S. Borkar, "An 80-tile sub-100-w teraflops processor in 65-nm cmos," *IEEE JSSC*, vol. 43, no. 1, pp. 29–41, 2008.

[4] U. Ogras, R. Marculescu, and D. Marculescu, "Variation-adaptive feedback control for networks-on-chip with multiple clock domains," in *Proc. of DAC 2008*, 2008, pp. 614–619.

[5] S. Pasricha, N. Dutt, and F. J. Kurdahi, "Dynamically reconfigurable on-chip communication architectures for multi use-case chip multiprocessor applications," in *ASP-DAC '09*, 2009, pp. 25–30.

[6] G. Liang and A. Jantsch, "Adaptive power management for the on-chip communication network," in *Proc. 9th EUROMICRO Conf. Digital System Design: Architectures, Methods and Tools DSD 2006*, 2006, pp. 649–656.

[7] W. Kim, M. Gupta, G.-Y. Wei, and D. Brooks, "System level analysis of fast, per-core DVFS using on-chip switching regulators," in *Proc. HPCA 2008*, 16–20 Feb. 2008, pp. 123–134.

[8] J. M. Chabloz and A. Hemani, "Distributed DVFS using rationally-related frequencies and discrete voltage levels," in *Proc. ACM/IEEE Int Low-Power Electronics and Design (ISLPED) Symp*, 2010, pp. 247–252.

978-1-4673-2895-1/12 $31.00 © 2012 IEEE

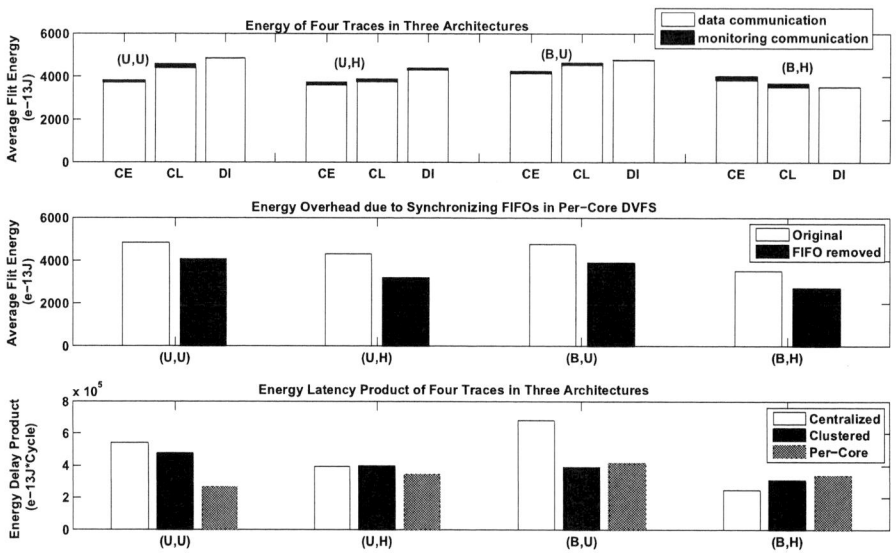

Fig. 8.   Simulation Results on Energy Consumption, Synchronization Overhead and Energy-Delay Product

TABLE I
SUMMARIZING COMPARISON OF CENTRALIZED, CLUSTERED AND PER-CORE DVFS IN ENERGY-PERFORMANCE TRADEOFF

Architecture	Temporal Adaptation	Spatial Adaptation	Monitoring Communication Overhead	Complexity	EDP
Centralized	slow	not addressing locality	congested monitoring network	low	high
Clustered	moderate (due to the VR delay)	addressing each cluster	very low energy overhead no congestion	moderate (due to VRs)	low
Distributed	fast (voltage switching in 10ns)	very fine-grained adaptation (with possible oscillation)	no global monitoring network	high (MPN technology needed)	low (Oscillation leads to performance penalties)

[9] U. Ogras, R. Marculescu, P. Choudhary, and D. Marculescu, "Voltage-frequency island partitioning for gals-based networks-on-chip," in *Proc. of DAC '07*, 2007, pp. 110–115.

[10] A. Yin, L. Guang, P. Liljeberg, P. Rantala, J. Isoaho, and H. Tenhunen, "Hierarchical agent based noc with dvfs techniques," *International Journal of Design, Analysis and Tools for Circuits and Systems*, vol. 1, no. 1, pp. 32–40, 2011.

[11] L. Fiorin, G. Palermo, and C. Silvano, "MPSoCs run-time monitoring through Networks-on-Chip," in *Proc. DATE '09. Design, Automation & Test in Europe Conference & Exhibition*, 2009, pp. 558–561.

[12] I. Miro Panades and A. Greiner, "Bi-synchronous fifo for synchronous circuit communication well suited for network-on-chip in gals architectures," in *Proc. of NOCS 2007*, 2007, pp. 83–94.

[13] P. Hazucha, G. Schrom, J. Hahn, B. Bloechel, P. Hack, G. Dermer, S. Narendra, D. Gardner, T. Karnik, V. De, and S. Borkar, "A 233-mhz 80%-87% efficient four-phase dc-dc converter utilizing air-core inductors on package," *IEEE Journal of Solid-State Circuits*, vol. 40, no. 4, pp. 838–845, 2005.

[14] D. Wentzlaff, P. Griffin, H. Hoffmann, L. Bao, B. Edwards, C. Ramey, M. Mattina, C.-C. Miao, J. Brown, and A. Agarwal, "On-chip interconnection architecture of the tile processor," *IEEE MICRO*, vol. 27, no. 5, pp. 15–31, 2007.

[15] J. Gjanci and M. Chowdhury, "A hybrid scheme for on-chip voltage regulation in system-on-a-chip (soc)," *IEEE Transactions on VLSI*, vol. 19, no. 11, pp. 1949 –1959, nov. 2011.

[16] J. Wibben and R. Harjani, "A high efficiency dc-dc converter using 2nh on-chip inductors," in *Proc. IEEE Symposium on VLSI Circuits*, 2007, pp. 22–23.

[17] K. Nose, A. Shibayama, H. Kodama, M. Mizuno, M. Edahiro, and

N. Nishi, "Deterministic inter-core synchronization with periodically all-in-phase clocking for low-power multi-core socs," in *Proc. of ISSCC. 2005*, 2005, pp. 296–599.

[18] L. Guang, E. Nigussie, J. Plosila, J. Isoaho, and H. Tenhunen, "HLS-DoNoC: High-level simulator for dynamically organizational NoCs," in *DDECS*, 2012, pp. 89–94.

[19] J. C. García, J. A. Montiel-Nelson, and S. Nooshabadi, "High performance cmos dual supply level shifter for a 0.5v input and 1v output in standard 1.2v 65nm technology process," in *ISCIT'09*. IEEE Press, 2009, pp. 963–966.

[20] A. Kahng, B. Li, L.-S. Peh, and K. Samadi, "Orion 2.0: A fast and accurate noc power and area model for early-stage design space exploration," in *Proc. DATE '09*, 2009, pp. 423–428.

[21] A. J. Stratakos, "High-efficiency low-voltage dc-dc conversion for portable applications," Ph.D. dissertation, University of California, Berkeley, 1998.

[22] M. Wang, T. Madhyastha, N. H. Chan, S. Papadimitriou, and C. Faloutsos, "Data mining meets performance evaluation: fast algorithms for modeling bursty traffic," in *Proc. 18th International Conference on Data Engineering*, 2002, pp. 507–516.

[23] Z. Lu, A. Jantsch, E. Salminen, and C. Grecu, "Network-on-chip benchmarking specification part 2: Microbenchmark specification version 1.0," OCP International Partnership Association, Inc., Tech. Rep., May 2008.

Author Index

Saleh M. Abdel-Hafeez
Jean-Thomas Acquaviva
Junwhan Ahn
Tapani Ahonen
Roberto Airoldi
Rachid Al-Khayat
Gabriel M. Almeida
Amir Amin
Gerd Ascheid
Prabhat Avasare
Nadine Azemard
David A. Bader
Amer Baghdadi
Jürgen Becker
Luca Benini
Davide Bertozzi
Shuvra S. Bhattacharyya
Martin Broich
Harald Bucher
Jeronimo Castrillon
Yi-Hsin Chang
Kiyoung Choi
Ilya Chukhman
Simone Corbetta
Loïc Cudennec
Jeroen Declerck
Andy Dewilde
Rolf Drechsler
Gilles Ducharme
Gökhan Erdogan
William Fornaciari
Miguel Glassee
Manfred Glesner
Ann Gordon-Ross
Vineeth Govind
Daniel Große
Liang Guang
Finn Haedicke
Andreas Herkersdorf
Waqar Hussain
Leandro Soares Indrusiak
Jouni Isoaho
Axel Jantsch
Yingtao Jiang
Michel Jézéquel
Mohammad Kakoee
Chih-Chen Kao
Torsten Kempf
Yuchen Kuo
Jan Kuper
Hoang M. Le
Imyong Lee
Rainer Leupers

Stephane Louise
Li Lu
Zhonghai Lu
Kun Lu
Jussara Marandola
Philippe Maurine
Antonio Miele
Leandro Moller
Fernando Moraes
Daniel Müller-Gritschneder
Abdul Naeem
Anja Niedermeier
Ethiopia Nigussie
Tobias G. Noll
Jari Nurmi
Luciano Ost
Martin Palkovic
Vladimir Petrovic
Christian Pilato
William Plishker
Juha Plosila
Roman Plyaskin
Praveen Raghavan
Jaan Raik
Simon Reder
Christoph Roth
Piia Saastamoinen
Oliver Sander
Ulf Schlichtmann
Donatella Sciuto
Mohammad Shatnawi
Gerard Smit
Alessandro Strano
Ilter Suat
Hervé Tatenguem
Hannu Tenhunen
Shiao-Li Tsao
Erik Umans
Anthony Van Herrewege
Bart Vanthournout
Ingrid Verbauwhede
Ling Wang
Zhen Wang
Thomas Wild
Di Wu
Zeqin Wu
Cheng-Kun Yu
Diandian Zhang
Davide Zoni

9781467328951